MACHINING AND FORMING TECHNOLOGIES

MACHINING AND FORMING TECHNOLOGIES

VOLUME 3

MACHINING AND FORMING TECHNOLOGIES

Additional books in this series can be found on Nova's website
under the Series tab.

Additional E-books in this series can be found on Nova's website
under the E-books tab.

MACHINING AND FORMING TECHNOLOGIES

MACHINING AND FORMING TECHNOLOGIES

VOLUME 3

J. PAULO DAVIM
EDITOR

Nova Science Publishers, Inc.
New York

Copyright © 2012 by Nova Science Publishers, Inc.

All rights reserved. No part of this book may be reproduced, stored in a retrieval system or transmitted in any form or by any means: electronic, electrostatic, magnetic, tape, mechanical photocopying, recording or otherwise without the written permission of the Publisher.

For permission to use material from this book please contact us:
Telephone 631-231-7269; Fax 631-231-8175
Web Site: http://www.novapublishers.com

NOTICE TO THE READER

The Publisher has taken reasonable care in the preparation of this book, but makes no expressed or implied warranty of any kind and assumes no responsibility for any errors or omissions. No liability is assumed for incidental or consequential damages in connection with or arising out of information contained in this book. The Publisher shall not be liable for any special, consequential, or exemplary damages resulting, in whole or in part, from the readers' use of, or reliance upon, this material. Any parts of this book based on government reports are so indicated and copyright is claimed for those parts to the extent applicable to compilations of such works.

Independent verification should be sought for any data, advice or recommendations contained in this book. In addition, no responsibility is assumed by the publisher for any injury and/or damage to persons or property arising from any methods, products, instructions, ideas or otherwise contained in this publication.

This publication is designed to provide accurate and authoritative information with regard to the subject matter covered herein. It is sold with the clear understanding that the Publisher is not engaged in rendering legal or any other professional services. If legal or any other expert assistance is required, the services of a competent person should be sought. FROM A DECLARATION OF PARTICIPANTS JOINTLY ADOPTED BY A COMMITTEE OF THE AMERICAN BAR ASSOCIATION AND A COMMITTEE OF PUBLISHERS.

Additional color graphics may be available in the e-book version of this book.

Library of Congress Cataloging-in-Publication Data

ISBN: 978-1-61324-787-7
ISSN: 2157-250X

Published by Nova Science Publishers, Inc. + *New York*

CONTENTS

Preface		ix
Editorial		xv
	A.M. Abrão and J. Paulo Davim	
Chapter 1	Contribution to the Modeling of Forces in Drilling with Twist Drills	**1**
	Angelo Marcos Gil Boeira, Friedrich Kuster, Konrad Wegener, Ricardo Knoblauch, Roger Margot, Rolf Bertrand Schroeter	
Chapter 2	Dimensional Accuracy of End Milled Slots in 7075–T7 Aluminium Alloy	**17**
	Mauro Paipa Suarez, Eder Silva Costa, Álisson Rocha Machado and Alexandre Mendes Abrão	
Chapter 3	Experimental Evaluation and Numerical Validation of Surface Residual Stresses after Hard Turning of Mechanical Components Made of Din 21NiCrMo2 (AISI 8620) Carburized Steel	**29**
	Adalto de Farias, Gustavo Henrique B. Donato, Sergio Delijaicov and Gilmar Ferreira Batalha	
Chapter 4	On the Fundamentals of Orthogonal Metal Cutting	**45**
	P. A. R. Rosa, V. A. M. Cristino, C. M. A. Silva and P. A. F. Martins	
Chapter 5	Thermal Analysis in TiN and Al_2O_3 Coated ISO K10 Cemented Carbide Cutting Tools Using Design of Experiment (DoE) Methodology	**61**
	Rogério Fernandes Brito, João Roberto Ferreira, Solidônio Rodrigues de Carvalho and Sandro Metrevelle Marcondes de Lima e Silva	
Chapter 6	Effect of the Loading Sequence on the Work-Hardening of Low Carbon Steel and CuZn34 Brass Sheets	**73**
	Wellington Lopes, Elaine Carballo Siqueira Corrêa, Haroldo Béria Campos, Maria Teresa Paulino Aguilar and Paulo Roberto Cetlin	

Chapter 7	Numerical and Experimental Analysis of Crimping Tubular Steel Components by Means of a Spherical Punch *João Emanuel Soffiati and Sérgio Tonini Button*	**85**
Chapter 8	Prediction of Local Necking Limit Curve in Sheet Metal Forming *José Divo Bressan*	**99**
Chapter 9	Study of an Automotive Component Production through the Utilization of the Forming Limit Strain and Stress Diagrams *S. M. R. Cravo, B. P. P. A. Gouveia and J. M. C. Rodrigues*	**115**
Chapter 10	The Assessment of Hot Forging Batches through Cooling Analysis *Edilson Guimarães de Souza, Wyser José Yamakami,* *Alessandro Roger Rodrigues, Miguel Ângelo Menezes,* *Juno Gallego, Vicente Afonso Ventrella* *and Hidekasu Matsumoto*	**135**
Chapter 11	3D Modeling of Precision High Speed Turning and Milling for the Prediction of Chip Morphology Using FEM *A. P. Markopoulos, K. Kantzavelos, N. Galanis* *and D. E. Manolakos*	**147**
Chapter 12	Finite Elements Modelling and Simulation of Precision Grinding *A. P. Markopoulos*	**161**
Chapter 13	Prediction of Thin Wall Surface Shape through Simulation of the Machining Process for Light Alloy Workpieces *Serge St-Martin, Jean-François Chatelain, René Mayer,* *Serafettin Engin and Stéphane Chalut*	**181**
Chapter 14	Experimental Study of Drilling Ceramic Matrix Brake Pad (CMBP) *K.L. Kuo and C.C. Tsao*	**199**
Chapter 15	Turning Hardened Steel Using Carbide Insert under High-Pressure Coolant Condition *Prianka B. Zaman, S. K. Dey and N. R. Dhar*	**211**
Chapter 16	Optimization of CNC Turning of AISI 304 Austenitic Stainless Steel Using Grey Based Fuzzy Logic with Multiple Performance Characteristics *C. Ahilan, S. Kumanan and N. Sivakumaran*	**227**
Chapter 17	Optimization of Roller Burnishing Process on Tool Steel Material Using Response Surface Methodology *M. R. Stalin John and B. K. Vinayagam*	**247**
Chapter 18	Parametric Optimization of AFM Process during Finishing of Al/5wt%SiC-MMC Cylindrical Surface *Harlal Singh Mali and Alakesh Manna*	**265**

Chapter 19	OIM and EBSD Studies on the Influence of Texture on the Mechanical Properties Developed in Cold Rolled Duplex Stainless Steels *K. Manikanda Subramanian, P. Chandramohan* *and S. Basavarajappa*	**281**
Index		**297**

PREFACE

Nowadays, machining is one of the most important of manufacturing processes in which a cutting tool or other techniques are used to remove excess material from a workpiece so that the remaining material is the desired part shape. Forming technologies includes a large group of manufacturing processes in which plastic deformation and other techniques are used to change the shape of workpieces. Machining and forming technologies can be applied to a wide variety of materials, namely, metals, polymers, ceramics, composites, biomaterials, nanomaterials. This book focuses on machining and forming technologies.

Chapter 1 - Drilling and turning are the two most used cutting operations in the metal industry, which were also the first machining operations carried out. The drilling process is much more complex and very difficult to characterize considering tool geometry and variable cutting speed along the cutting edge.

Chapter 2 - Dimensional accuracy and surface roughness are key elements for a competitive production due to the tight tolerances demanded in high technology application. This is particularly critical when human lives are involved. Among the available alternatives for the milling process, end milling is probably one of the most used in the mould and die industry. Additionally, end mill cutters are largely employed in the manufacture of aircraft structural parts. The use of end mills has increased in the last decades owing to their ability to replace previously employed machining operation providing higher productivity.

Chapter 3 - Often, as a result of mechanical processing, the component surface can be left with a residual stress state after conventional machining operations. Although limited to a surface layer, such stress state can have a significant influence on the component performance and lifetime. Generally, compressive residual stresses are preferred, as they tend to reduce the effect of external applied tension fields, tending to close (or to reduce the opening tendency of) surface cracks. Furthermore, the reduction in the equivalent stress fields strongly favors fatigue life of components subjected to cyclic loads. This beneficial effect of compressive stresses on cyclically loaded mechanical components is a common sense for many authors who have investigated issues related to residual stress conditions and component performance.

Chapter 4 - In a recent paper, Astakhov (2005) discussed the major drawbacks of the single shear plane model for metal cutting, showing that this model cannot be utilized in the development of the predictive metal cutting theory as well as in the development and utilization of finite element computer programs. As initially claimed by Cook et al. (1954) and later by Atkins (2003), the single shear plane model is inadequate for real metal cutting because it cannot operate in plane strain plasticity at constant plastic volume without 'new surface' formation at the tip of the tool. In connection to this, Astakhov (1999) points out that there is a major difference between metal cutting and other metal working processes in that there must be physical separation of the layer to be removed from the work material and that the process of separation forms new surfaces.

Chapter 5 - Machining processes generate enough heat to deform the materials involved: the chip and the tool. This level of heat is a factor that strongly influences tool performance. Friction wear and heat distribution affect the temperature on both the chip and tool. In order to increase tool life, its surface can be coated with materials having thermal insulation features that reduce tool wear. Hence, the influence of coatings on heat transfer and friction wear is an area of study that deserves an in-depth investigation.

Chapter 6 - Sheet metal forming employs relatively simple tooling and can produce parts with a variety of shapes and dimensions under economically attractive conditions. The success of this process hinges, however, on the formability of the material being formed, involving variables such as the material initial conditions, ductility, strength and anisotropy. In addition to that, the process parameters (strain rate, temperature, friction between the material and the tools, etc.) are also of importance.

Chapter 7 - The increasing necessity to supply products with low cost and high level of quality has been forcing Manufacturing Engineering to develop and search alternatives in the fabrication and assembling elements of machines and auto-parts. Mechanical elements for fixtures such as screws, rings and rivets have been extensively used. However depending on the application of these elements they tend to become expensive and not trustworthy throughout the time.

Chapter 8 - Sheet metal blanking and forming technologies are the principal fabrication processes in the automotive, aeronautic, kitchen utensils, metal packaging and other metal-mechanic sectors. The usual sheet metals utilized in these industries are steel, aluminum, brass and titanium alloys. The success of these fabrication processes are due to various factors which make them attractive and competitive as for example: good surface finishing, low weight, flexibility to change dies, production of parts with complex shapes, near net shape forming process, and sometimes is of low cost owing to its high mass production. However, a set of dies and stamping presses costs in the automotive industry is in the order of million dollars.

Chapter 9 - During design of sheet metal forming tools, the knowledge on sheet metal formability limits plays an important role, since the amount of plastic deformation in the sheet metal part is limited by the occurrence of undesirable phenomena such as localized necking and fractures. To study and prevent these formability problems during stamping operations, strain-based forming limit curves have been widely used. However, its utilization is restricted to linear or near linear loading conditions, leading to erroneous assessments on the analysis of sheet metal formability under complex strain paths such as multi-step forming. Recently, stress based forming limit diagrams and the respective curves have been successfully applied under these circumstances.

Chapter 10 - Several research works concerned with hot forging followed by controlled cooling have been performed. Other works correlated to this subject, such as developments metallurgic of alloys, that permit the direct cooling, and the application of the thermomechanical treatment have allowed to obtain more adequate microstructures and mechanical properties required in forging applications, thus enabling the reduction of production cycle and energy costs.

Chapter 11 - High Speed Machining (HSM) refers to processes with cutting speed or spindle rotational speed substantially higher than some years before or also than the still common and general practice, according to a definition by Tlusty (1993) in a keynote paper of CIRP. In this definition, a spectrum of speeds or a lower value above which a machining process is characterized as HSM is avoided, because the speed involved in high speed machining depends upon the actual cutting process, the cutting tool and mainly on the workpiece material; for instance aluminum alloys can be machined at significantly higher speeds than stainless steels or titanium alloys.

Chapter 12 - Grinding is a precision manufacturing process, traditionally used as a finishing operation because of its ability to produce high workpiece surface quality. Improvements in its performance have allowed for the use of grinding in bulk removal of metal, maintaining at the same time its characteristic to be able to perform precision processing, thus opening new areas of application in today's industrial practice. The ability of the process to be applied on metals and other difficult to machine materials such as ceramics and composites is certainly an advantage of this manufacturing method. In a keynote paper by CIRP, a thorough discussion on the applications of grinding, especially in the automotive industry, as well as the limitations and research conducted towards the improvement of the process was presented, justifying the great importance of grinding and the amount of researchers that show interest in analyzing its characteristics.

Chapter 13 - High quality product manufacturers largely benefit from the increase of high performance machining technology. This is particularly true for the manufacturing of airframe and complex aircraft parts that feature large amount of milled pockets. The machining process of these parts becomes delicate at the finishing step, as the occurrence of thin wall between pockets demands additional caution to avoid inaccuracies and form errors to the components. For one part, finish machining is normally used to mill components to their designed dimensions within tolerances. It includes the cutting of the material in excess that have been left on the surfaces during the high removal rate machining due to tool static deflection. For low rigidity components such as thin walls, the local deflection of the workpiece as well has to be considered in order to reach dimensional and geometrical objectives. Secondly, the low tool immersion process will have a significant impact on the thin wall dynamic behaviour. At best, this will have an effect on the surface finish quality; at worst chatter could arise and generate cracking or breakage. In order to achieve dimensional accuracy of the component right on the first attempt, it is imperative to predict the surface profile generated by presently used cutting strategies. Thus, one is able to optimize the key parameters that will permit the compensation of the additional sources of inaccuracies inherent to thin wall machining.

Chapter 14 - Most developed countries have prohibited importing traditional brakes pads owing to the presence of asbestos in those products. The merits of the present brake pads, e.g. ceramic matrix brake pads, iron fiber brake pads and fiber magnesium brake pads, possess the high temperature durability, wear resistance and corrosion resistance. These brake pads are made of composite materials and have better mechanical properties than those made with traditional asbestos and cast iron. Moreover, the effective lifespan of the present brake pads can even reach 300,000 kilometers, which approaches the life expectancy of most vehicles. To use these brake pads, accurate, precise high quality holes need to be drilled to ensure proper and durable assemblies. Conventional drilling with twist drill still remains one of the most economical and, therefore, commonly adopted machining processes for drilling holes in structural parts. However, the geometry and the velocity of cutting edge for twist drill are not constant, but vary along the cutting edge. During drilling these brake pads, the problem of delamination often occurs at the entrance plane and the exit plane of brake pad. The drilling induced-delamination has been a number of studies with different methods of measurement.

Chapter 15 - Hard turning of harder material differs from conventional turning because of its larger specific cutting forces requirements. Typically, in the machining of hardened steel materials, no cutting fluid is applied in the interest of low cutting forces and low environmental impacts. Higher temperatures are generated in the cutting zone, and because cutting is typically done without coolant, hard turned surfaces can exhibit thermal damage in the form of microstructural changes and tensile residual stresses. The potential economic benefits of hard turning can be offset by rapid tool wear or premature tool failure if the brittle cutting tools required for hard turning are not used properly. Even, progressive tool wear can result in significant changes in cutting forces, residual stresses, and microstructural changes in the form of a rehardened surface layer.

Chapter 16 - In engineering industries, turning is one of the important and extensively used machining processes. The cutting conditions such as cutting speed, feed rate and depth of cut, features of tools, work piece materials affects the process efficiency and performance characteristics. Performance evaluation of CNC turning is based on the performance characteristics like surface roughness, material removal rate (MRR), tool wear, tool life and power consumption. Surface quality is an important performance to evaluate the productivity of machine tools as well as machined components. Hence achieving desired surface quality is of great importance for the functional behavior of the mechanical parts. Surface roughness is used as the critical quality indicator for the machined surfaces. Very few research attempts have been done to estimate the significance of energy required for the machining process. Recent increase in energy demand and constraints in supply of energy becomes a priority for the manufacturing industry. Now in manufacturing industries, special attention is given to surface finish and power consumption. Austenitic stainless steels are a high work hardening rate, low thermal conductivity and resistance to corrosion. Stainless steels are known for their resistance to corrosion but their machinability is more difficult than the other alloy steels due to reasons such as having low heat conductivity, high BUE tendency and high deformation hardening. Work to date has shown that little work has been carried on the determination of optimum machining parameters when machining austenitic stainless steels. The influences of cutting fluids on tool wear and surface roughness during turning of AISI 304 are investigated. The optimum cutting speed for turning of AISI 304 austenitic stainless steel based on tool wear and surface roughness are determined and it needs further investigation. The high cost of CNC machine tools compared to their conventional counterparts, there is an economic need to operate these machines as efficiently as possible in order to obtain the required payback. The desired cutting parameters are determined based on experience or by use of a handbook which does not guarantee optimal performance. It is mandatory to select the most appropriate machining settings in order to improve cutting efficiency, process at low cost and produce high-quality products.

Chapter 17 - Conventional machining process such as turning has inherent irregularities and defects like tool marks and scratches that cause energy dissipation and surface damage. Conventional finishing processes such as grinding, honing and lapping are used to overcome these defects. But these methods essentially depend on chip removal to attain the desired surface finish and also skill of the workers. To resolve these problems, burnishing process is applied for better surface finish on the post machined components due to its chip-less and relatively simple operations.

Chapter 18 - The aluminium alloy reinforced with discontinuous silicon carbide (SiC) particulates (Al/SiC-MMC) is rapidly replacing conventional materials in various industries mainly automotive and aerospace due to considerable weight saving. Despite superior physical and mechanical properties, particulate reinforced metal matrix composites are not widely used in industry because of their poor machinability. Wide spread engineering application is resisted due to its poor machining characteristics particularly excessive tool wear and poor surface finish. The hard SiC particles of Al/SiC-MMC, which intermittently come into contact to the tool surface, act as small cutting edges like those of a grinding wheel on the cutting tool edge which in due course is worn out by abrasion, resulting in the formation of poor surface finish during machining. Hence, proper identification of a cost effective surface finishing process to achieve the required quality of surfaces on Al/SiC-MMC jobs is a challenge to the manufacturing engineers. Abrasive flow machining (AFM) is an advanced finishing process that can be used for cleaning and fine finishing of Al/SiC-MMC. This process can also be used for deburring, polishing, radiusing, removing recast layers, producing compressive residual stresses of difficult to reach areas. Application of abrasive flow finishing (AFF) processes has been reported on Al/6063 workpieces and optimization of AFF parameters carried out.

Chapter 19 - Duplex Stainless Steels (DSS) contain both phases, i.e. ferrite (Alpha-BCC) and austenite (Gamma-FCC) in almost equal proportions. Therefore, to make a study of texture in this cold rolled alloy, it is necessary to report few literatures related to the deformation mechanisms, texturing and orientations that exist in cold rolled DSS alloys.

Versions of these chapters were also published in *Journal of Machining and Forming Technologies,* Volume 3, Numbers 1-4, edited by J. Paulo Davim, published by Nova Science Publishers, Inc. They were submitted for appropriate modifications in an effort to encourage wider dissemination of research.

EDITORIAL

*A.M. Abrão[*1] and J. Paulo Davim[2]*

[1]University of Minas Gerais
Av. Antônio Carlos, 6627, Pampulha, Belo Horizonte MG,
31.270-901, Brazil
[2]University of Aveiro
Campus Santiago, 3810-193 Aveiro, Portugal

In recent years, Brazil has experienced a steady growth in the manufacturing engineering field with an industrial sector of great technological expression as well as economic and social relevance. The Brazilian Conference on Manufacturing Engineering is undoubtedly the principal forum in the country where academy and industry congregate to discuss recent advanced in manufacturing technology as well as issues of mutual interest. During the fifth edition of the conference, held in Belo Horizonte, Minas Gerais from the 14th to the 17th April 2009, 178 papers were presented during 33 technical sessions. The public attending the technical sessions represented 57 institutions, including universities, research centers and companies.

The best papers presented during the conference were selected for publication in this special issue of the Journal of Machining and Forming Technologies after having their content extended and being peer reviewed. These papers are divided into two blocks: in the first block can be found five papers concerned with various aspects of metal cutting. Another five papers related to the forming process are presented in the second block. In both cases, fundamental issues related to the phenomena involved in cutting and forming are addressed, as well as the technological aspects of cutting and forming. We trust the selected papers will provide useful information for those interested in manufacturing.

Finally, on behalf of the organizing committee we would like to thank Nova Science Publishers for providing the opportunity to publish this book.

---

[*] Email: abrao@ufmg.br

In: Machining and Forming Technologies, Volume 3
Editor: J. Paulo Davim

ISBN: 978-1-61324-787-7
© 2012 Nova Science Publishers, Inc.

Chapter 1

CONTRIBUTION TO THE MODELING OF FORCES IN DRILLING WITH TWIST DRILLS

*Angelo Marcos Gil Boeira[1], Friedrich Kuster[1], Konrad Wegener[1], Ricardo Knoblauch[1], Roger Margot[1], Rolf Bertrand Schroeter[2]**

[1]Institute of Machine Tools and Manufacturing (IWF), ETH Zentrum, CLA, Tannenstrasse 3, Zürich

[2]Laboratório de Mecânica de Precisão (LMP), UFSC, Caixa Postal-476 EMC, CEP: Florianópolis - SC, Brazil

ABSTRACT

Specific cutting force and specific thrust force in drilling vary according to changes in the main cutting edge and in the cutting angles. Considering the geometry of the main cutting edge of a twist drill, angles such as the working orthogonal rake angle, the tool orthogonal clearance angle and the working cutting edge inclination angle vary along the main cutting edge and the chisel edge. In this paper a new method for prediction of the cutting forces in drilling is presented. This method is based on Kienzle force model, using results obtained in turning operations and applying correcting factors for variations in the tool geometry. Finally, the simulation results of cutting and thrust forces in drilling using Kienzle constants were compared with experimental forces.

Keywords: Drilling, Force Modeling, Twist Drills, Simulation of Drilling

1. INTRODUCTION

Drilling and turning are the two most used cutting operations in the metal industry, which were also the first machining operations carried out. The drilling process is much more complex and very difficult to characterize considering tool geometry and variable cutting speed along the cutting edge (Hsieh and Lin, 2002; Castillo, 2005).

---

* Email: rolf@emc.ufsc.br

In drilling, the most used tools are twist drills, mainly due to their universal application. Although this type of tool has been applied over the past 200 years and even with all of the efforts to improve the efficiency of its operation during this period, twist drills are typically the bottleneck of machining operations. Thus, a better understanding of this process is necessary to enable its optimization (Bork, 1995; Hsieh and Lin, 2002).

One of the several possibilities to improve the efficiency of drilling is the optimization of the tool geometry. Many variations of tool geometry have been proposed, in order to reduce machining forces, increase tool life, reduce machining time and improve hole quality, among other improvements (Stephenson and Agapiou, 1992; Bork, 1995; Chen *et al.*, 1996; Paul *et al.*).

While the characteristics of other machining processes are relatively easy to analyze in real time, monitoring the phenomena occurring during the drilling process is very complex, because they occur within the piece to which access is blocked by the tool. Only the thrust force and torque can be measured on the axis of the tool. The chip flow and conveying in the flutes and the material compression produced by the minor cutting edge complicate the interpretation of results. Furthermore, the cutting speed varies along the main edge (Castillo, 2005; König and Klocke, 2002).

In order to better understand machining processes, the use of computational tools is necessary to get reliable and accurate result from simulation replacing the scientifically proven empirical data used in the shop floor. Specifically in the case of drilling, it is important to know the behavior of the drilling forces associated with the twist drill, for different tool geometries, substrate materials, coatings and other tool characteristics.

In drilling optimization, the techniques for modeling and simulating are important due to the fact that there is a complex relation between the input variables and the results, making certain process phenomena difficult to evaluate. To experimentally evaluate the relationship between these variables and their influences, many experimental tests would be required, taking into account multiple cutting parameters.

Ascertaining the magnitude and direction of the forces during machining is fundamental in determining the cutting parameters, evaluating the machine tool precision under specified working conditions (deformation of the machine and the tool), recognizing the chip formation processes and explaining the wear mechanisms (König and Klocke, 2002). The modeling and simulation of machining forces provides a simplified alternative to predict cutting forces with good accuracy (Kim and Ahn, 2005). In this study, a drilling model is applied that allows the use of the Kienzle equation coefficients obtained from turning tests, a process with significantly less difficulties compared to drilling. Based on these models, the drilling forces can be simulated for different tool geometries, since the relations between forces on the edge of the drill and their geometrical characteristic are parameterized.

2. THEORETICAL BACKGROUND

As in other areas of engineering, in the recent years a great effort has been done to develop models and simulate machining processes. The models are developed based on specific parameters, such as geometry of the cutting tool, the workpiece material, the cutting parameters, the static and dynamic characteristics of the tool, workpiece and machine.

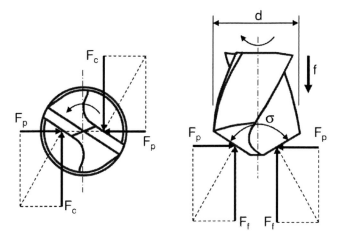

Figure 1. Decomposition of machining forces on the drill lips.

The study of cutting forces in drilling has been carried out classically by models based on forces measured in the practical experiment and a mathematical function fitted in, depending on the characteristics of the process and the tool geometry. In the case of drilling some difficulties due to the complex geometry of the drill occurs. In general, empirical equations are used to predict the force and torque, using the drill diameter and the feed rate as input data for the calculations, generating approximated equations that do not take into account all of the parameters involved in the process (Campos, 2004; Castillo, 2005).

The machining force F on the drill cutting edge can be decomposed into components such as cutting force F_c, thrust force F_f and passive force F_p as shown in Figure 1. The cutting force F_c is associated with the cutting resistance of the workpiece material. This force is predominantly responsible for the torque M_z generated in the process (König and Klocke, 2002; Grote and Feldhusen, 2004; Castillo, 2005).

According to Witte (1980), the thrust force F_f and the torque M_z in full drilling can be determined using the following equations:

$$F_c = k_c \cdot \frac{f \cdot d}{4} \qquad (1)$$

$$F_f = k_f \cdot \frac{f \cdot d \cdot \sin\left(\frac{\sigma}{2}\right)}{2} \qquad (2)$$

$$M_z = k_c \cdot \left(\frac{f \cdot d^2}{8000}\right) \qquad (3)$$

where:
- k_c - Specific cutting force [N/mm^2];
- k_f - Specific thrust force [N/mm^2];
- f - Feed rate [mm];

d - Drill diameter [mm];
σ - Drill point angle [°].

Equations (1), (2) and (3) are based on the Kienzle model (1951) *apud* König and Klocke (2002), which describes the behavior of machining forces and uses a potential function (Eq. (4)), developed for turning processes in general, to provide a non-linear relationship between the specific cutting force and the cutting width. Thus, this model predicts the cutting forces and power, acting on the machine-tool, the workpiece and the tool, under several working conditions. It requires the experimental determination of the Kienzle equation constants $k_{c1.1}$ and ($1-m_c$) (Rocha, 1984; König and Klocke, 2002; Grote and Feldhusen, 2004).

$$F_c = k_{c1.1} \cdot b \cdot h^{(1-m_c)}$$

(4)

where:

$k_{c1.1}$ - Specific cutting force for 1×1 mm cross-section ($b \times h = 1$ mm^2);
b - Chip width [mm];
h - Chip thickness [mm];
$1-m_c$ - Kienzle exponent.

Based on the work of Kienzle, Witte (1980), Rocha (1984), Bach *et al.* (2004) and other researchers have proposed equations and methods, that use known turning parameters for the prediction of drilling forces. These equations use correction factors, which consider the influence of dependent and independent factors of the drilling process. However, due to the complexity of this problem, these models are difficult to implement given factors, such as the large chip deformations occurring in the region of the chisel edge, low cutting speeds, strongly negative tool geometries and the difficulty in disposing of the chips. Thus, efforts are still underway to improve the accuracy of the simulation models and facilitate the prediction of the force involved in the drilling processes under different working conditions.

3. MODEL-BUILDING METHODOLOGY

The model of drilling forces proposed in this paper aims to allow the simulation of forces associated with twist drills, which act on the chisel edge and on the main cutting edge. The data available on such forces originates from turning tests. The machining force components, especially the cutting and the thrust force, must be simulated for different working conditions, tool geometries and workpiece materials, *inter alia*.

From longitudinal orthogonal turning tests the specific cutting $k_{c1.1}$ and thrust $k_{f1.1}$ forces are determined, along with the coefficients ($1-m_c$) respectively ($1-m_f$) of the Kienzle equation. For both forces, cutting and thrust, a model is developed which allows the prediction of the forces for a given cutting condition in the process of drilling through correction factors related to differences at the detailed geometry along the cutting edge between turning and drilling tool.

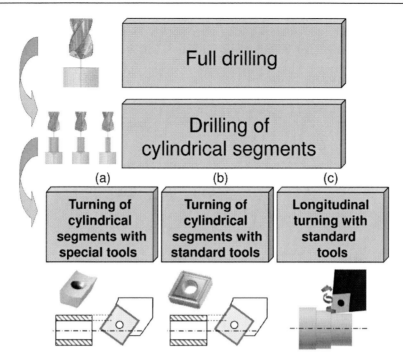

Figure 2. Description of the method used in this study.

Figure 2 shows the correlation obtained between the processes of turning and drilling, which was used in this proposed model. This radial segmentation seeks to facilitate, in drilling, a better definition and understanding of the drill local geometry along its radius and the influence that these geometric variations have on the machining force components, for further elaboration of an expanded Kienzle model through correction coefficients. Based on cutting and thrust forces from turning tests, this model adjusts the specific cutting and thrust forces, allowing the simulation of the machining forces in drilling. Indexable inserts with constant geometry, as shown in Figure 2b, were used to turn the segments.

For the drill main cutting edge, Figure 3a, the cutting and thrust forces in drilling are directly related to each segment (2, 3 and 4) in turning, representing the same cutting speed for both processes. However, to evaluate the chisel edge, Figure 3b, it is necessary to subtract the pre-drilling forces from the full drilling data. For the turning test, segment (3), used for the main cutting edge, is machined using a cutting speed corresponding to that of the chisel edge. This solution was adopted due to the impossibility of turning a cylinder, corresponding to the chisel edge, with the tool nose radius. Because of the eccentric cutting force, torsion and bending load will cause overload and break away the small cylinder.

As previously mentioned, the proposed model is based on observations of the geometric differences between turning and drilling tools. In turning, the use of indexable inserts with constant geometry implies that the angles remain fixed throughout the experiment. But, in the drilling process, the geometric and kinematic parameters of cutting vary along the cutting edge, which has consequences in terms of the component forces, as shown in Figure 4. In the case of the main cutting edge (GP), the rake angle γ_n (which is measured in the plane normal to the edge (P_n)) and the lateral lead angle λ_s vary as a function of the drill radius, and therefore they are directly related to the cutting speed. For the chisel edge (GT), because it is aligned with the center of the drill, the lateral lead angle is zero and the rake and clearance angles remain constant, making them easy to measure directly on the drill. Thus, the only variation is in the cutting speed and thus the effective clearance angle.

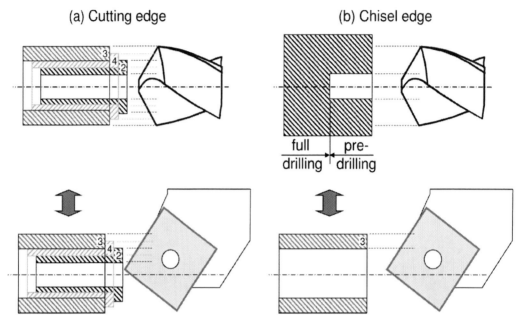

Figure 3. Workpiece segmentation in drilling and turning for the main cutting (a) and chisel (b) edges.

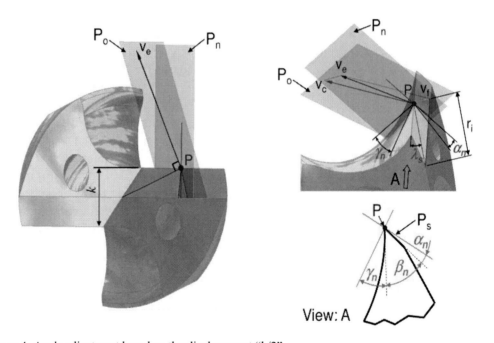

Figure 4. Angle adjustment based on the displacement "k/2".

The rake angle γ_n, measured in the plane normal to the main cutting edge (P_n), and the lateral lead angle λ_s, measured in the plane of the edge (P_s), are shown in Figure 5.

4. EXPERIMENTAL PROCEDURE

The drilling force model is based on steel Ck45 as standard workpiece material and a standard twist drill with diameter of 10 mm as a tool, both widely used in industry. The tests were done with twist drills without relief and coated with TiAlN. The turning tests were carried out with rhombic carbide inserts type CNMM-120408 with standard geometry, without chip breaker, coated with TiN, because for turning TiAlN coating are not available, and a rake angle of 6°, constant throughout the edge. The used tool holder resulted in a back rake angle (κ_r) of 93° and in a rake angle of zero (0°).

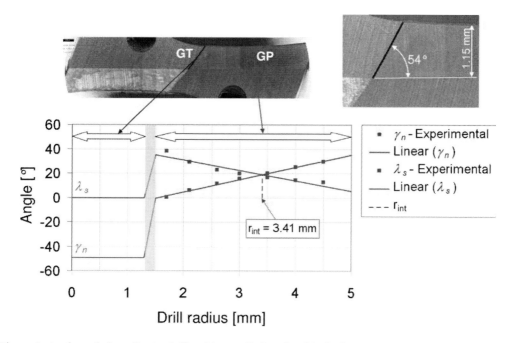

Figure 5. Angle variation of twist drills without relief on the chisel edge.

4.1. Tests Bench

Turning and drilling tests were performed on a Schaublin CNC machining center for turning and milling, model 42L, equipped with a test bench for turning and drilling (Figure 6). Both benches are equipped with Kistler force dynamometers, type 9121A5 for turning and type 9271A for drilling.

4.2. Cutting Parameters

The cutting parameters selected are based on conditions used in industry, where the cutting speed, for a carbide drill coated with TiAlN, was set at 120 m/min. The segment of largest diameter, but without influence of minor cutting edge of the drill, was selected as reference for the turning tests, together with the cutting speed for the chisel edge region and those of the chisel edge. For determining the parameters at the chisel edge a work-piece with a pre-drilled hole of 3 mm has been used because the chisel edge has a length of 2.8 mm.

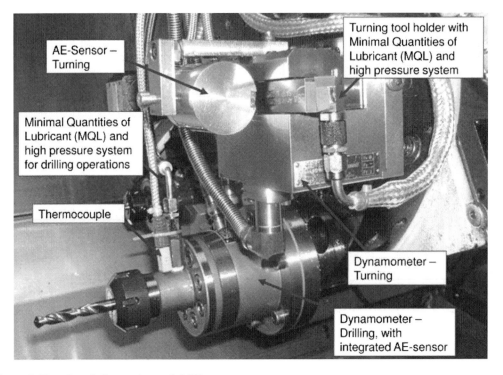

Figure 6. Tests bench for turning and drilling.

Table 1 shows the cutting parameters used in turning and drilling tests. All the experiments were performed without coolant.

Table 1. Cutting parameters of drilling and turning tests

Parameter	Turning	Drilling
Segments of cutting lip GP	D9	D5, D6, D7, D8, D9
Segments of chisel edge GT	D9	GP, FC
Cutting speed for GP [m/min]	108 (segment D9)	120
Cutting speed for GT [m/min]	24 (segment D9)	36
Feed rate [mm]	0.05; 0.1; 0.15; 0.2	0.1; 0.15; 0.2; 0.25

5. DRILLING FORCE MODEL

Kienzle model determines the specific force required to remove a predetermined section of material during its machining. The model proposed in this study simulates the thrust and cutting forces in drilling using as a basis the adaptation of specific forces originating from turning tests.

For the main cutting edge, turning and drilling tests are performed. The turning tests are performed for only one of the segments (usually that with the largest diameter), using the respective cutting speed and sweeping the feed rates in order to enable the calculation of the Kienzle equation. The values of specific cutting force $k_{c1.1}$ and the Kienzle coefficient $(1-m_c)$ thus determined are taken as the reference for the other segments that cover the main cutting edge.

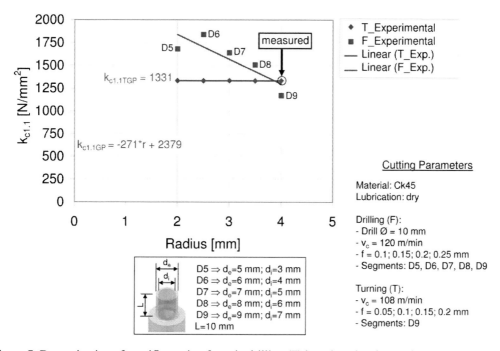

Figure 7. Determination of specific cutting force in drilling (F) based on that for turning (T).

In drilling, due to variations in the geometry and cutting speed as a function of the drill radius, machining tests were carried out on each segment with their respective cutting speeds. Figure 7 shows the results obtained for the specific cutting force at machining Ck45 steel, both in the drilling and turning of segments, for the main cutting edge. The turning tests were performed only under the conditions that represent segment D9, and this value of specific cutting force obtained was used as the reference for the other segments.

5.1. Modeling of Forces on the Main Cutting Lip

The adequacy of the application of the Kienzle equation obtained from turning to drilling is achieved through the application of correction factors that take into account the geometric variations between the cutting tools of the two processes. Depending on the geometric region of the drill (main cutting or chisel edge), the specific force is corrected for the variation of the tool orthogonal rake angle γ_n and the lateral lead angle λ_s. The same adjustment is made to the Kienzle coefficients $(1-m_c)$ and $(1-m_f)$ used in the model.

Based on the variation of the rake angle γ_n, as shown in Fig (5), a rotation of the line fitted to the values for the specific cutting forces associated with turning is performed, making it parallel to the line fitted to the points related to the drilling tests, as a function of the radius. After this rotation, a translation is performed using the values for the lateral lead angle, so that the fitted line for turning coincides with that for drilling. To make this displacement, the value of λ_s was determined at the point of intersection r_{int}, which corresponds to the radius at which the angles γ_n and λ_s have the same value.

Using Equation (6), by the application of correction coefficients $C_{c,GP(\gamma)}$ for γ_n and $C_{c,GP(\lambda)}$ for λ_s, the specific cutting force for drilling can be calculated from the specific cutting force of turning, as a function of the drill radius r_i.

$$k_{c1.1GP}(r) = k_{c1.1TGP} \cdot \{[1 + ((\gamma_{nT} - \gamma_n(r)) \cdot C_{c,GP(\gamma)})] + (\lambda_{rint} \cdot C_{c,GP(\lambda)})\} = A_c \cdot r + B_c \tag{6}$$

where:

$k_{c1.1GP}(r)$ - Specific cutting force adjusted as a function of drilling radius;

$k_{c1.1TGP}$ - Specific cutting force obtained from tests on orthogonal turning;

γ_{nT} - Normal rake angle used in orthogonal turning [°];

$\gamma_n(r)$ - Normal rake angle determined as a function of drill radius [°];

$C_{c,GP(\gamma)}$ - Rake angle adjustment factor for the specific cutting force;

λ_{rint} - Lateral lead angle for the point of intersection between $\gamma_{n(r)}$ and $\lambda_{s(r)}$ [°];

$C_{c,GP(\lambda)}$ - Lateral lead angle adjusting factor for the specific cutting force;

The same procedure is applied to the specific thrust force $k_{f1.1GT}$, Eq. (7), with its respective correction coefficients $C_{f,GP(\gamma)}$ and $C_{f,GP(\lambda)}$.

$$k_{f1.1GP}(r) = k_{f1.1TGP} \cdot \{[1 + ((\gamma_{nT} - \gamma_n(r)) \cdot C_{f,GP(\gamma)})] + (\lambda_{rint} \cdot C_{f,GP(\lambda)})\} = A_f \cdot r + B_f \tag{7}$$

The exponents of the Kienzle equation $(1-m_c)$, for the cutting force, and $(1-m_f)$, for the thrust force, are adjusted applying the same methodology used in Equations (6) and (7), respectively. The correction coefficients for the cutting direction are $C_{mc,GP(\gamma)} = 0.02$ and $C_{mc,GP(\lambda)} = 0.025$ and for the feed direction $C_{mf,GP(\gamma)} = 0.008$ and $C_{mf,GP(\lambda)} = 0.025$.

Figure 8 shows the fitting of the curves, based on the angles γ_n (short-dashed line) and λ_s (long-dashed line), which allows the determination of the specific cutting force for drilling based on data originating from turning tests. The curve equation obtained for the drilling case (long-dashed line) is then used to calculate the forces along the drill main cutting edge.

Because the specific force decreases as a function of the radius, the calculation of cutting and thrust forces is performed by integration of Equation (10) over the main cutting edge, as shown schematically in Figure 9. Based on the variation of specific cutting force $k_{c1.1GP}(r)$ as a function of the radius, according to Equation (8), and using an infinitesimal width for the cutting edge db, Equation (9), the cutting force in drilling $F_{cGP}(r)$ can be determined applying this expanded Kienzle equation, resulting in a force as a function of the radius, as shown in Equation (10).

$$k_{c1.1GP}(r) = A_c \cdot r + B_c \tag{8}$$

$$db = \frac{\sqrt{\dfrac{r^2}{r^2 - \left(k/2\right)^2}}}{\sin(\kappa_r)} \cdot dr \tag{9}$$

$$F_{cGP} = \int_{r_{GT}}^{r} k_{c1.1GP}(r) \cdot h^{(1-m_{cGP})(r)} \cdot \frac{\sqrt{\dfrac{r^2}{r^2 - \left(k/2\right)^2}}}{\sin(\kappa_r)} \cdot dr \tag{10}$$

where:
- db — Infinitesimal cutting depth [mm];
- dr — Infinitesimal cutting depth adjusted in radial direction [mm];
- κ_r — Side cutting edge angle [°];
- F_{cGP} — Cutting force related to cutting lip [N];
- r_{GT} — Radius of chisel edge [mm];
- k — Web Thickness [mm].

The same procedure is applied to determine the specific thrust force, resulting in a $C_{f,GP(\gamma)}$ = 0.04 and $C_{f,GP(\lambda)}$ = 0.013. In the case of the Kienzle exponents for the feed direction, the correction coefficients $C_{mf,GP(\gamma)}$ = 0.076 and $C_{mf,GP(\lambda)}$ = 0.039, respectively.

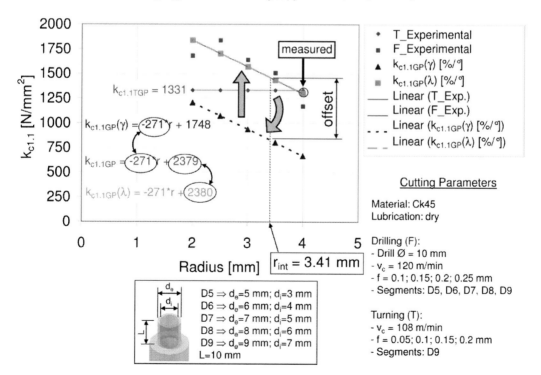

Figure 8. Determination of specific cutting force in drilling (F) based on that for turning (T).

5.2. Modeling of Forces on the Chisel Edge

Considering that this is a tool region where the cutting occurs at extremely negative orthogonal rake angles, as well as the fact that the cutting speed tends to zero, an understanding of the phenomena occurring on the chisel edge becomes more complex. According to Risse (2006), among others, the contribution of the chisel edge to the total drilling cutting force is small, what also can be explained by the small torque and torque radius, but for the thrust force, its contribution may reach high levels. These values are estimated to be between 65 and 75% of the total drill thrust force, while the main edge is responsible for 17 to 25% and the other parts of the drill for less than 10% of the force.

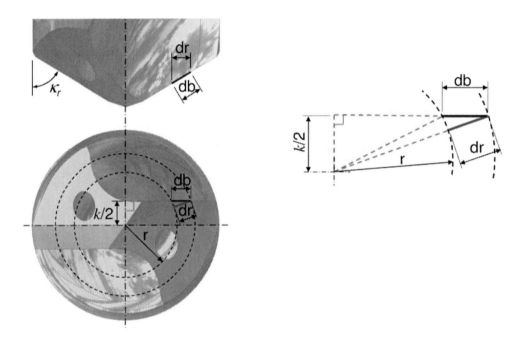

Figure 9. Integration of forces along the cutting lip of the drill.

There are many methods to calculate the forces at the chisel edge. Among them, it is possible to distinguish those based on the principles of deformation, yield, friction of the material and shear stress. Because in practical experiments chips formation at the chisel edge has been observed, the forces at the chisel edge are modeled according to the Kienzle model. Ohbuchi and Obikawa (2003), Fang (2004) observed in the machining of steel, even with rake angles as negative as -60°, chip formation too. In drilling processes, at the chisel edge region the drill has a rake angle in the order of -50°. As a result, the procedure adopted in this proposition to obtain the specific cutting and thrust forces at the chisel edge region is similar to that used for the main cutting edge. However, due to the big difficulties of segmentation in this region, the values obtained for the specific cutting and thrust forces, respectively, as well as those obtained for the exponents $(1-m_c)$ and $(1-m_f)$, for the turning and drilling, were directly correlated through the correction of the rake angle, using Eq. (6) and (7).

5.3. Simulation of Forces in Drilling

Based on the values of the specific cutting and thrust forces adjusted for application to the drilling process, it is possible to simulate the cutting and thrust forces by the integral for the main cutting edge as well as the direct application of the Kienzle equation for the chisel edge.

In the feed direction, the two (2) components of the thrust force F_{fFC} can be directly added, Eq. (11), as they have the same axial direction as the drill feed, Fig. (10).

The residual cutting force is calculated from Eq (13) as the force, that has equivalent torque effect, using the lever arms r_{AGP} for the main cutting edge, r_{AGT} for the chisel edge, r_{AGS} for the minor cutting edge and r_{AFC} for the full drilling. The length of the lever arms r_{AGP}, r_{AGT}, r_{AGS} and r_{AFC}, were determined for a Standard carbide twist drill, TiNAl coated, based on the geometric method presented by Spur (1960).

$$F_{fFC} = F_{fGP} + F_{fGT} + F_{fGS} \tag{11}$$

$$M_{zFC} = M_{zGP} + M_{zGT} + M_{zGS} \tag{12}$$

$$F_{cFC} = \frac{r_{AGP} \cdot F_{cGP} + r_{AGT} \cdot F_{cGT} + r_{AGS} \cdot F_{cGS}}{r_{AFC}} \tag{13}$$

Figure (11) compares the simulation of the cutting force corresponding to the cutting edge (GP) and the sum of main cutting edge (GP), chisel edge (GT) and the minor cutting edge (GS), which correspond to the full drilling (FC) of Ck45 steel.

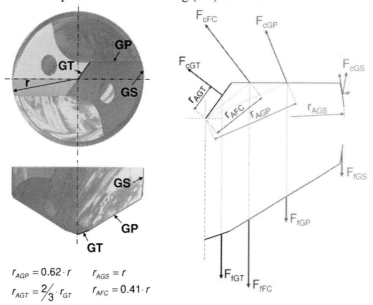

Figure 10. Schematic representation of the thrust forces on the main and secondary cutting and chisel edges.

The simulated values for the cutting force have a confidence interval of 95%. The simulated cutting force for the chisel edge (GT) corresponds to approximately 54% of the total force for the full drilling.

Similarly, Figure 12 shows the simulation of the thrust force corresponding to the full drilling (FC) of Ck45 steel.

It is possible to verify that the thrust force simulated for the chisel edge cumulates values up to 82% of the total value of the full drilling, as shown by the specific literature, validating, thus, the expanded Kienzle model.

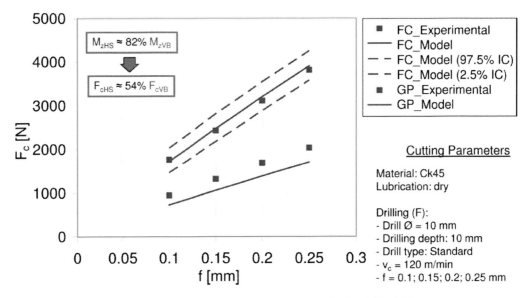

Figure 11. Cutting force for drilling Ck45 steel with 10 mm standard carbide drill.

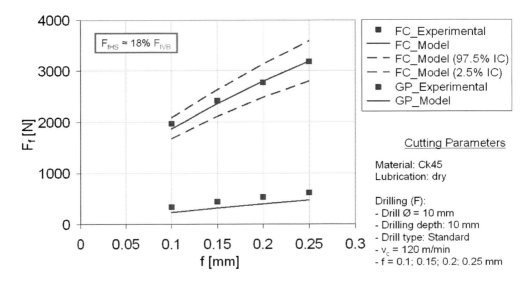

Figure 12. Thrust force for drilling Ck45 steel with 10 mm standard carbide drill.

CONCLUSION

This paper aimed to model drilling forces using the experimental data obtained through turning tests. For the main cutting edge, the adapted Kienzle model proved to be satisfactory, i.e., the values for the specific cutting and thrust forces corresponding to those for turning transformed into values for drilling, thus enabling the calculation of the thrust forces and torque of the drilling process. The values obtained, in percentage terms, are to be expected and are described in the literature. In the region of the chisel edge, the results obtained with the model were also shown to be satisfactory, but with a greater discrepancy in the simulated values when compared with the main cutting edge. This resulted in a confidence range of around ± 11% for both the cutting and thrust forces. A better adaptation of the geometric correction from turning, with a zero rake angle, to drilling, with an extremely negative rake angle, is required along with a better understanding of the chip formation mechanism.

The use of cylindrical segments is demonstrated to be efficient and easy to apply in both the drilling and turning tests. Thus, this could become a standard procedure to find the correction factors to calculate the drilling forces based on turning tests for different drill geometries, diameters, materials and coatings. Furthermore, with the segmentation along the main and chisel edges, the regions with greater mechanical forces were identified. With this information it is then possible to propose changes in the geometry of the drill tip to minimize these forces, for example, the application of special sharpening features for the chisel edge relief, the application of other materials for the cutting tool and new coatings. Another possibility that the model allows is the application of data taken directly from books and tables, from which the Kienzle coefficients can be obtained.

ACKNOWLEDGMENTS

The authors thank Schweizer Förderagentur für Innovation KTI as well as Swiss Steel AG, Blaser Swisslube AG, Sphinx Werkzeuge AG, Laubscher Präzision AG, Äschlimann AG and Fiber Optic PandP AG for financial support, and also CAPES for the granting of doctoral and post-doctoral scholarships.

REFERENCES

Bach, Fr. -W., Koehler, W., Schäperkötter, M. and Weinert, K., 2004, "Kontaktflächenanalyse beim Hochgeschwindigkeits-Bohren", VDI-Z - Integrierte Produktion, *Springer VDI Verlag*, pp. 30-32.

Bork, C. A. S., 1995, "Otimização de Variáveis de Processo para a Furação do Aço Inoxidável Austenítico DIN 1. 4541", *Mechanical Engineering Master's thesis*, Universidade Federal de Santa Catarina, Florianópolis, Brazil.

Campos, D. D. V., dezembro de 2004, "Análise Teórico-experimental da Deflexão de Ferramentas no Fresamento de Topo", *Mechanical Engineering Master's thesis*, Universidade Federal de Santa Catarina, Florianópolis, Brazil.

Castillo, W. J. G., 2005, "Furação Profunda do Ferro Fundido Cinzento GG25 com Brocas de Metal-duro com Canais Retos", *Mechanical Engineering Master's thesis,* Universidade Federal de Santa Catarina, Florianópolis, Brazil, 134 p.

Chen, W., Fuh, K., Wu, C. and Chang, B., 1996, "Design optimization of a split-point drill by force analysis", *Journal of Materials Processing Technology*, Vol. 58, Issue 2, pp. 314–322.

Fang, N., February 2005, "Tool-chip friction in machining with a large negative rake angle tool", *Wear*, Vol. 258, Issues 5-6, pp. 890-897.

Grote, K. H. and Feldhusen, J., 2005 "Dubbel - Taschenbuch für den Maschinenbau", 22. *Auflage*. Springer-Verlag Berlin Heidelberg, Berlin, Germany, 1856 p.

Hsieh, J. and Lin, P. D., 2002, "Mathematical Model of Multiflute Drill Point", *International Journal of Machine Tools and Manufacture,* Nr. 42, pp. 1181-1193.

Kim, K. W. and Ahn, T. K., 2005, "Force Prediction and Stress Analysis of a Twist Drill from Tool Geometry and Cutting Conditions", *International Journal of Precision Engineering and Manufacturing*, Vol. 6, No. 1, pp. 65–72.

König, W. and Klocke, F., 2002, "Fertigungsverfahren 1: Drehen, Fräsen, Bohren", 7. *korrigierte Auflage,* Springer-Verlag Berlin Heidelberg, Aachen, Germany, 471 p.

Ohbuchi, Y. and Obikawa, T., July 2003, "Finite Element Modeling of Chip Formation in the Domain of Negative Rake Angle", *Cutting Journal of Engineering Materials and Technology,* Vol. 125, Issue 3, pp. 324-332.

Paucksch, E., 1992, "Zerspantechnik 11., überarbeitete Auflage", Viewegs Fachbücher der Technik, Braunschweig, Germany, 404 p.

Paul, A., Kapoor, S. G., DeVor, R. E., 2005, "Chisel Edge and Cutting Lip Shape Optimization for Improved Twist Drill Point Design", International Journal of Machine Tools and Manufacture, Vol. 45, pp. 421-431.

Risse, K., May 2006, "Einflüsse von Werkzeugdurchmesser und Schneidkantenverrundung beim Bohren mit Wendelbohrern in Stahl", *Doctoral thesis*, RWTH, Aachen, Germany, 137 p.

Rocha, A. S., 1984, "Determinação de um Modelo de Força de Usinagem para a Furação a Partir do Modelo de Força de Usinagem no Torneamento", *Mechanical Engineering Master's thesis,* Universidade Federal de Santa Catarina, Florianópolis, Brazil, 117 p.

Spur, G., 1960, "Beitrag zur Schnittkraftmessung beim Bohren mit Spiralbohrern unter Berücksichtigung der Radialkräfte", *Doctoral thesis,* Technische Hochschule Braunschweig, Braunschweig, Germany, 132 p.

Stephenson, D. A. and Agapiou, J. S., 1992, "Calculation of Main Cutting Edge Forces and Torque for Drills with Arbitrary Point Geometries", *International Journal of Machine Tools and Manufacture*, Vol. 32, No. 4, pp. 521-538.

Witte, L., 1980, "Spezifische Zerspankräfte beim Drehen und Bohren", *Doctoral thesis,* RWTH, Aachen, Germany, 143 p.

In: Machining and Forming Technologies, Volume 3
Editor: J. Paulo Davim

ISBN: 978-1-61324-787-7
© 2012 Nova Science Publishers, Inc.

Chapter 2

DIMENSIONAL ACCURACY OF END MILLED SLOTS IN 7075–T7 ALUMINIUM ALLOY

*Mauro Paipa Suarez[1], Eder Silva Costa[1], Álisson Rocha Machado[1] and Alexandre Mendes Abrão[2]**

[1]Universidade Federal de Uberlândia, Faculdade de Engenharia Mecânica,
Av. João Naves de Ávila, 2.121, Uberlândia-MG, 38.408-100, Brazil
[2]Universidade Federal de Minas Gerais,
Departamento de Engenharia Mecânica, Av.Antônio Carlos,
Belo Horizonte-MG, Brazil

ABSTRACT

Heat generated during machining is dissipated through the components involved in the process (chip, tool, workpiece and environment) and it is widely known that the percentage distribution depends on the materials involved as well as on the cutting conditions. The relationship between the heat generated and the heat dissipated in machining is known as thermal energy balance. If the amount of heat dissipated through the workpiece is reduced, thermal expansion of the component will also be diminished and tighter tolerances will be achieved. The aim of this work is to assess the influence of the cutting parameters (cutting speed, feed rate, depth of cut, distance between the slots and cutting fluid) on the dimensional accuracy of slots produced by end milling 7075–T7 aluminium-zinc alloy with uncoated cemented carbide tools. In order to verify the influence of the cutting parameters on the workpiece temperature, an infrared sensor was used when dry cutting. The machining tests followed a design of experiment approach and the results were evaluated using the analysis of variance with a significance level of 5%. The results indicated that tighter tolerances were obtained using cutting fluid (either minimal quantity lubrication or flooding). Increasing cutting speed, feed rate and depth of cut resulted in wider tolerances. The use of the Surface Response Method allowed the identification of the cutting conditions responsible for the highest temperature at the bottom surface of the slots, which represented the worst condition when the smallest

---

* E-mail: juniorpaipa@hotmail.com

thermal expansion of the workpiece and, consequently, the tightest dimensional accuracy are targeted.

Keywords: End-milling; Aeronautic aluminium; Thermal energy balance; Dimensional accuracy; Cutting fluid

INTRODUCTION

Dimensional accuracy and surface roughness are key elements for a competitive production due to the tight tolerances demanded in high technology application (Carrino et al., 2002). This is particularly critical when human lives are involved. Among the available alternatives for the milling process, end milling is probably one of the most used in the mould and die industry. Additionally, end mill cutters are largely employed in the manufacture of aircraft structural parts. The use of end mills has increased in the last decades owing to their ability to replace previously employed machining operation providing higher productivity (Drozda and Wick, 1983).

In general, aluminium alloys present excellent machinability with respect to tool life, surface finish, cutting forces and temperature, unless abrasive reinforcing materials such as particulate silicon carbide are added. Nevertheless, when machinability is assessed in terms of dimensional accuracy, aluminium alloys present poor behaviour. Being of particular interest to the present work, 7075-T7 aluminium-zinc alloy is widely used in the aircraft and aerospace industries due to its superior strength in addition to its good corrosion resistance and ductility.

In metal cutting almost all the consumed energy is converted into heat, which is generated by plastic deformation at the primary and secondary shear zones and, to a lesser extent, friction between chip and tool at the rake face and between tool and machined surface at the clearance face. Heat generation is affected by the cutting parameters. Heat dissipation takes place through the chip, cutting tool, workpiece and environment (air or cutting fluid). Longbottom and Lanham (2005) present a survey with the estimates of various authors for the amount of heat dissipated through the chip, tool and workpiece. In general, between 75 and 90% of the heat is dissipated through the chip, 5-20% trough the cutting tool and 5-10% through the workpiece.

The relationship between the amount of heat generated at the shear zones and the heat dissipated is known as thermal energy balance and the fraction of heat dissipated into the workpiece is responsible for its temperature elevation and, to some extent, for the dimensional accuracy of the machined part. In order to increase the accuracy of the machined component the amount of heat dissipated through the workpiece should be minimized.

High precision, powerful and stiff machine tools are essential for the production of components with high dimensional accuracy. Furthermore, the tool and workpiece clamping systems must be equally rigid and the blank must not present internal defects. Finally, cutting tools dimensionally accurate and made of a material that can withstand the severe conditions imposed during the operation are essential for the success of the operation.

Even when these requirements are fulfilled, machining aluminium alloys offers difficulties associated with their high thermal expansion coefficient (approximately 23×10^{-6} $1/^\circ$C against 11×10^{-6} $1/^\circ$C for carbon steels). When the workpiece is heated as a consequence of chip formation (shear and friction), dimensional variation takes places due to the thermal expansion, resulting in overcutting. In order to compensate such overcut, modelling can be used to calculate the correct cutting geometry, nevertheless, this is a time consuming and costly task owing to the large number of variables involved and the lack of knowledge on the thermal behaviour of the machine-tool-workpiece system. In contrast, when milling quenched and tempered steels, temperature elevation may cause the annealing of the layers below the cutter path, thus impairing the mechanical properties of the component.

Stephenson and Ali (apud Lazoglu and Altintas, 2002) compared the temperature behaviour when interrupted cutting an aluminium alloy with tungsten carbide inserts using experimental and numerical techniques. A steady state temperature of 255°C was found for the experimental work, while the temperature obtained numerically was 250°C.

When pocket milling, the cutter can dwell at the pocket corner as the tool changes its direction, promoting higher temperatures in comparison with straight cuts, as well as longer contact times that allow heat transfer from the tool to the workpiece at dwell points (Longbottom and Lanham, 2005). In order to minimize deflection error in corner sections, Law et al. (1999) recommends the diagonal cutting method (diagonal cuts followed by straight slots on the sides) to be used instead of the conventional inside-out spiral tool path.

Richardson et al. (2006) proposed a model to predict the workpiece temperature in peripheral milling of an aerospace aluminium alloy. The findings indicated that the workpiece temperature is drastically affected by the machining conditions. For instance, increasing cutting speed from 300 to 3000 m/min reduces the proportion of energy heating the workpiece by a factor of 6, while increasing feed rate from 0.1 to 0.3 mm/rev/tooth reduces the proportion of heat conducted into the workpiece by 64%. Increasing the radial depth of cut can also reduce the amount of heat conducted to workpiece.

Dewes et al. (1999) report that the maximum cutting temperature is never superior to the melting point of the work material, consequently there is no limit to the cutting speed when high speed machining aluminium alloys owing to the fact that the melting point of these alloys is approximately 660°C, which is below the temperature required to make cemented carbide and ceramic tools to loose their strength and wear resistance. In contrast, when cutting ferrous alloys and other high melting point metals, it is necessary to control cutting speed accordingly to the tool material employed.

According to Budak and Altintas (1994), form errors are generated in the presence of static deflections, while dynamic movements are responsible for the surface roughness quality. In order to improve the quality of the machined surface, in general the material removal rate is reduced, abdicating of the power and torque available in the machine tool.

A cutting strategy which maintains cutting forces stable is critical for the production of moulds and dies with a uniform quality due to the fact that at a constant average cutting force, the static cutter deflection remains unaltered (Toh, 2005). Furthermore, Kim et al. (2003) investigated tool deflection and surface form error of a half cylinder workpiece when ball end milling and noticed that the form error in upward ramping is larger than in downward ramping, due to the fact that the length of the engaged edge of the cutter is larger in the former strategy. On the other hand, Ratchev et al. (2004) propose a flexible force model that takes into account part deflection when end milling low-rigidity components.

The cutting tool slenderness ratio, i.e., the ratio between tool diameter and length is crucial in order to assure an acceptable quality to the machined surface when finish cutting

(López de Lacalle et al., 2002). High slenderness ratio values lead to tool vibration, cutting edge degradation and poor surface finish.

The present work aims to study the influence of the machining parameters on the dimensional accuracy of slots produced in 7075-T7 aeronautic aluminium alloy under various cutting conditions. Workpiece temperature was monitored in order to better understand its influence on the dimensional tolerancing of the slots.

Experimental Procedure

End milling tests were conducted on bars of 7075–T7 aeronautic aluminium (Al-Zn alloy) with average hardness of 150 HBN using uncoated cemented carbide milling cutters. The bars presented square section with the following dimensions: 400x100x100 mm. The alloying elements added are expected to improve mechanical strength and machinability compared with commercially pure aluminium, promoting chip breakability under severe cutting conditions without galling and adhesion of the material onto the milling cutter surfaces. Owing to its application, particularly in the aircraft and aerospace industries, research on machining of this alloy has been frequently developed, especially focused on high speed machining (Balkrishna and Yung, 2001). The 7075-T7 alloy is an overaged alloy, obtained through precipitation followed by solution heat treatment and quenching (Davis, 1993).

Prior to the machining trials the bars were face milled. Figure 1 shows a schematic diagram of the distribution of the slots. Three tests were conducted for each cutting condition. The distance between the slots (d) is also depicted in Figure 1.

The tools used were ISO grade K10 cemented carbide milling cutter with two teeth manufactured by OSG-Tungaloy, see Figure 2a. The nose radius was 0.5 mm and the geometry indicated for the machining of aluminium alloys. Although the nominal diameter of the cutter is 10 mm, the actual average diameter was 9.942 mm. The diameter is an important feature due to its effect on the width of the slots. Figure 2b presents a photograph of the rake face at the end of the experimental programme and shows that tool wear was negligible.

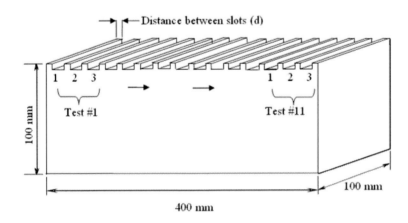

Figure 1. Schematic diagram of the workpiece and the sequence of tests.

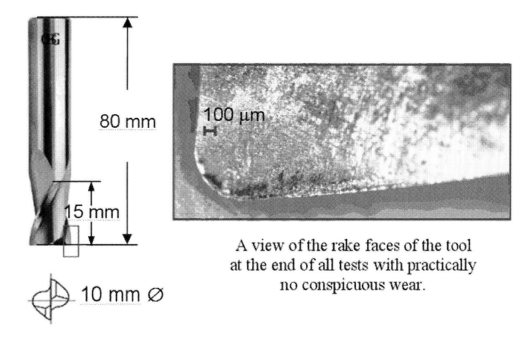

Figure 2. (a) K10 cemented carbide end mill cutter and (b) Rake face of the cutter.

A Tesa Micro Hite coordinate measuring machine (CMM) with resolution of 1 μm was used to determine the width of the machined slots, see Figure 3. A ruby probe with 2 mm diameter was used (Figure 3b). The smallest depth of cut used in the experimental work was 2 mm to match the probe diameter. The value of the slot width considered for analysis for each cutting condition tested was the average of 15 widths (5 equidistant measurements for each of the 3 slots generated under the same cutting condition).

 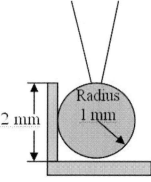

Figure 3. (a) Coordinate measuring machine and (b) Probe used in the CMM.

A Raytek Thermalert TX infrared camera was used to measure the temperature at the bottom surface of the workpiece (see Figure 4). A clamping system was devised in order to keep the distance between the camera and measuring spot constant at 76 mm. The target diameter was 2 mm. The input parameters (factors) studied were cutting speed, feed rate, depth of cut, distance between the slots and cutting fluid (dry, minimum quantity lubrication and flooding) and the output parameter was average slot width. The tests followed a fractional factorial design and the levels for each factor are presented in Table 3. Tests conducted under minimum quantity lubrication (MQL) used vegetable neat oil (Vascomill) supplied in an air stream with 4 bar pressure, whereas tests under flooding were carried out with vegetal based soluble oil (Vasco) diluted in water in a concentration of 9%. Both fluids were supplied by Blaser Swisslube.

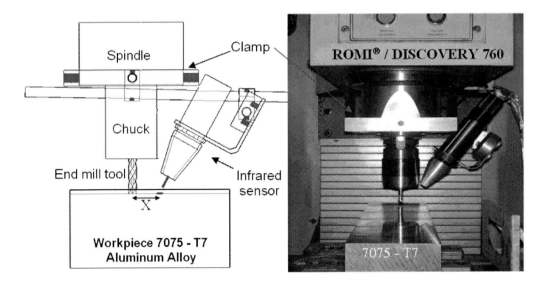

Figure 4. Setup for the temperature measurements.

Temperature measurement tests were performed under dry machining using the cutting conditions indicated in Table 4. A full factorial design was used, thus resulting in 40 tests. Each test was repeated three times and the average temperature was calculated from the maximum value recorded during each test.

Table 3. Cutting conditions used for slot width measurement

Factor	Level		Unit
	-1	+1	
Cutting speed (v_c)	70	265	[m/min]
Feed rate (f)	0.05	0.12	[mm/rev]
Depth of cut (a_p)	2	4	[mm]
Distance between the slots (d)	1.5	2.5	[mm]
Cutting fluid	Dry	MQL Flooding	20 [ml/h] 1200 [l/h]

Table 4. Cutting conditions for temperature measurement

Factor	Level					Unit
	1	2	3	4	5	
Cutting speed (v_c)	10	86	162	238	314	[m/min]
Feed rate (f)	0.01	0.055	0.1	0.145		[mm/rev]
Depth of cut (a_p)	2	4				[mm]

RESULTS AND DISCUSSION

The analysis of variance for slot width indicated that cutting fluid was the only factor which statistically affected the dimensional accuracy of the slots produced with a significance level of 5%, i.e., using either MQL or flooding resulted in slot width values closer to the cutter diameter in comparison to dry milling.

Table 5. Influence of factors variation on slot width

Factor variation	Level -1 Slot width (mm)	Level +1 Slot width (mm)	Percentage variation
v_c=70 → 265 m/min	9.983	9.999	↑ 19.2%
f=0.05 → 0.12 mm/rev	9.985	9.988	↑ 3.5%
a_p=2 → 4 mm	9.986	9.987	↑ 1.2%
d=1.5 → 2.5 mm	9.984	9.988	↑ 4.8%
Cutting fluid			
Dry → MQF at 20 ml/h	9.989	9.983	↓ 6.7%
Dry → Flooding at 1200 l/h	9.989	9.984	↓ 5.6%

Table 5 presents the influence of each factor on the average slot width. In addition to that, the percentage variation between the slot width and the end mill cutter diameter (9.942 mm) is given. The arrows ↑ and ↓ indicate, respectively, whether slot width is farther or closer to the actual cutter diameter when each factor is changed from the lower level (-1) to the upper level (+1).

Table 5 shows that, except for cutting fluid, an increase in any parameter results in wider dimensional deviation from the cutter diameter, worst results being found for cutting speed, which elevation from 70 to 265 m/min resulted in the largest slot width (9.999 mm). The reason for that may reside in the fact that an increase in cutting speed will promote temperature elevation at the chip-tool interface together with the reduction in the chip thickness. As a consequence, the amount of heat conducted through the chip is reduced, increasing the workpiece temperature and its thermal expansion. Additionally, tool vibration due to the increase in spindle speed may have contributed to a wider deviation.

Increasing feed rate and depth of cut resulted in wider sloth width values, albeit the effect was not as evident as that observed for cutting speed (percentage variations of 3.5 and 1.2%, respectively). The elevation of feed rate results, on the one hand, in higher plastic strain, which elevates the temperature in the cutting zone. On the other hand, however, higher feed rate means higher shear plane area, thus allowing more heat to be conducted through the chip and promoting a milder increase in the workpiece temperature. In addition to that, friction between the tool and workpiece is reduced as feed rate is elevated, thus diminishing the influence of temperature on the workpiece dimensional deviation.

A similar analysis can be carried out for the effect of depth of cut, nevertheless, owing to the fact that the ability of the cutting fluid to penetrate in the cutting zone is reduced when depth of cut is elevated, one would expect a more pronounced influence of depth of cut on the width deviation, which was not observed.

Surprisingly, increasing the distance between the slots from 1.5 to 2.5 mm resulted in higher dimensional deviation (4.8%). Theoretically, increasing the cross section of the workpiece results in higher volume of material available to dissipate heat, however, thinner walls produced smaller deviation than larger ones.

As previously mentioned, cutting fluid was the only factor which significantly affect the dimensional deviation of the slot width, i.e., tighter dimensional tolerancing was obtained using either minimum quantity lubrication (6.7%) or flood cooling (5.6%) in comparison with dry machining. These findings suggest that both cooling systems are effective in transferring heat away from the cutting zone and, consequently, in reducing the dimensional deviation.

Figure 5 shows the influence of the cooling system on slot width. When MQL is applied, see Figure 5a, the difference between slot width and cutter diameter was reduced in 6μm (from 32 μm when dry cutting to 26 μm with MQL). Figure 5b shows that a smaller difference was obtained when milling under flood cooling (3μm), nevertheless, MQL provided slot width closer to the cutter diameter. The dashed areas in Figure 5 represent the tool radial run-out measured with a dial gauge.

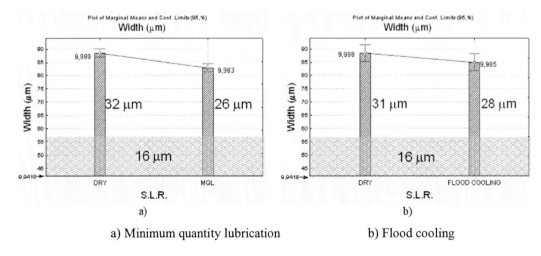

Figure 5. Effect of cooling system on slot width: a) Minimum quantity lubrication and b) Flood cooling.

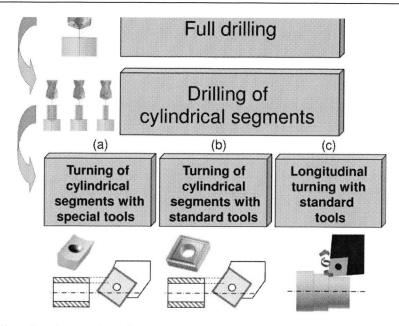

Figure 6. Effect of cutting speed and feed rate on the temperature at the bottom surface of the slot.

The results suggest that the mist produced by the MQL system is more effective in penetrating the chip-tool interface, acting as lubricant and thus reducing cutting forces. The vegetable based oils used are claimed to present higher lubricant action when compared to mineral base oils due to their composition. While the mineral oil possesses only hydrocarbons in its formulation, the vegetable oil has got functional groups with oxygen. This element provides small polar charges which work similarly to magnets and make the vegetable oil lubricant effective and capable of bearing heavier loads (Woods, 2005).

After conducting the temperature measurement tests indicated in Table 4, the Surface Response Method was employed in order to determine the cutting parameters which are responsible for highest temperature and, consequently, adversely affect the dimensional accuracy due to the workpiece thermal expansion. The influence of cutting speed and feed rate on the temperature at the bottom of the slot is presented in Figure 6. It can be noticed that the highest temperature is recorded when dry end milling with a cutting speed of 265 m/min and a feed rate of 0.06 mm/rev.

These results are in agreement with those given in Table 5, i.e., an elevation in cutting speed results in a steep increase in cutting temperature, thus leading to wider slot widths due to the thermal expansion of the work material. Nevertheless, increasing feed rate does not result in a drastic increase in temperature, especially when higher cutting speed are used, probably due to the reduction in friction between the cutter teeth and workpiece and the increase in the shear plane area, which enhances heat conduction through the chip and, therefore, reduces the temperature at the bottom surface of the workpiece.

Figure 7 shows the slot temperature against the time required to machine a complete slot (100 mm long) under different cutting conditions using each depth of cut indicated in Table 4. The results indicate that, in general, higher slot temperatures are recorded when milling with the highest depth of cut and that the difference between the temperatures obtained at depths of cut of 2 and 4 mm varies with cutting speed and feed rate, the largest gradient being observed when milling at v_c=162 m/min and f=0.055 mm/rev.

These findings suggest that attaching the infrared camera to the machine tool spindle is not recommended when the effects of cutting speed and feed rate on temperature are investigated owing to the fact that they result in distinct feed speeds. As a consequence, the temperature at the bottom surface of the slot is obtained after different periods, which may bias the results, especially in the case of work materials possessing high thermal conductivity values.

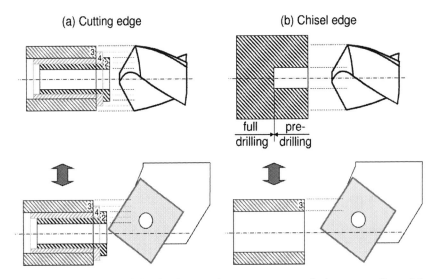

Figure 7. Effect of cutting time and depth of cut on the temperature at the bottom surface of the slots.

CONCLUSION

After end milling 7075-T7 aluminium alloy with a plain tungsten carbide cutter, the following conclusions can be drawn:

- Cutting fluid applied either as minimum quantity lubrication or flood cooling was the only factor which statistically affected the dimensional accuracy when compared with dry milling. A reduction in the dimensional deviation of 6.7% was obtained when minimum quantity lubrication was used, whereas flooding allowed a reduction in the slot width deviation of 5.6%. Cutting speed, feed rate, depth of cut and the distance between the slots were not relevant at a significance level of 5%.
- Increasing cutting speed resulted in wider dimensional deviation, probably due to both the elevation in the cutting temperature and the reduction in the cross section area of the swarf, which makes heat conduction through the chip more difficult and promotes the thermal expansion of the workpiece.
- The effect of increasing feed rate and depth of cut on the elevation of the slot width deviation was marginal, probably owing to both the facts that the cutting temperature did not increase as drastically as when cutting speed was elevated and that an elevation in these factors resulted in larger shear plane areas. The larger the shear plane area, the higher the amount of heat flowing through the chip and the lower the workpiece temperature. Consequently, tighter tolerances are obtained.

- The temperature measured at the bottom surface of the slot using an infrared camera indicated that cutting speed is the principal factor influencing temperature, however, instead of measuring temperature at a fixed distance from the cutter, the measurement should have been conducted at a constant period when distinct cutting speeds and feed rates are employed.

ACKNOWLEDGMENTS

The authors are grateful to CAPES, CNPq and FAPEMIG for financial support, to the Universities of Uberlândia and Minas Gerais for providing laboratory facilities and to EMBRAER, Blaser Swisslube do Brasil Ltda. and OSG-Tungaloy Sulamericana de Ferramentas Ltda. for providing consumables.

REFERENCES

Balkrishna, R. and Yung, C. S. (2001), "Analysis on high-speed face-milling of 7075-t6 aluminum using carbide and diamond cutters", *International Journal of Machine Tools & Manufacture,* 41, 1763–1781.

Budak, E. and Altintas, Y. (1994), "Perpheral milling conditions for improved dimensional accuracy", *International Journal of Machine Tools & Manufacture,* 34(7), 907-918.

Carrino, L., Giorleo, G., Polini, W. and Prisco, U. (2002), "Dimensional errors in longitudinal turning based on the unified generalized mechanics of cutting - Part I: Three-dimensional theory", *International Journal of Machine Tools & Manufacture* 42, 1509–1515.

Davis, J. R. (1993*), "Aluminum and aluminum alloys",* Davis & Associates, ASM International. Handbook Committee, USA, 784 p.

Dewes, R. C. Ng, E., Chua, K. S., Newton, P. G. and Aspinwall, D. K. (1999), "Temperature measurement when high speed machining hardened mould/die steel", *Journal of Materials Processing Technology,* 92-93, 293-301.

Drozda, T. J. and Wick, C. (1983), "Tool and Manufacturing Engineers Handbook – Machining", *Dearborn: Society of Manufacturing Engineers,* 4(1), 1. 1-1. 66, 10. 1-10. 76.

Kim, G. M., Kim, B. H. and Chu, C. N. (2003), "Estimation of cutter deflection and form error in ball-end milling processes", *International Journal of Machine Tools & Manufacture,* 43, 917–924.

Law, K. M. Y., Geddam, A. and Ostafiev, V. A. (1999), "A process-design approach to error compensation in the end milling of pockets", *Journal of Materials Processing Technology*, 89–90, 238–244.

Lazoglu, I. and Altintas, Y. (2002), "Prediction of tool and chip temperature in continuous and interrupted machining", *International Journal of Machine Tools & Manufacture,* 42, 1011–1022.

López de Lacalle, L. N., Lamikiz, A., Sánchez, J. A. and Arana, J. L. (2002), "Improving the surface finish in high speed milling of stamping dies", *Journal of Materials Processing Technology,* 123, 292–302.

Ratchev, S., Liu, S., Huang, W. and Becker, A. A. (2004), "Milling error prediction and compensation in machining of low-rigidity parts", *International Journal of Machine Tools & Manufacture*, 44(15), 1629-1641.

Richardson, D. J., Keavey, M. A. and Dailami, F. (2006), "Modelling of cutting induced workpiece temperatures for dry milling", *International Journal of Machine Tools & Manufacture* 46, 1139–1145.

Toh, C. K. (2005), "Design, evaluation and optimisation of cutter path strategies when high speed machining hardened mould and die materials", *Materials and Design* (26), 517–533.

Woods, S. (2005), "Going green", *Cutting Tool Engineering Magazine*, 57(2), 11-13.

In: Machining and Forming Technologies, Volume 3
Editor: J. Paulo Davim

ISBN: 978-1-61324-787-7
© 2012 Nova Science Publishers, Inc.

Chapter 3

EXPERIMENTAL EVALUATION AND NUMERICAL VALIDATION OF SURFACE RESIDUAL STRESSES AFTER HARD TURNING OF MECHANICAL COMPONENTS MADE OF DIN 21NiCrMo2 (AISI 8620) CARBURIZED STEEL

Adalto de Farias[1], Gustavo Henrique B. Donato[2,], Sergio Delijaicov[2] and Gilmar Ferreira Batalha[1]*

[1]Department of Mechatronics and Mechanical Systems Engineering,
(PMR-EPUSP), Manufacturing Engineering Laboratory,
University of São Paulo, São Paulo, SP 05508-900 – Brazil
[2]Department of Mechanical Engineering, Ignatian Educational
Foundation (FEI). São Bernardo do Campo, SP 09850-901 – Brazil

ABSTRACT

Residual stresses are commonly found in machined components and represent a key issue for performance and integrity assessment of mechanical critical components. The aim of the present work is to evaluate the residual stresses resulting from the hard turning process of mechanical components made of DIN 21 NiCrMo 2 (AISI 8620) case hardened steel (58-62 HRC). An experimental evaluation of surface residual stress fields for six samples was conducted based on the incremental hole drilling method, in which a small hole is machined on the surface of the component and the strain relaxation is quantified through special strain gage rosettes. In order to validate the experimental procedure and obtain confidence in the measured residual stresses, the results served as a basis for the development of refined finite element models representing the stressed area found in the real samples. The experimental residual stress measurements revealed a compressive stress state in all the analyzed samples and the comparison of results with the numerical predictions showed an excellent agreement between measured and predicted relieved strains. In addition to that, both methods presented the same trends of

---

[*] E-mail: gdonato@fei.edu.br

30 Adalto de Farias, Gustavo Henrique B. Donato, Sergio Delijaicov et al.

residual stresses as a function of depth from the surface, thus providing confidence in the results and encouraging the application of current studies to additional cases of machining processes and manufacturing control.

Keywords: residual stress, incremental hole drilling method, finite element method

1. INTRODUCTION

Often, as a result of mechanical processing, the component surface can be left with a residual stress state after conventional machining operations. Although limited to a surface layer, such stress state can have a significant influence on the component performance and lifetime. Generally, compressive residual stresses are preferred, as they tend to reduce the effect of external applied tension fields, tending to close (or to reduce the opening tendency of) surface cracks (Griffiths, 2001). Furthermore, the reduction in the equivalent stress fields strongly favors fatigue life of components subjected to cyclic loads (Schwach and Guo, 2005). This beneficial effect of compressive stresses on cyclically loaded mechanical components is a common sense for many authors who have investigated issues related to residual stress conditions and component performance (Toenshoff, 2000; Delijaicov, 2004; Abrão, 2005 and Umbrello *et al.*, 2007).

These residual stresses are produced when a component undergoes localized plastic deformation or non-homogeneous elastic deformation. They can be classified as macro or micro-residual stresses, depending on the scale at which they are distributed. Their effects may be beneficial or detrimental to the component, depending on their nature (tension or compression), magnitude and distribution. The most common application of residual stresses is to improve the performance of materials against the demands of the external environment and to reduce the occurrence of fatigue failures by means of compressive stress fields, as already mentioned (Guo and Barkey, 2004; Schwach and Guo, 2005 and Umbrello *et al.*, 2007). However, in manufacturing processes, residual stress imposition may also generate component bending and distortion, making it necessary to introduce a further finishing process, thus raising leading times and processing costs.

Therefore, in order to achieve a relevant improvement in the design, quality control and performance of mechanical components, it is necessary to incorporate more information about the residual stresses on manufacturing practices and to develop reliable methods for its practical determination. Dahlman et al. (2004) presented some results which indicate that the control of the parameters involved in the hard turning process allows controlling (and predicting) the levels and distribution of the residual stresses during the component manufacturing.

The aim of this study is to experimentally evaluate the distribution and levels of residual stresses resulting from hard turning of mechanical components made of DIN 21 NiCrMo 2 (ABNT 8620) case hardened steel (with 700HV, 58-62HRC). To achieve these goals, an experimental evaluation of superficial residual stress fields for six samples was carried out based on the incremental hole drilling method, in which a small hole is machined on the surface of the component and the strain relaxation is quantified through special strain gage rosettes. In order to validate the experimental procedure and obtain confidence in the measured residual stresses, the results served as a basis for the development of refined finite element models representing the stressed area found in the real samples.

2. RESIDUAL STRESS STATE DETERMINATION BY THE BLIND HOLE METHOD

Currently, the blind hole technique is considered as an important methodology for determining residual stress levels and distribution in layers near the surface of mechanical components. The experimental procedure is based on standardized practices, such as the ASTM E837 (ASTM, 2008), and consists of making a small incremental hole (1 to 5 mm diameter) on the tested material surface. The hole generates a redistribution of stresses due to stressed material removal and the respective strain relaxation can be measured by using special strain gage rosettes. These strain gages are usually electrical resistance rosettes comprehending three gages placed around the area to be drilled. For each incremental hole depth, strains are measured by the extensometers and, by using numerically calibrated coefficients (recommended by the ASTM E837), the residual stresses can be calculated, as will be described in the sequence.

A restriction on the application of this technique concerns the magnitude of measured residual stresses, since the effect of the notch imposed by the hole leads to stress concentration and could strain harden the root of the hole. As a biaxial state of stress around the hole is assumed, the material strain hardening sensitivity at the root of the notch is in the range of $0.6 \cdot \sigma_{ys-L}$ to $0.6 \cdot \sigma_{ys-L}$ (Nobre *et al.*, 2000), where σ_{ys-L} represents the localized current yield stress of the material. This localized yield stress is higher than the uniaxial tensile value due to plasticity and strain hardening effects near the hole root. Consequently, depending on the sensitivity of the material to strain hardening, the localized current yield stress can be significantly increased.

Nobre *et al.* (2006), proposed a criterion for calculating the localized current yield stress (σ_{ys-L}), based on Vickers micro indentation hardness profiles from the processed surface in the form:

$$\sigma_{ys-L} = \sigma_{ys} \cdot \left[1 + \alpha \, \frac{\Delta HV}{HV_0} \right] \tag{1}$$

where σ_{ys} is the material original uniaxial yield stress, ΔHV is the Vickers micro indentation hardness variation ($HV-HV_0$) across the thickness, HV_0 is the bulk material hardness and α is the factor that considers the plastic effects of the Vickers micro indentation probe.

A common standard for profile measuring of non-uniform residual stresses along the thickness of a solid body has not been well established until the present days. However, many methods based on the incremental blind hole technique have been proposed: Method of Equivalent Uniform Stress, Method of Power Series, Integral Method and others (Nobre *et al.*, 2000). In these techniques, the hole drilling is divided into several incremental steps, in which the strain relaxation is measured by the rosette fixed around the hole, as already mentioned. By using coefficients obtained by elasticity considerations and the finite element method (Schajer, 1988b), and based on a chosen method for the calculations, one can obtain the values of residual stresses in the whole studied profile. For strongly non-uniform residual stress fields, the Integral Method has been most indicated. In this method, the total contributions of relaxed strains, measured in all increments, are considered simultaneously, as will be presented in the sequence.

The Integral Method mathematical basis (Schajer, 1988a and 1988b) can be summarized as follows: the transformed strains ($\varepsilon(h)$), which were relaxed due to the total hole depth h, are the integral sum of the transformed stresses $(\sigma(H))$ in the infinitesimal depths in the range $0 \leq H \leq h$ where there are components of relaxed strains. Zuccarello (1999) proposed the following expression to describe the methodology:

$$\varepsilon_x(h) = \frac{1}{2E} \int_0^h \left\{ (1-v)\hat{A}(H,h)\left[\sigma_x(H) + \sigma_y(H)\right] + \hat{B}(H,h)\left[\sigma_x(H) - \sigma_y(H)\right] \right\} d(H)$$

$$(0 \leq H \leq h) \tag{2}$$

where $\sigma_x(H)$ and $\sigma_y(H)$ are the residual stresses contributions along the principal orthogonal directions x and y, $\hat{A}(H,h)$ and $\hat{B}(H,h)$ are the influence functions that can be determined by using numerical methods, E is the elastic modulus and v is the Poisson's ratio.

For conceptual simplicity, if the problem has been limited to a simple equal biaxial field, expression (2) can be written in the form:

$$\varepsilon_x(h) = \frac{1}{E} \int_0^h (1+v)\hat{A}(H,h)\left[\sigma_x(H) + \sigma_y(H)\right] d(H) \qquad (0 \leq H \leq h) \tag{3}$$

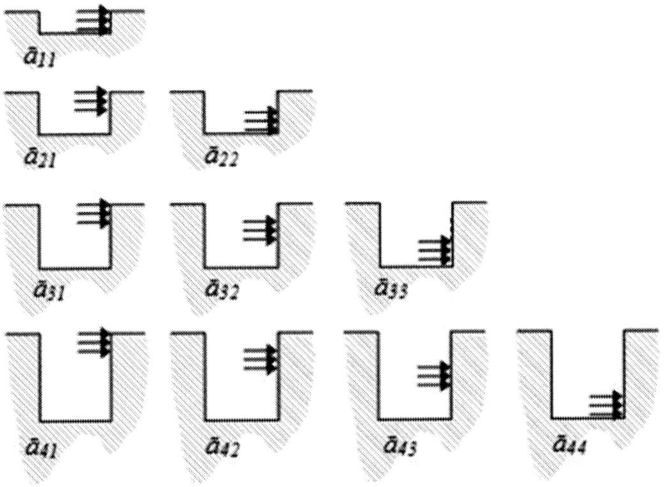

Figure 1. Stress loadings corresponding to the coefficients $\overline{a}_{i,j}$ of matrix $\overline{a}$ (ASTM, 2008).

Experimental Evaluation and Numerical Validation ... 33

However, since the experimental technique is incremental, the discrete form of the above expression can be applied, which in matrix representation is stated as:

$$\overline{a}\,\sigma = E\varepsilon\,/(1+v) \qquad (1 \le j \le i \le n) \tag{4}$$

where $\overline{a} = [\overline{a}_{ij}] = \int_{H_{j-1}}^{H_j} \hat{A}(H,h_i)d(H)$ represents the relaxed strain due to unit stress within increment j of a hole of instantaneously i depth increments. These values are represented by a lower triangular matrix where n is the total number of hole depth increments. Figure 1 shows the physical interpretation of coefficients $\overline{a}_{ij}$ from matrix $\overline{a}$, which enhances the methodology understanding.

In this work, due to the interest in the profile characterization and the incremental features of the experimental program, the Integral Method was used for the residual stresses evaluation, in conjunction with the computational software H-Drill (Schajer, 2009). The equipment used to drill the holes was the RS-200 and the strain indicator was the P-3, both from Vishay Micro-Measurements Group (Vishay, 2007).

3. EXPERIMENTAL PROCEDURE

3.1. Hard Turning Experiments

Investigations on six specimens made of DIN 21 NiCrMo 2 (ABNT 8620) case hardened steel were conducted, with hardness of approximately 700 HV (58-62 HRC) and 1.0 mm average case depth. The CBN insert for turning operations was a TNGX110308S-WZ, with low CBN content, tip radius of 0.8 mm and wiper geometry. Insert tool holding used was CTJNL2525M11 with the following tool angles: position angle = 93 °; cutting edge inclination angle = - 6 °; rake angle = - 6 °. Details can be found on Seco (2003).

The machine used was a INDEX MC400 CNC lathe (20 kW power) and all tests were conducted without cutting fluid. For each change in the pair V_c (cutting speed) and f (feed rate), the insert cutting edge was changed, so the tool wear was neglected for all tests. Table 1 presents the cutting conditions for cutting speed (V_c - in m/min), feed rate (f - in mm/rev) and constant depth of cut (a_p - in mm).

Table 1. Cutting conditions for the analyses matrix

	Vc	f	ap
	[m/min]	[mm/rev]	[mm]
Condition 1	180	0.05	0.18
Condition 2	180	0.08	0.18
Condition 3	180	0.12	0.18
Condition 4	200	0.05	0.18
Condition 5	200	0.08	0.18
Condition 6	200	0.12	0.18

3.2. Residual Stress Evaluation with the Blind Hole Method

The experimental residual stresses have been evaluated in circumferential and axial directions by the incremental blind hole method using the RS-200 Milling equipment from Vishay Micro-Measurements Group, with incremental manual step for the acquisition of the corresponding strain values at each step in the three known strain gage directions. The strain values were processed and analyzed using the H-Drill software based upon the integral method (Calle, 2004; Schajer, 1988a and 2006). Figure 2 shows one representative sample and the measuring equipment consisting of high speed pneumatic spindle and strain indicator. The strain gage rosette installed in the sample can be viewed in Figure 2(a) and the sintered tungsten carbide drill with Ø 1.8 mm is shown in detail in Figure 2(c).

The drill used has 1.80 mm diameter and the strain gage applied to the case was the 062RE manufactured by Vishay Micro-Measurements (Vishay, 2007). By using the device micrometer and controlling the depth of cut, the drill is set close to the extensometer surface; at this point, all the P3 indicator channels are set to zero and the incremental drilling process begins. The hole drilling is conducted carefully and slowly, using cutting oil and high-speed (up to 400,000 rpm). At each 20 μm penetration depth, the relieved strain values shown in the indicator were recorded in a spreadsheet. These data were implemented into the software H-Drill for the numerical computations by using the integral method previously described (Schajer, 1988a and 2006).

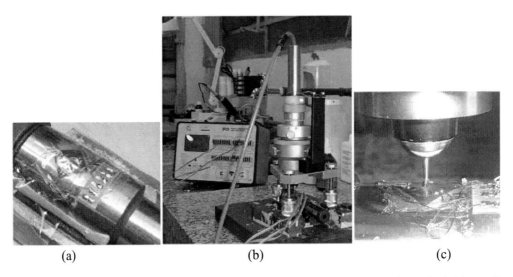

(a) (b) (c)

Figure 2. Sample, strain gage installation scheme and equipment for the blind hole method. (a) Details of the strain gage 062RE rosette installed in the sample, (b) equipment for residual stress measurement - high speed pneumatic spindle and strain indicator and (c) detail of cemented carbide drill with Ø 1.8 mm.

3.3. Experimental Results

Table 2 presents the strain gage readings for each drill penetration depth on the surface samples machined with the cutting conditions presented by Table 1. Gage number 1 represents the circumferential direction, gage number 3 represents the axial direction and gage number 2 represents the intermediate direction (45 °), see Figure 2.

Table 2. Strain results for different samples and corresponding tested cutting depths (machining conditions were presented in Table 1)

	Hole depth	Gage 1 strain	Gage 2 strain	Gage 3 strain		Hole depth	Gage 1 strain	Gage 2 strain	Gage 3 strain
	[mm]	$[10^{-6}]$	$[10^{-6}]$	$[10^{-6}]$		[mm]	$[10^{-6}]$	$[10^{-6}]$	$[10^{-6}]$
Condition 1	0	0	0	0	Condition 2	0	0	0	0
	0.04	6	-5	1		0.04	3	12	4
	0.06	10	-7	0		0.06	8	20	11
	0.08	20	-12	-3		0.08	10	32	17
	0.10	22	-7	6		0.10	12	40	23
	0.12	26	0	12		0.12	15	53	31
	0.14	30	4	17		0.14	17	59	37
	0.16	33	7	20		0.16	18	64	43
Condition 3	0	0	0	0	Condition 4	0	0	0	0
	0.04	5	6	4		0.04	6	2	3
	0.06	18	19	14		0.06	17	8	12
	0.08	26	27	20		0.08	32	15	21
	0.10	30	32	23		0.10	43	20	29
	0.12	34	37	25		0.12	48	22	32
	0.14	41	52	32		0.14	55	29	37
	0.16	45	58	34		0.16	58	33	40
	0.18	47	66	37		0.18	63	35	43
Condition 5	0	0	0	0	Condition 6	0	0	0	0
	0.04	1	1	3		0.04	7	2	3
	0.06	4	2	5		0.06	13	5	6
	0.08	6	3	9		0.08	21	8	10
	0.10	7	3	12		0.10	27	11	13
	0.12	8	3	13		0.12	29	12	15
	0.14	9	3	14		0.14	---	---	---

Each of the six specimens was measured once, and the strain results presented above were then processed by the H-Drill software using the integral method in order to evaluate the residual stress profiles. No filters or fitting parameters were applied to the results obtained. Figures 3(a) and 3(b) show the profiles obtained from H-Drill for residual stresses in axial and circumferential directions respectively. It can be realized that all measurements indicate compressive stress fields with maximum values below the surface (depths between 0.05 and 0.1 mm in most cases). Exactly the same compressive trend is observed for overall equivalent stress fields, which are suppressed for objectiveness.

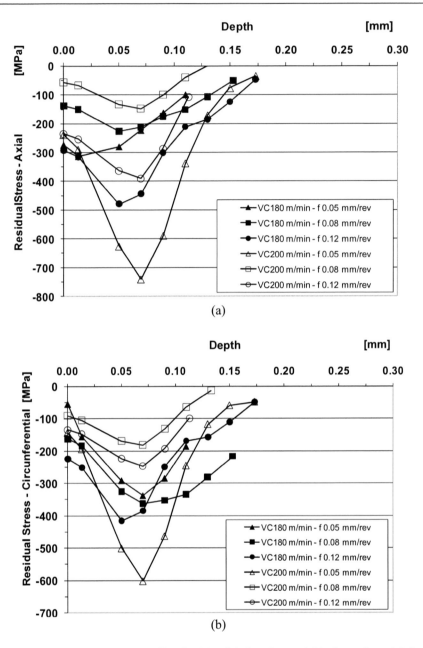

Figure 3. Experimental residual stress profiles for (a) axial direction and (b) circumferential direction.

4. VALIDATION USING THE FINITE ELEMENT METHOD

In order to validate the experimental measurements and gain confidence in the residual stress profiles previously shown, several refined finite element models where developed to characterize the measured strain fields near the evaluated areas. Due to the reduced dimensions of the strain gage rosettes (whose gages are disposed in a 5 mm radius – see Figure 2 – Vishay, 2007) and to the component axisymmetry, the finite element model could be simplified to a plane plate with main dimensions relatively large if compared to the hole diameter and maximum depth. This strategy simulates a semi-infinite media with adequate boundary conditions applied far from the hole, as will be presented in the sequence, and did not represent any deleterious influence on the obtained results.

Figure 4 shows the geometrical features and main dimensions of the developed models. To mitigate possible boundary effects on the strain measurements, several validation models were built and after the tests the reference dimensions were conservatively set as: *width=length=30.hole diameter; thickness=70.maximum drilling depth*. The same figure indicates the planes and respective boundary conditions applied to the model, in order to guarantee unstressed boundaries.

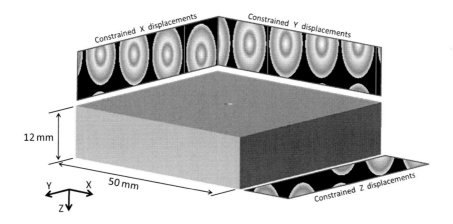

Figure 4. Geometrical features, main dimensions and boundary conditions of the developed models.

4.1. Finite Element Models

In addition to that, Figure 5(a) shows the structured refined mesh built for the finite element models, where improved and increasing refinement can be noted in the direction of the hole.

Figure 5(b) represents a detail of the hole area for one illustrative depth. Here, the different experimental depths were simulated by the incremental exclusion of element layers. The meshes possess 36 radial element sectors on the XY plane near the hole area and each drilling step comprehends 4 element layers in the Z direction, in order to improve accuracy.

Figure 6, in its turn, shows a top view of the mesh configuration near the hole area. It can be realized from the figure that the mesh complies with the real strain gage rosette configuration (and physical dimensions), in order to allow the integration of superficial displacement fields along all the strain gage sensitive area during post-processing (Schajer, 1988b).

Here, the strain gages were assumed to be uniformly sensitive over its active area, and to have negligible transverse sensitivity, which showed no effect on the computations and is in accordance with other researchers such as Schajer (1988b).

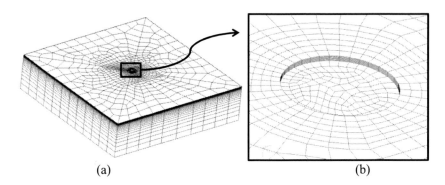

Figure 5. (a) Finite element model developed for the hole drilling simulation and (b) detail of the hole area. Different depths are simulated by the incremental exclusion of element layers.

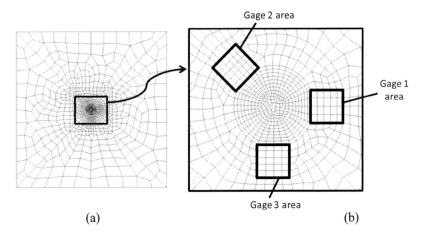

Figure 6. (a) Top view of the structured finite element mesh and (b) detail of the hole area where the mesh pattern respects the real strain gage rosette configuration and dimensions.

The material considered for the computation obeys a simple linear-elastic constitutive model with $E = 200\ GPa$ and $v = 0.3$. Pre and post-processing were conducted using software MSC Patran 2008r1 (MSC, 2009), while the finite element solutions were processed using MD Nastran R3b (MSC, 2009). The mesh shown in Figures 5 and 6 was developed using approximately 54,000 nodes and 52,000 3-D 8-node hexahedric isoparametric elements. The computations were conducted separately for each specimen and hole depth considering small chance in geometry (SCG) setting.

For each model, loads to be applied were computed based on the experimental Cartesian stresses obtained from H-Drill software (see Figures 3a and 3b for examples) for the different depths evaluated in the real experimental program. Based on the normal (axial and circumferential) and shear stresses provided by the software for each hole depth, conventional stress transformations were applied to describe the whole stress fields in hole walls using cylindrical coordinates in the form

$$\sigma_{x'} = \frac{\sigma_x + \sigma_y}{2} + \frac{\sigma_x - \sigma_y}{2}\cos 2\theta + \tau_{xy} \sin 2\theta \qquad (5)$$

$$\tau_{x'y'} = -\frac{\sigma_x - \sigma_y}{2}\sin 2\theta + \tau_{xy}\cos 2\theta \qquad (6)$$

where σ and τ represent, respectively, normal and shear stress components.

Figure 7 shows the loading strategy for one illustrative model. Basically, normal ($\sigma_{x'}$) and shear ($\tau_{x'y'}$) stress fields computed by the coordinate transformations are imposed to the hole walls obeying equilibrium considerations (Schajer, 1988a, ASTM, 2008), according to the integral method approach (see Figure 1). The computational implementation of the described strategy is based on spatial fields (using PCL functions) in MSC Patran (MSC, 2009). In this way, the relaxed stresses due to material removal can be virtually re-established for each drilling step to assess (and consequently validate) the experimentally measured strains.

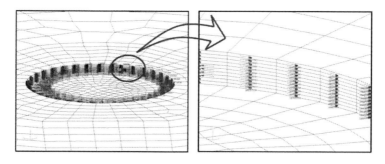

Figure 7. Normal ($\sigma_{x'}$) and shear ($\tau_{x'y'}$) stress fields application to the hole walls. Stresses were computed using the coordinate transformations presented by Equations (4) and (5).

Figure 8(a) shows a top view of the loading scheme using stress fields and Figure 8(b) shows, for illustration purposes only, the corresponding equivalent von Mises superficial strain field. All the evaluated models and also the stress field results follow very similar trends and will not be presented due to space considerations. Based on the strain fields presented, each gage predicted strain was calculated from the model by the integration of surface displacement field along all the strain gage sensitive area. These resulting strains (denoted here FEM) were compared to the experimental measurements for validation purposes, as described in the next section.

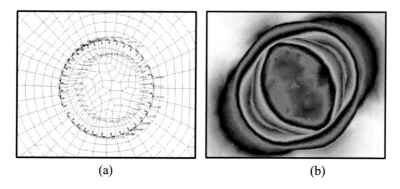

Figure 8. (a) top view of the loading scheme using stress fields and (b) illustrative equivalent von Mises surface strain field.

4.2. Numerical Predictions and Comparison

Figures 9 to 12 show selected results for the comparison between experimentally and numerically obtained strains. For illustrative purposes, four representative cases with different cutting speeds and feed rates are reported, with the corresponding three strains evaluated for each depth. The comparison is based on measured and predicted strains, since they represent the direct experimental measurements and are free of any method dependent assumption (see Table 1 for experimental values and conditions). It can be realized that in all cases, regardless of the gage direction, the numerical results corroborate the experimental findings with excellent agreement, with mean deviation of 0,4 % with respect to the experimental results.

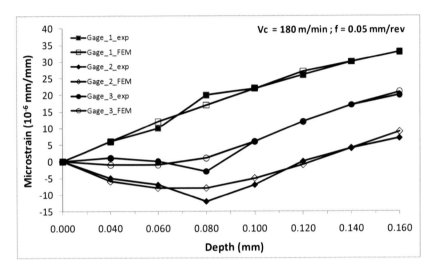

Figure 9. Relieved microstrains comparison for the three measuring strain gages as a function of hole depth for Vc = 180 m/min and f = 0.05 mm/rev.

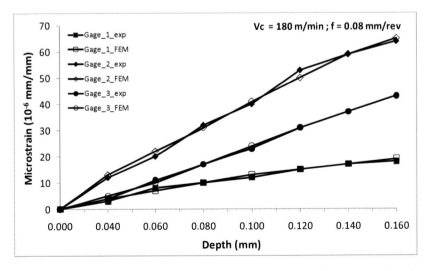

Figure 10. Relieved microstrains comparison for the three measuring strain gages as a function of hole depth for Vc = 180 m/min and f = 0.08 mm/rev.

It can also be realized from the results that the numerical predictions present smoother curves, which can be used for the identification of experimental inaccuracies or problems, calling the attention to unreliable or censored values and leading to better global results.

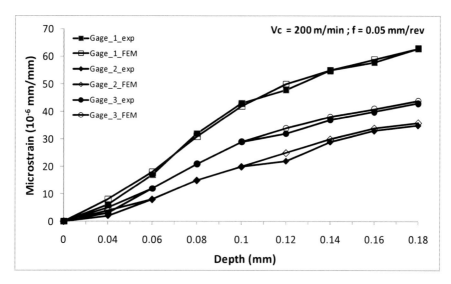

Figure 11. Relieved microstrains comparison for the three measuring strain gages as a function of hole depth for Vc = 200 m/min and f = 0.05 mm/rev.

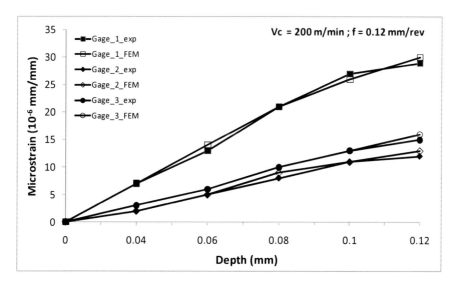

Figure 12. Relieved microstrains comparison for the three measuring strain gages as a function of hole depth for Vc = 200 m/min and f = 0.12 mm/rev.

In general, the results encourage the use of the incremental blind hole method as an accurate technique for measuring residual stresses on machined case hardened steels. In addition, finite element models provided good agreement between predicted and measured strains, which indicates that the correlation between relieved residual stresses on the hole wall and the strain gage readings are properly described by the integral method. The influence coefficients of that method, which are dependent on material properties and must be numerically calibrated, can be used as recommended in the current practices (ASTM, 2008) for the studied cases. For other manufacturing conditions and material properties, they can represent a potential source of deviations and deserve, therefore, further investigation. For these investigations, the methodology presented here can be of great interest for the validation of experimental measurements.

CONCLUSION

The experimental evaluation of residual stresses on hard turned case hardened AISI 8620 steel indicated, for all studied cases, compressive stress fields both in longitudinal and circumferential directions, with maximum values below the surface (depths between 0.05 and 0.1 mm in most cases). These results are in agreement with the literature and (due to the assessed magnitudes) call the attention to the importance of studying residual stresses due to hard turning as a means of improving performance and lifetime of mechanical components.

The numerical core results obtained from this study indicate the feasibility of modeling and validation, using finite element method, of the residual stress distribution from experimental results obtained by the incremental blind hole method in hard turned case hardened steels with linear elastic behavior. The conducted methodology is able to represent the experimental phenomenon with good accuracy and average deviation under 0.5 %, which is considered negligible from an engineering viewpoint. For the specific case of hardened components, focus of this investigation, additional testing in other samples with different machining parameters should be implemented in order to verify the model's ability to recreate the results with higher levels of residual stresses and plasticity development, extending its applicability through the use of elastic-plastic finite element models.

As a potential contribution to experimental evaluation of residual stresses, the present investigation showed that the numerical predictions can be applied to other isotropic materials and residual stress profiles, and in case of poor agreement due to plasticity or experimental procedures, can be an important tool to identify inaccuracies/problems or inadequate direct application of standardized recommendations for the integral method (and its coefficients), which are a function of the material and should, in these cases, be numerically calibrated. The conducted validation practice can indicate potentially inaccurate results, thus avoiding failures and improving manufacturing and quality control.

ACKNOWLEDGMENTS

The authors would like to acknowledge Polimold Industrial S/A for providing the work material samples, Seco Tools Brazil for the machine tool and tooling and the Department of Mechanical Engineering of the Ignatian Educational Foundation (FEI) for the personnel and facilities used for the experimental and numerical procedures.

REFERENCES

Abrão, A. M. (2005), *"Steel Hard Turning, Advanced Manufacturing Technologies"*, Chapter 5, Novos Talentos Press, 89-103 (in Portuguese).

American Society for Testing and Materials (2008), "Standard Test Method for Determining Residual Stresses by the Hole-Drilling Strain-Gage Method", *ASTM E-837-08*, Philadelphia, USA.

Calle, M. A. (2004), "Numerical and Experimental Analysis of the Residual Stresses Induced by Shot Peening in Automotive Springs", Master Dissertation, University of Sao Paulo - Escola Politecnica. Sao Paulo, Brazil.

Dahlman, P., Gunnberg, F., Jacobson, M. (2004), "The Influence of Rake Angle, Cutting Feed and Cutting Depth on Residual Stresses in Hard Turning", *Journal of Materials Processing Technology*, 147(2), 181–184.

Delijaicov, S. (2004), "Residual Stresses Experimental Evaluation after Hard Turning of 100CrMn6 Steel and its Correlation to Cutting Loads", Dr. Eng. *Thesis*, University of Sao Paulo - Escola Politecnica, Sao Paulo, Brazil.

Guo, Y. B., Barkey, M. E. (2004), "Modeling of Rolling Contact Fatigue for Hard Machined Components with Process-induced Residual Stress", *International Journal of Fatigue*, 26 (6), 605-613.

Griffiths, B. (2001), *"Manufacturing Surface Technology"*, Penton Press, London, UK.

MSC Software Corporation (2009), website available at http://www.mscsoftware.com/. Visited in July 24, 2009.

Nobre, J. P., Dias, A. M., Gibmeier, J., Kornmeier, M. (2006), "Local Stress-Ratio Criterion for Incremental Hole-Drilling Measurements of Shot-Peening Stresses", Transactions of the ASME - *Journal of Engineering Materials and Technology*, 128(2), 193-201.

Nobre, J. P., Kornmeier, M., Dias, A. M., Scholtes, B. (2000), "Use of the Hole-drilling Method for Measuring Residual Stress in Highly Stressed Shot-peened Surfaces" *Experimental Mechanics*, 40(3), 289-297.

Rech, J., Moisan, A. (2003), "Surface Integrity in Finish Hard Turning of Case-hardened Steels", *International Journal of Machine Tools and Manufacture*, 43(5), 543-550.

Schajer, G. S. (1988a), "Measurement of Non-uniform Residual Stresses Using the Hole Drilling Method. Part I – Stress Calculation Procedures", Transactions of the ASME - *Journal of Engineering Materials and Technology*, 110(4), 338-343.

Schajer, G. S. (1988b), "Measurement of Non-uniform Residual Stresses Using the Hole Drilling Method. Part II – Practical Application of the Integral Method", *Journal of Engineering Materials and Technology*, 110(4), 344-342.

Schajer, G. S. (2006), *"H-Drill 3.0 - Hole-Drilling Residual Stress Calculation Program"*, website available at http://www.schajer.org/ . Visited in July, 2009.

Schwach, D. W., Guo, Y. B. (2005), "A Fundamental Study on the Impact of Surface Integrity by Hard Turning on Rolling Contact Fatigue", *Transactions of NAMRI/RI/SME*, 33, 541-548.

Seco (2003), "Secomax PCBN Technical Guide", *Seco Tools AB*, Sweden, Grafiska byran/db Oerebro.

Töenshoff, H. K., Arendt, C., Amor, R. (2000), "Cutting of Hardened Steel", *CIRP Annals - Manufacturing Technology*, 49(2), 547-66.

Umbrello D., Ambrogio G., Filice L., Shivpuri R. (2007), "An ANN Approach for Predicting Subsurface Residual Stresses and the Desired Cutting Conditions During Hard Turning", *Journal of Material Processing Technology*, 189(6), 143-52.

Vishay-Measurements Group (2007), website available at http://www.schajer.org/ . Visited in December, 2007.

Zuccarello, B. (1999), "Optimal Calculation Steps for the Evaluation of Residual Stresses by the Incremental Hole-Drilling Method", *Experimental Mechanics*, 39(2), 117-124.

In: Machining and Forming Technologies, Volume 3
Editor: J. Paulo Davim

ISBN: 978-1-61324-787-7
© 2012 Nova Science Publishers, Inc.

Chapter 4

ON THE FUNDAMENTALS OF ORTHOGONAL METAL CUTTING

P. A. R. Rosa[*], V. A. M. Cristino, C. M. A. Silva and P. A. F. Martins

IDMEC, Instituto Superior Tecnico, TULisbon, Av. Rovisco Pais,
Lisboa, Portugal

ABSTRACT

This paper is based on the proposition by Atkins (2003) that the physics behind the separation of material at the tip of the tool is of great importance for understanding the mechanics of chip formation. Knowledge on how material separates along the parting line to form the chip and how to include the significant fracture work involved in the formation of new surfaces, as well as the traditional components of plastic flow and friction, are presently the most disturbing issues of the fundamental metal cutting research. This work attempts to provide answers to these issues by means of a combined numerical and experimental investigation based on the interaction between finite elements and modern ductile fracture mechanics. The overall presentation is supported by specially designed orthogonal metal cutting experiments that were performed on lead test specimens under laboratory-controlled conditions. Comparisons between theoretical predictions and experimental results comprise a wide range of topics such as material flow, primary shear plane angle, cutting force and specific cutting pressure. The paper shows that numerical modelling of metal cutting exclusively based on plasticity and friction is suitable for estimating material flow, chip formation and the distribution of the major field variables whereas the contribution of ductile fracture mechanics is essential for obtaining good estimates of cutting forces and of the specific cutting pressure.

Keywords: Metal cutting, Experimentation, Finite element method, Ductile fracture

---

[*] Corresponding author: Fax: +351-21-8419058, E-mail: pedro.rosa@ist.utl.pt

1. INTRODUCTION

In a recent paper, Astakhov (2005) discussed the major drawbacks of the single shear plane model for metal cutting, showing that this model cannot be utilized in the development of the predictive metal cutting theory as well as in the development and utilization of finite element computer programs. As initially claimed by Cook et al. (1954) and later by Atkins (2003), the single shear plane model is inadequate for real metal cutting because it cannot operate in plane strain plasticity at constant plastic volume without 'new surface' formation at the tip of the tool. In connection to this, Astakhov (1999) points out that there is a major difference between metal cutting and other metal working processes in that there must be physical separation of the layer to be removed from the work material and that the process of separation forms new surfaces.

As recently pointed out by Atkins (2005), there are two different views of metal cutting mechanics. The commonly accepted view considers that the energy required for cutting is mainly due to plasticity and friction which far exceed the energy required for the formation of new surfaces (Shaw, 1984). This view, which hereafter we refer to as 'plasticity and friction only' (PFO), gives support to the initial work of Ernst and Merchant (1941) and is implicit in most of the major contributions to the understanding of the process made by Shaw (1984), Zorev (1966), Oxley (1989) and many others.

The non-traditional, and sometimes controversial, view of metal cutting states that the energy to form new surfaces at the tip of the tool is not negligible and ought to be at kJ/m^2 levels, rather than at the few J/m^2 level of the chemical surface free energy that was employed in the calculations (Shaw, 1984) which purported to show that surface work should be negligible. First suggestion that fracture plays an important role in metal cutting goes back to Reuleux (1900), but the understanding of its overall influence in the process was only recently provided by Atkins (2003) after re-examining metal cutting in the light of modern ductile fracture mechanics. Atkins claims that quantitative explanations for longstanding problems related to material separation and chip formation mechanisms require a proper interaction between plasticity and ductile fracture by means of the inclusion of a new parameter related to fracture toughness in addition to plastic flow and friction.

In connection to this it is not surprising that analytical and numerical modelling of metal cutting based on the PFO approach may experience difficulties at estimating the cutting forces for practical metal cutting. In a recent study Bil et al. (2004) performed a comprehensive assessment of the estimates provided by three different commercial finite element computer programs with experimental data and concluded that although individual parameters (such as the cutting force, the thrust force and the shear angle) may be made to match experimental results, none of the models was able to achieve a satisfactory correlation with all the measured process parameters all of the time. Possible reasons for mismatches between finite element simulations and experimentation were identified as lack of information about flow stresses and friction at the rates and temperatures experienced in practical cutting, both of which can drastically affect the predictions. Indeed this has always been the case, even for algebraic models of machining. It was noted (Bil et al., 2004) that tuning of friction yields good agreement only for some variables in the range: smaller values of friction provided good estimates for the cutting force, whereas higher values of friction were required to make theory and experiment agree for other variables such as the thrust force and shear angle. This is probably the reason why Astakhov (2005), after analysing the work of several researchers, discovered that the results of finite element modelling of metal cutting always seemed to be in good agreement with the experimental results regardless of the particular value of the friction coefficient selected for modelling (probably because some authors adjusted the rate and temperature dependent flow stress to fit, without independent determinations of the flow stress under comparable conditions).

From what was mentioned before it is possible to conclude that finite element computer programs currently employed in metal cutting generally experience limitations in providing good estimates of the experimental measurements. In addition, the majority of the simulative works reported in the literature are limited to examples that had been previously experimentally observed and, therefore, computations are not being employed predictively.

The proposed paper is directed at re-examining the fundamentals of orthogonal metal cutting and proposing new modelling and experimental strategies to ensure a proper interaction between finite elements and modern ductile fracture mechanics. The overall presentation is supported by especially designed orthogonal metal cutting experiments that were performed on lead test specimens under laboratory-controlled conditions. Comparisons between theoretical predictions and experimental results comprise a wide range of topics such as material flow, primary shear plane angle, cutting force and specific cutting pressure.

2. Experimental Background

The experiments were performed in orthogonal metal cutting conditions using a testing apparatus that was installed in a computer-controlled hydraulic press after being instrumented for monitoring the displacements and the forces acting on the tool face. The flow curve, friction coefficient and fracture toughness of the work material were determined by means of special purpose tests that were performed in the range of strain rates to match those obtained in the metal cutting experiments.

2.1. Material and Tribological Characterization

The choice of material is of prime importance if metal cutting investigation is expected to be performed in near-ideal orthogonal conditions. In principle, it is important to select a material that is capable of providing a narrow zone of intense plastic shearing (shear plane as required by the ideal rigid plastic analytical theories) under low cutting speeds. The utilization of low speeds is also crucial for eliminating the possible influence of temperature (Pugh, 1958).

Unfortunately, commonly used engineering materials are unable to fulfil the above mentioned requirements in low speed cutting conditions because of strain-hardening during plastic deformation. Therefore, the shear plane usually broadens into a plastic shearing zone denoted as primary shear zone. Technically-pure lead is probably one the best material choices because of its very low strain hardening and of its ability for being able to model the plastic deformation of engineering materials (e.g. steels) at higher strain-rates and temperatures (Altan, 1970).

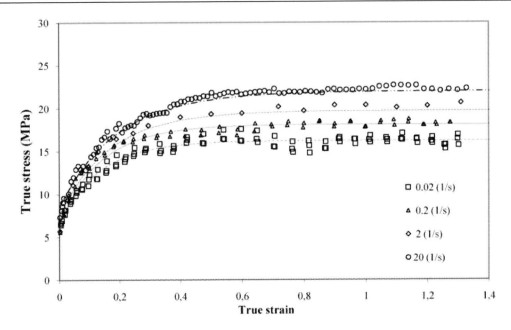

Figure 1. Stress-strain curve of technically-pure lead obtained by means of compression tests for a range of strain-rates that match those obtained in the metal cutting experiments.

The material characterization of technically-pure lead was performed by means of compression tests carried out at room temperature on cylindrical specimens using four different strain-rate conditions $\dot{\varepsilon}$ = 0.02, 0.2, 2 and 20 s^{-1}. The specimens with a height to diameter ratio, $h/d = 1.5$ were manufactured from the supplied raw material and lubricated with PTFE (polymer based lubricant) for ensuring homogeneous deformation. The mathematical fitting of the experimental data resulting from the compression tests (Figure 1) by means of the following mixed plastic/viscoplastic material law is given by:

$$\bar{\sigma} = 19.275\, \dot{\varepsilon}^{0.043} - 11.874\, \dot{\varepsilon}^{0.047} \exp^{(0.019\,\dot{\varepsilon} - 6.652)\varepsilon} \quad \text{MPa} \qquad (1)$$

The coefficient of friction μ at the contact interface between the cutting tool and the material was determined directly from the ratio $\mu = F_t/F_c$ between the experimental values of the thrust F_t and cutting F_c forces in metal cutting experiments performed by means of a cutting tool with a rake face angle $\alpha = 0°$. Under this particular tool geometry, the cutting F_c and thrust F_t forces become the normal F_N and frictional F_f forces, respectively, allowing the coefficient of friction μ to be directly obtained from the ratio between the experimental measured values of the cutting and thrust forces (Figure 2).

The coefficient of friction μ was determined using lubrication and surface quality conditions similar to those employed in the orthogonal metal cutting tests and, as seen in Figure 2, the coefficient of friction is not influenced by the cutting velocities (0.05 to 4 m/min) that were utilized in the experiments. The value $\mu = 0.4$ of the coefficient of friction resulting from Figure 2 is within the range of values that are commonly found in metal cutting processes.

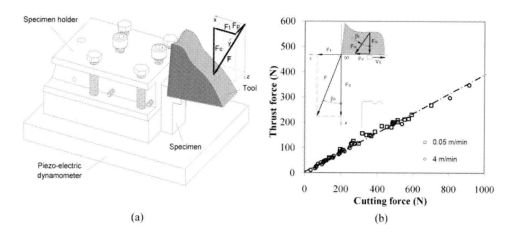

Figure 2. a) Schematic representation of the metal cutting testing apparatus and b) Experimental data for the evaluation of the coefficient of friction μ in orthogonal metal cutting conditions with a rake face angle $\alpha = 0°$ (lubricated conditions with different cutting velocities of 0.05 and 4 m/min).

2.2. Fracture Toughness Characterization

The characterization of fracture toughness was performed by means of experimental tests carried out at room temperature on double-notched cylindrical specimens loaded in shear (Figure 3). The geometry of the specimens was optimized by means of finite element analysis so that the crack running ahead of the punch propagates under plastic conditions similar to those taking place in orthogonal metal cutting at the tip of the tool and in the adjacent primary shear zone.

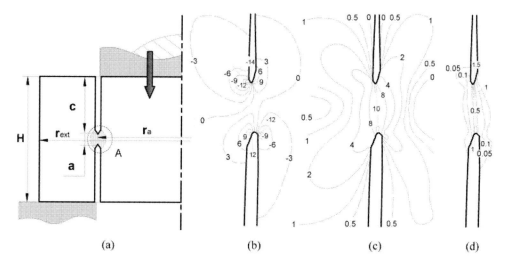

Figure 3. Finite element modelling of a double-notched cylindrical test specimen with $H = 12.5$ mm, $r_a = 7.5$ mm and $r_{ext} = 15.0$ mm after 0.9 mm punch displacement. Results show the computed predicted distribution of b) the average stress (MPa), c) the shear stress (MPa) and d) the effective strain-rate (s^{-1}) at the onset of cracking.

The presence of stresses of opposite sign at the vicinity of the notches gives rise to a narrow bounding strip of pure shear starting at the tip of the crack and to a separation mechanism typical of a sliding opening mode (mode II of fracture mechanics). As will be shown later (refer to section 4) this opening mode is compatible with the formation of the crack ahead of the cutting tool edge, making characterization of fracture toughness by means of double-notched cylindrical specimens loaded in shear very appropriate for understanding the importance of the energy required for the formation of new surfaces at the tip of the cutting tool. The double-notched cylindrical tests were performed with specimens where started cracks have been introduced to various depths 'c' before loading (Figure 3) and the experiments consist on determining the punch shearing load-displacement evolution from which fracture toughness of the material may be deduced. The maximum length of the ligament between notches was limited to $a = 2.5$ mm so that plastic deformation is confined to a small region in between the notches. If $a > 2.5$ mm the plastic deformation extends outside this zone and the experimental measurements are no longer adequate for calibrating fracture toughness. The energy W to initiate crack propagation is calculated by direct integration of each punch shearing load-displacement curve up to the maximum punch shearing load and the specific work of surface formation (fracture toughness) R is:

$$R = \frac{W}{2 \pi r_a a} \qquad (2)$$

where, r_a is defined in Figure 4. The shearing load depends directly on the length of the ligament between notches but fracture toughness R remains approximately constant and independent from 'a' ($R \cong 13.3$ kJ/m^2 is taken from Figure 4).

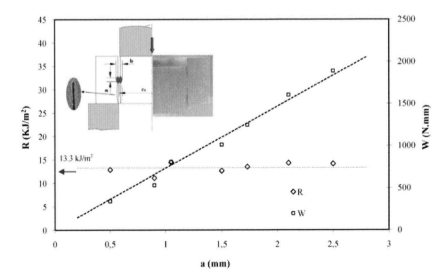

Figure 4. Fracture toughness R (left axis) and energy W (right axis) to initiate crack propagation as a function of the length of the ligaments a of the double-notched cylindrical test specimens.

The corresponding separation work per volume of the plastically deformed region in-between the notches U is written as (refer to Figure 4):

$$U = \frac{W}{2\pi r_a ab} = \frac{R}{b} \cong 10 - 13.5 \ (MJ/m^3) \qquad (3)$$

where b denoted as the plastic shear band width, is defined in Figure 4. This factor is of major importance in the forecoming sections of the paper and its value was obtained from finite element simulations after assuming the boundary of the plastically deformed region to be characterized by low values of the effective strain-rate. In the present case, $\bar{\varepsilon}_{lim} = 0.05 \ s^{-1}$ i.e. approximately 20 times below the average values of effective strain rate within the plastic region.

2.3. Orthogonal Metal Cutting Experiments

The orthogonal metal cutting experiments were performed on a special tooling apparatus installed on a 500 kN computer-controlled hydraulic press (Figure 2a). The cutting forces were monitored by a piezo electric multi-component force (x,y,z) plate transducer (Kistler, type 9257B with a range ±10 kN, attached to a charge amplifier type 5011B). The transducer is linear across its entire range and measures forces with an accuracy of 0.1%. The displacement of the cutting tool was measured using a micro-pulse position transducer (Balluff, range 600 mm) with an overall accuracy below 6 μm. An electronic board (National Instruments, type PCI-6070E) and a windows based software (Labview) were utilized for performing data acquisition directly from the previous mentioned equipments.

The experiments were designed in order to cope with the major process parameters that govern low speed orthogonal metal cutting; (i) the rake angle α, (ii) the uncut chip thickness t_0, and (iii) lubrication. A clearance angle of 5° was maintained throughout the experiments and cutting speeds were taken as 0.05 m/min and 3.5 m/min. The tests were carried out on parallelepiped specimens with 100 x 49 x 24 mm that were manufactured from the supplied raw material.

3. THEORETICAL BACKGROUND

In what follows, authors start by reviewing the fundamentals of the finite element computer program I-Cut that has been utilized and extensively validated against experimental measurements of metal cutting (Rosa et al., 2007) and conclude by presenting a new simulative approach based on the interaction between finite elements and ductile fracture mechanics.

3.1. Fundamentals of the Computer Program I-Cut

The finite element computer program I-Cut is based on the flow formulation that considers the cutting specimen as an isotropic plastically deformable body occupying an open bounded domain V with a sufficiently smooth boundary. The boundary consists of several portions; S_u, is the part of the boundary S on which the velocity $\bar{u}_i$ is prescribed, while S_t and S_f are the remaining parts on which traction t_i and frictional stress τ_f are applied. The unknown variables are the velocity u_i and the stress field σ_{ij} resulting from the following set of equations (Kobayashi et al., 1989):

$$\sigma_{ij,j} = 0 \tag{4}$$

$$\dot{\varepsilon}_{ij}(u) = \frac{1}{2}(\frac{\partial u_i}{\partial x_j} + \frac{\partial u_j}{\partial x_i}) = \frac{1}{2}(u_{i,j} + u_{j,i}) \tag{5}$$

$$\sigma'_{ij} = \frac{2\bar{\sigma}}{3\dot{\bar{\varepsilon}}}\dot{\varepsilon}_{ij} \tag{6}$$

where, $\sigma'_{ij} = \sigma_{ij} - \sigma_m \delta_{ij}$ is the deviatoric stress tensor, $\sigma_m = \frac{1}{3}\sigma_{kk}$ is the average stress, $\bar{\sigma} = \sqrt{\frac{3}{2}\sigma'_{ij}\sigma'_{ij}}$ is the effective stress, $\dot{\varepsilon}_{ij}(u)$ is the strain-rate tensor, δ_{ij} is the Kronecker delta and $\dot{\bar{\varepsilon}}(u) = \sqrt{\frac{2}{3}\dot{\varepsilon}_{ij}(u)\dot{\varepsilon}_{ij}(u)}$ is the effective strain-rate.

In addition, it is expected that the solution of metal cutting problems should generally meet the following boundary conditions:

$$u_i = \bar{u}_i \text{ on } S_u \tag{7}$$

$$(u - u_{tool})_i n_i = 0 \text{ on } S_f \tag{8}$$

$$\sigma_{ij} n_j = t_i \text{ on } S_t \tag{9}$$

$$\sigma_{ti} = \sigma_{ij} n_j - ((\sigma_{kl} n_l) n_k) n_i = \tau_{fi} = -\left|\tau_f\right| \frac{\tilde{u}_i}{\left|\Delta u_f\right|} \text{ on } S_f \tag{10}$$

where, σ_{ti} denotes the tangential component of stress at the contact interface, S_f, between material and tooling and, $\left|\Delta u_f\right| = \sqrt{\tilde{u}_i \cdot \tilde{u}_i}$ is the norm of the relative slipping velocity, $\tilde{u}_i = (u - u_{tool})_i$ along S_f (where friction shear stresses τ_{fi} are applied and $\left|\tau_f\right|$ is its norm).

Under these circumstances, the variational principle associated with the finite element flow formulation considers the admissible velocity field $\mathbf{u}(\mathbf{x}) \in H^1$, ($H^1$ the is Soblev space of degree one), to be the solution of the problem only if it corresponds to the absolute minimum of the total potential energy rate $E(\mathbf{u})$:

On the Fundamentals of Orthogonal Metal Cutting

$$E(\mathbf{u}) \cong \int_V \left(\int \overline{\sigma}(\mathbf{u}) d\overline{\dot{\varepsilon}}(\mathbf{u}) \right) d\Omega \; + \; \frac{k}{2} \int_V \dot{\varepsilon}_v^2 \, dS \; - \; \int_{S_t} \mathbf{t}\,\mathbf{u}\,dS \; + \; \int_{S_f} \left| \tau_f \right| \cdot \left| \Delta u_f \right| dS \tag{11}$$

where the unknown variables are the velocity u_i and the stress σ_{ij} fields and k is a large positive constant penalising the volumetric strain-rate component, $\dot{\varepsilon}_v = \dot{\varepsilon}_{ii}$, in order to enforce incompressibility. The utilisation of the finite element flow formulation based on the penalty function method offers the advantage of preserving the number of independent variables, because the average stress can be computed after the solution is reached through:

$$\sigma_m = k\,\dot{\varepsilon}_v \tag{12}$$

The simplest procedure to satisfy the equilibrium Equation (4) and to ensure the boundary conditions (8-10) is to express the weak form of the variational principle (11) entirely in terms of an arbitrary variation in the velocity $\mathbf{v} \in H^1$:

$$\int_V \sigma_{ij}(\mathbf{u})\dot{\varepsilon}_{ij}(\mathbf{v})\,d\Omega + k \int_V \dot{\varepsilon}_{ii}(\mathbf{u})\dot{\varepsilon}_{jj}(\mathbf{v})\,dS - \int_{S_t} t_i v_i\,dS - \int_{S_f} \sigma_n v_i\,dS = 0 \quad \forall\,\mathbf{v} \in H^1 \tag{13}$$

The enforcement of the essential boundary conditions (7) is straightforward because finite element shape functions are interpolants ($\mathbf{N}_i(\mathbf{x}_j) = \delta_{ij}$) while the contact requirements (8) are dealt by means of an explicit direct contact algorithm that determines the increment of time, Δt, for the material to reach the surface of the cutting tool by integrating the trajectories of the nodal points placed on the boundary of the cutting specimen (Alves et al., 2004).

By employing a discretization procedure that considers the cutting specimen to be represented by M finite elements linked through N nodal points, it is possible to rewrite equation (13) as the following set of non-linear equations:

$$\sum_{m=1}^{M} \left\{ \int_{V^m} \mathbf{B}^T (\frac{\overline{\sigma}}{\overline{\dot{\varepsilon}}}\mathbf{D} + k^m \mathbf{C}\mathbf{C}^T)\mathbf{B}\,dV^m - \int_{S_t^m} \mathbf{N}^T\,\mathbf{T}\,dS^m + \int_{S_f^m} \left| \tau_f \right| \frac{\mathbf{I}}{\left| \Delta u_f \right|} \mathbf{N}^T \mathbf{N}\,dS^m - \int_{S_f^m} \mathbf{N}^T \left| \tau_f \right| \frac{\mathbf{u}_{tool}}{\left| \Delta u_f \right|} dS^m \right\} = 0 \tag{14}$$

where, $\mathbf{B}$ denotes the strain-rate matrix, $\mathbf{C}$ represents the vectorial form of the Kronecker delta and $\mathbf{D}$ is a matrix relating stresses with strain rates. The numerical evaluation of the integrals included in equation (14) is performed through the utilization of quadrilateral elements, as shown in Figure 5.

In order to ensure the incompressibility requirements of the plastic deformation of metals, both complete and reduced Gauss point integration schemes are utilised for the calculation of the domain integral. The numerical integration of the boundary integrals is performed by means of a 5 Gauss point quadrature. Further details on the finite element discretization of equation (14) can be found elsewhere (Alves et al., 2004).

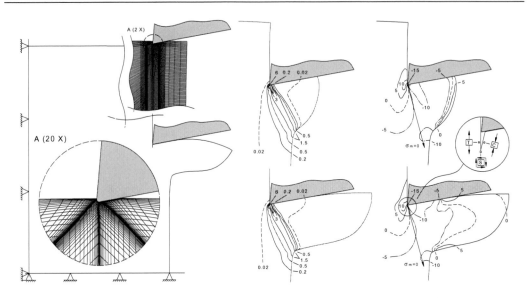

Figure 5. Numerical simulation of orthogonal metal cutting. a) Finite element model of the cutting specimen showing two details of the mesh, b) Computed distribution of the effective strain-rate $\dot{\bar{\varepsilon}}$ (s^{-1}) and c) Computed distribution of the average stress σ_m from transient to steady-state metal cutting conditions (lubricated, with $\alpha = 10°$ and $t_0 = 0.5$ mm for tool displacements of 1.5 and 4.4 mm).

The computer implementation of the finite element flow formulation is based on an updated Lagrangian description of coordinates in which the mesh follows the material and deforms as its shape changes. As a consequence, a material separation criterion is necessary to cope with the singularity at the sharp tip of the tool and to allow the tool to move. The overall strategy implemented in the finite element computer program makes use of a chip separation criterion based on a critical distance between the tip of the tool and the nodal point placed immediately ahead. Whenever this critical distance is reached, the computer model automatically rezones the mesh in order to insert of a new nodal point and allow the chip to separate from the cutting specimen. Details can be found elsewhere (Rosa et al., 2007).

3.2. Interaction between Finite Elements and Ductile Fracture Mechanics

The evaluation of the fracture work involved in the formation of new surfaces in metal cutting may be achieved by means of a 'decoupled approach' that neglects the influence of fracture work in the yield surface of the material being cut. Material is assumed to be continuous and isotropic throughout the cutting process and the conventional constitutive equations for rigid-plastic/viscoplastic material behaviour are utilised. In other words, the major field variables such as strain, stress and strain rate are to be directly calculated from PFO calculations, while the cutting force F_c is to be determined by summing up afterwards the solution derived from PFO finite element modelling with the resistance to crack propagation derived from experimental measurements of the specific work of surface formation (refer to section 2.2):

$$F_c = \left(F_c\right)_{FEM} + U\,ws \tag{15}$$

where ' s ' is the length of the shear plane (in close analogy to the length ' a ' of the ligament between notches of the fracture test specimen) and $\left(F_c\right)_{FEM}$ is the cutting force calculated by finite elements. Because the specific work of surface formation expressed as a work per volume U can also be expressed as a work per area R (fracture toughness) it follows from Equations (3) and (15) that:

$$F_c = \left(F_c\right)_{FEM} + U\,ws = \left(F_c\right)_{FEM} + \frac{Rw}{Q^*} \qquad Q^* = \frac{b}{s} = \frac{b}{t_0}\sin\phi \tag{16}$$

The evaluation of the cutting force F_c by means of Equation (16) presents very close resemblance with the analytical solution proposed by Atkins (2003). In fact, the evaluation of the cutting force F_c by means of the proposed decoupled approach (16) can be seen as a modified version of the equation proposed by Atkins in which the correction factor Q is to be replaced by a new factor Q^*. The new factor Q^* (dimensionless parameter) differs from Q by the fact that adjustment is made in terms of the width ' b ' of the plastically deformed region placed in between the notches of the fracture specimen, as well as in terms of the length ' s ', of the primary shear zone, instead of friction.

To conclude it is worth noticing that the overall values of the length ' s ' of the primary shear zone lie within the range of values of the length ' a ' of the ligament between the notches of the fracture test specimens. This fact justifies the utilization of $b = 0.8$ mm (refer to section 2.2) in the forecoming sections of the paper.

4. RESULTS AND DISCUSSION

4.1. Physics behind Material Separation at the Tip of the Tool

Figure 5c shows the finite element distribution of the hydrostatic (mean) stress from the instant of time when the tool starts to advance into the cutting specimen until reaching steady-state conditions. As seen, the material adjacent to the rake surface is under compression, whereas the material in the vicinity of the relief (flank) surface of the tool is pulled under tensile stresses (positive values means tensile mean stress).

The presence of stresses of opposite sign at the vicinity of the cutting edge gives rise to a narrow bounding strip of pure shear starting at the tip of the tool that proves new surfaces to be formed by shear. The implication that physics behind separation of material to take place in shear is experimentally corroborated by means of scanning electron microscope observations of the material adjacent to the tip of the tool (Figure 6) showing a kink perpendicular to the plane of the figure — a clear indication that the crack formed ahead of the cutting edge is closed and growth occurs in shear (mode II of fracture mechanics).

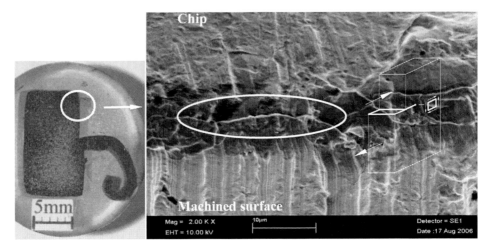

Figure 6. Scanning electron microscopic observation of the material adjacent to the tip of the tool showing a crack being formed by shear. The observation refers to steady-state metal cutting conditions.

4.2. Shear Plane Angle

Figure 7 shows the experimental and numerical evolution of the shear plane angle ϕ as a function of the uncut chip thickness t_0, obtained from orthogonal metal cutting with a rake angle $\alpha = 10°$, $t_0 = 1.0$ mm and $v_c = 0.05$ m/min. Two different patterns of behaviour can be easily distinguished; i) a first zone in which the shear plane angle ϕ is affected by the uncut chip thickness and ii) a second zone where the shear plane angle ϕ is less sensitive and remains approximately constant.

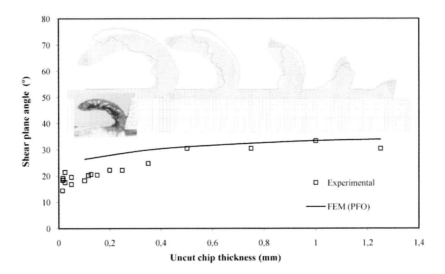

Figure 7. Variation of the shear plane angle ϕ with the uncut chip thickness t_0. Insets show details of the mesh and of the experimental profile of the chip at different stages of the chip formation (lubricated conditions with $\alpha = 10°$, $t_0 = 1.0$ mm and $v_c = 0.05$ m/min).

The first zone corresponds to the typical range of metal cutting processes where $t_0 < 0.2$ mm while the second zone is outside common metal cutting processing but still within possible working area.

As shown by the insets included in Figure 7 finite element modelling was further assessed by comparing the predicted curling of the chip with the experimentally measured profile. The shaded profile placed on top of the finite element mesh helps visualize the differences between the experimental and the numerical prediction of chip curling. In general terms results compare reasonably well and confirm that finite elements are capable of successfully modelling the plastic deformation mechanics of orthogonal metal cutting. The key point in the following sections of the paper is to check if the aforementioned conclusions will remain when finite element estimates of the cutting force and specific cutting pressure are to be compared with experimental data derived from orthogonal metal cutting tests.

4.3. Cutting Force

Figure 8 presents the steady-state cutting forces F_c for different values of the uncut chip thickness t_0 obtained from finite element modelling and experimental testing with rake angles $\alpha = -5°$ and $10°$.

As shown, PFO finite element estimates, of the cutting forces vs. uncut chip thickness, pass through the origin and considerably underestimate the experimental measurements. The failure of finite element modelling seems to suggest that PFO analysis of orthogonal metal cutting is not adequate for calculating the forces in actual cutting processes.

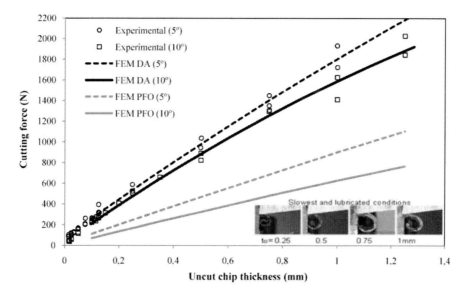

Figure 8. Theoretical and experimental variation of the cutting force F_c with the uncut chip thickness t_0, after reaching steady-state conditions (lubricated conditions with $\alpha = 10°$ and $\alpha = -5°$ and $v_c = 0.05$ m/min). Insets show details of the formed chip.

As discussed earlier, the missing value in the cutting force is related to the fracture work involved in the formation of new cut surfaces. Confirmation of this conclusion is given by the results supplied by the proposed decoupled finite element-fracture mechanics approach, where the effect of fracture work on the cutting force F_c is added on afterwards (refer to the FEM (DA) in Figure 8). In fact, the proposed decoupled approach is capable of providing very good correlations between theory and experimentation and it is interesting to notice that the trend curves containing the numerical estimates 'bend down' at small values of the uncut chip thickness t_0 in agreement with the experimental measurements.

4.4. Specific Cutting Pressure

The specific cutting pressure k_s can be defined as the ratio between the cutting force F_c and the cross sectional area $t_0 w$ of the uncut chip:

$$k_s = \frac{F_c}{t_0 w} \tag{17}$$

and it is commonly utilized for performing approximate calculations of the cutting forces. Its value is available in tables or calibration curves derived from experimentation.

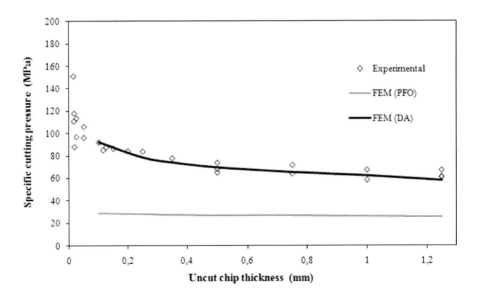

Figure 9. Variation of the specific cutting pressure k_s with the uncut chip thickness t_0 (lubricated conditions with $\alpha = 10°$ and $v_c = 0.05$ m/min).

Standard text books of metal cutting often claim that the specific cutting pressure k_s is not a stress even though it has the dimensions of stress and looks like a nominal normal stress. A possible explanation for this is due to the fact that a normal stress $\sigma = k_s$ would not be consistent with the stress-strain/strain-rate behaviour of the material and with its corresponding state-of-stress. Taking Figure 9 as an example, it follows that lead would have to provide normal compressive stresses up to 60 to 120 MPa in order to allow the specific cutting pressure k_s to be considered as a normal stress σ.

However, taking a first glimpse at the specific cutting pressure k_s obtained by finite element analysis it is possible to conclude that (i) PFO modelling underestimates its value and that (ii) fracture work is once again required to end up with a good correlation with experimentation (refer to FEM (DA) in Figure 9).

This conclusion is in close agreement with what was previously mentioned and allows rewriting the specific cutting pressure as a summation of two terms (with different physical meanings):

$$F_c = k_s t_0 w = \left(\sigma + \frac{R}{b \sin \phi} \right) t_0 w \Rightarrow k_s = \left(\sigma + \frac{R}{b \sin \phi} \right) \tag{18}$$

The first term of k_s results from the normal compressive stress acting in the cross-sectional area of the uncut chip while the second term is due to the fracture work involved in the formation of new surfaces.

The above relationship (18) provides the explanation for a longstanding incompatibility of the metal cutting theory.

CONCLUSION

The paper demonstrates that traditional finite element approaches are capable of predicting quite well the experimentally-observed chip flow fields in metal cutting but are unable to correctly estimate the values of the cutting forces. In fact, they always underestimate cutting forces when independently-determined mechanical properties (flow stress, toughness and friction), at the correct rates and temperatures, are employed. The deficiency in forces is connected directly to not including work of surface formation in the analyses.

The paper resolves the aforementioned shortcomings by introducing a decoupled finite element approach that allows toughness work to be included in the analysis and demonstrates how, from independent determinations of mechanical properties, cutting forces can be successfully predicted.

The physics behind separation of material at the tip of the tool is also comprehensively investigated and it is proved that it is shear deformation that causes formation of new surfaces, not tensile deformation.

ACKNOWLEDGMENTS

The authors would like to acknowledge the support provided by Prof. Anthony G. Akins during the investigation and the financial support received from POCI-2010.

REFERENCES

Altan T., Henning H. J. and Sabroff A. M. (1970), "The use of model materials in predicting forming loads in metalworking", *Journal Engineering for Industry – Transactions of ASME*, 92, 444-452.

Alves M. L., Rodrigues J. M. C. and Martins P. A. F. (2004), "Three-dimensional modelling of forging processes by the finite element flow formulation", *Journal of Engineering Manufacture*, 218, 1695-1708.

Astakhov V. (1999), "*Metal cutting mechanics*", CRC Press, Boca Raton, Florida.

Astakhov V. (2005), "On the inadequacy of the single-shear plane model of chip formation", *International Journal of Mechanical Sciences*, 47, 1649-1672.

Atkins A. G. (2003), "Modelling metal cutting using modern ductile fracture mechanics: quantitative explanations for some longstanding problems, *International Journal of Mechanical Sciences*", 45, 373-396.

Atkins A. G. (2006), "Toughness and oblique machining", *Journal of Manufacturing Science and Technology – Transactions of ASME*, 128, 775-786.

Bil H, Kiliç S. E. and Tekkaya A. E. (2004), "A comparison of orthogonal cutting data from experiments with three different finite element models", *International Journal of Machine Tools and Manufacture*, 44, 933-944.

Cook N. H., Finnie I. and Shaw M. C. (1954), "Discontinuous chip formation", *Transactions of ASME*, 76, 153-162.

Ernst H. and Merchant M. E. (1941), "Chip formation, friction and high quality machined surfaces", *Transactions of ASME*, 29, 299.

Kobayashi S., Oh S. I. and Altan T. (1989), "*Metal forming and the finite element method*", Oxford University Press.

Oxley P. L. B. (1989), "*Mechanics of machining: An analytical approach to assessing machinability*", John Wiley and Sons, New York.

Pugh H. (1958), "Mechanics of metal cutting process", *Proceedings IME Conference Technology Engineering Manufacture*, London, 237-254.

Reuleaux F. (1900), "Über den taylor whiteschen werkzeugstahl verein sur berforderung des gewerbefleissen in preussen", *Sitzungsberichete*, 79, 179-189.

Rosa P. A. R., Martins P. A. F. and Atkins A. G. (2007), "Revisiting the fundamentals of metal cutting by means of finite elements and ductile fracture mechanics", *International Journal of Machine Tools and Manufacture*, 47, 607–617.

Shaw M. C. (1984), "*Metal cutting principles*", Clarendon Press, Oxford.

Zorev N. (1966), "*Metal cutting mechanics*", Pergamon Press, Oxford.

In: Machining and Forming Technologies, Volume 3
Editor: J. Paulo Davim

ISBN: 978-1-61324-787-7
© 2012 Nova Science Publishers, Inc.

Chapter 5

THERMAL ANALYSIS IN TiN AND AL_2O_3 COATED ISO K10 CEMENTED CARBIDE CUTTING TOOLS USING DESIGN OF EXPERIMENT (DoE) METHODOLOGY

Rogério Fernandes Brito[1], João Roberto Ferreira[1], Solidônio Rodrigues de Carvalho[2] and Sandro Metrevelle Marcondes de Lima e Silva[1]*

[1]Mechanical Engineering Institute - IEM, Federal University of Itajubá - UNIFEI, Campus Prof. José Rodrigues Seabra, Av. BPS, 1303, bairro Pinheirinho, CEP: 37500-903, Itajubá, MG, BRAZIL

[2]School of Mechanical Engineering - FEMEC, Federal University of Uberlândia - UFU, Campus Santa Mônica, Av. João Naves de Ávila, 2160, bairro Santa Mônica, CEP: 38400-902, Uberlândia, MG, BRAZIL

ABSTRACT

This paper is concerned with the effect of variations in the thickness of the tool coating on the heat transfer in cemented carbide tool substrates (ISO grade K10). Titanium nitride (TiN) and aluminum oxide (Al_2O_3) with thickness values of 1 and 10 µm were used as coatings. In order to increase tool life and reduce costs, the thermal parameters of the turning operation are investigated aiming a more uniform temperature distribution in the cutting zone. Boundary conditions by convection and heat flux are known, as well as the thermophysical properties of the tool and coating involved in the numerical analysis. Two commercial softwares were used and the proposed methodology was validated experimentally under controlled conditions. Design of experiments (DoE) was used to identify the optimal parameters in order to obtain the maximum temperature difference (ΔT) between the tool substrate and the coating. The cutting tool temperature distribution is discussed and a thermal analysis on the influence of the coating is presented. Finally, the results are discussed and compared with data available in the published literature.

---

* E-mail: rogbrito@unifei.edu.br

Keywords: TiN and Al_2O_3 coatings, cutting tool, DoE, finite volume method, heat transfer

1. INTRODUCTION

Machining processes generate enough heat to deform the materials involved: the chip and the tool. This level of heat is a factor that strongly influences tool performance. Friction wear and heat distribution affect the temperature on both the chip and tool. In order to increase tool life, its surface can be coated with materials having thermal insulation features that reduce tool wear. Hence, the influence of coatings on heat transfer and friction wear is an area of study that deserves an in-depth investigation.

A review on the literature suggests that most orthogonal metal cutting simulations are designed for uncoated cemented carbide tools. In recent years, however, an opposite trend has emerged, using single and multiple coatings. In Marusich *et al.* (2002), for example, the authors carry out a simulation using a numerical model based on the Finite Element Method (FEM). The Thirdwave AdvantEdge® software is used to simulate the chip-breakage of coated and uncoated tools. With multi-layered coatings, one result shows a temperature reduction, for the tool substrate, of 100 °C. Grzesik (2003) studied the cutting mechanisms of several coated cemented carbide tools and found that the tool-chip contact area and the average temperature of the tool-workpiece interface change according to the coating. Grzesik did not explain, however, whether the coatings are able to thermally insulate the substrate.

Yen *et al.* (2003) and Yen *et al.* (2004) offer the first comprehensive assessments, using FEM, of an orthogonal cutting model for multi-layer coated cemented carbide tools. In their model, the authors analyze, both individually and as a group, the thermal properties of the following three layers: titanium carbide, (TiC), aluminum oxide (Al_2O_3), and titanium nitride (TiN). They consider a layer with equivalent thermal properties. Their results indicate that the fine width coatings of an Al_2O_3 intermediary layer do not significantly alter the temperature gradients for a steady state between the chip and the tool substrate. Rech *et al.* (2004) and Rech *et al.* (2005) worked with the qualification of the tribological system "work material-coated cemented carbide cutting tool-chip." Their aim was to better understand the heat flux generated during the turning operation. Their methodology, applied to several coatings deposited on cemented carbide inserts, showed that the coatings possess no significant influence on the substrate thermal insulation.

Kusiak *et al.* (2005) analyzed how several coatings on a cutting tool influenced heat transfer. They performed this analysis with an analytical model of their own. For actual cutting conditions, the authors carry out an experimental work on turning AISI 1035 steel to examine the behavior of different coated inserts. Their results showed that, while the other coatings fail to significantly modify the thermal field, the Al_2O_3 coating yielded a slight reduction on the heat transferred to the tool.

Coelho *et al.* (2007) used FEM to simulate the performance of polycrystalline cubic boron nitride (PCBN) tools when turning AISI 4340 steel, Coelho use titanium aluminum nitride (TiAlN) and aluminum chromium nitride (AlCrN) coated and uncoated cutting tools. The simulations performed indicate that, regardless of the coating, the temperature on the tool-chip interface was approximately 800 °C with an flank wear was absent.

Sahoo (2009) reports an experimental study on the wear characteristics of electroless nickel-phosphorus (Ni-P) coatings sliding against steel. Sahoo (2009) optimized the coating process parameters aiming minimum wear. The optimization based on L_{27} Taguchi (Ross, 1995; Taguchi, 1986) orthogonal design, takes into account four process parameters: bath temperature, concentration of nickel source solution, concentration of reducing agent, and annealing temperature. The author observed that the two most significant factors influencing the wear characteristics of electroless Ni-P coating were the annealing and bath temperatures.

The aim of the present work is to numerically analyze how the coatings on cutting tools influence heat transfer during the cutting process. It is intended to verify the thermal and geometrical parameters of the coated tool, striving for a more adequate temperature distribution in the cutting region. In order to obtain the cutting tool temperature field, ANSYS CFX® Academic Research software v.12 was used. Additionally, a cutting tool with a single coating layer was used, as reported by Rech *et al.* (2005).

In this work, eight cases were analyzed, all with cutting tools having a single layer coating, varying in thickness (h) from 1 μm to 10 μm. Different coating materials were investigated with two types of heat fluxes used on the tool-chip interface.

The design of experiments (DoE) is used owing to the fact that it is the most economical and accurate method for performing process optimization. The DoE accelerates the understanding on the influence of the process parameters by determining which variables are critical to the process and at which level. This investigation required the evaluation of the effects of three variables (Montgomery, 2000). To ascertain the key relationships among them, DoE was used to find the best studied case of each simulation carried out. The temperature fields on the cutting tools were thus obtained. Finally, a numerical analysis of the thermal influence of these coatings is presented.

2. PROBLEM DESCRIPTION

Figure 1 presents the thermal model for heat conduction in a cutting tool and the regions for imposing boundary conditions. The tool geometry, within the computational domain, is represented respectively by Ω_1 and Ω_2, the coating solids of height h, the cutting tool substrate of height H, and interface C between the coating and the substrate. Only one type of material was considered for the cutting tool with dimensions 12.7x12.7x4.7 mm, with a nose radius R=0.8 mm and heat flux region S_2 with an area of approximately 1.424 mm^2. The coating thickness values adopted were: h=1 and 10 μm.

At room temperature, the thermal parameters of the materials investigated (substrate and coating) were as follows: ISO K10 cemented carbide tool with density ρ=14,900 kg.m^{-3} (Engqvist *et al.*, 2000), specific heat capacity C_p=200 J.kg^{-1}.K^{-1} and thermal conductivity k=130 W.m^{-1}.K^{-1} at 25 °C (Engqvist *et al.*, 2000), TiN coating with ρ=4,650 kg.m^{-3} (Yen *et al.*, 2004), C_p=645 J.kg^{-1}.K^{-1} (Yen *et al.*, 2004) and k=21 W.m^{-1}.K^{-1} at 100 °C (Yen *et al.*, 2004), Al$_2$O$_3$ coating with ρ=3,780 kg.m^{-3} (Yen *et al.*, 2004), C_p=1,079 J.kg^{-1}.K^{-1} (Yen *et al.*, 2004) and k=28 W.m^{-1}.K^{-1} (Yen *et al.*, 2004).

Figures 2a and 2b show one of the meshes formed by hexahedral elements and used in the numerical simulation. Figure 2c shows a typical contact area (A) on the tool-chip interface and the area used in the numerical simulation of the present work (A=1.4245 mm^2). From

Carvalho *et al.* (2006), the following cutting conditions were used: cutting speed of v_c=209.23 m/min, feed rate of f=0.138 mm/rot, and cutting depth of p=3.0 mm.

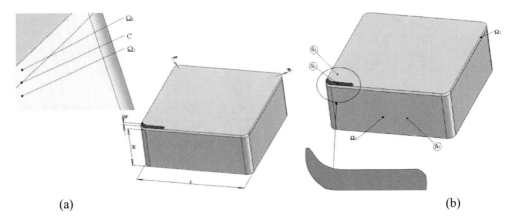

(a)   (b)

Figure 1. Coated cutting tool: (a) interface detail and (b) heat flux region.

(a) Typical hexahedral mesh used.

(b) Partial detail of the S_2 heat flux region with A area in red color.

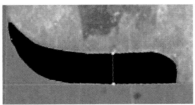

(c) Video image of the S_2 contact area on the chip-workpiece-tool interface (Carvalho *et al.*, 2006).

Figure 2. Non-structured mesh (a), mesh detail (b), image of the flux area (c).

2.1. Boundary Conditions

The present analysis assumed the following hypotheses: three-dimensional geometrical domain; transient regime; absence of radiation models; thermal properties, such as ρ, k, and C_p are uniform and the temperatures are independent for the coating layer and the substrate body; there is a perfect thermal contact and no thermal resistance contact between the coating layer and the substrate body; the boundary conditions of the heat flux $q''(t)$ are uniform and the time is variable; the boundary conditions of the heat transfer coefficient h and room temperature T_∞ are constant and also known; there is internal heat generation neither on the coating layer nor on the substrate body.

The heat diffusion equation is subject to two types of boundary conditions: imposed time-varying heat flux in S_2 and constant convection in S_1 of the cutting tool. The initial temperature conditions are described for the thermal states of the substrate and coating solids as $T_i=29.5\ °C$.

3. NUMERICAL METHOD

The solution of the continuity, momentum, and energy equations uses the Fluid Dynamics Calculus using the Finite Volume Method (FVM) with Eulerian scheme for the spatial and temporal discretization of the physical domain, using a finite number of control volumes (Versteeg and Malalasekra, 2007 and Löhner, 2008).

Through this method, the control volume elements follow the Eulerian scheme with unstructured mesh (Barth and Ohlberger, 2004). Using this approach, the transport equations may be integrated by applying the Gauss Divergence Theorem, where the approximation of surface integral is done with two levels of approximation. Firstly, the physical variables are integrated into one or more points on the control volume faces. Secondly, this integrated value is approximated in terms of nodal values.

This approximation represents, with second order accuracy, the average physical quantity of all the control volume (Shaw, 1992).

More details on the concepts involved in FVM may be found in Barth and Ohlberger (2004), where the authors explore discretization techniques, integral approximation techniques, convergence criteria, and calculus stability.

4. NUMERICAL VALIDATION

The commercial software used here was validated extensively by comparing this work's results of with those obtained in experimental and numerical investigations. For example, we compared our software's numerically obtained temperatures with those obtained, both numerically and experimentally, by Carvalho *et al.* (2006). The largest deviation was 6.07 %. In most of the simulated cases, the number of nodal points was 501,768 and the number of hexahedral elements was 481,500.

5. RESULTS AND DISCUSSION

Eight cases were selected to investigate the temperature distribution for a time interval t. The main goal is to study the influence of heat flux and variations in the coating thickness on the heat flux.

Table 1. Numerical results obtained from the temperature values after 63.14 s

Case	Coating/ Thickness μm	Heat flux W.m^{-2}	Chip-Tool Temperature T_{CT} °C	Coating-Substrate Temperature T_{CS} °C	Temperature Difference $T_{CT} - T_{CS}$ °C
1	TiN/1	$q_1''(t)$	86.71	86.53	0.18
2	TiN/1	$q_2''(t)$	601.72	599.86	1.86
3	TiN/10	$q_1''(t)$	87.28	86.46	0.82
4	TiN/10	$q_2''(t)$	607.37	599.17	8.20
5	Al$_2$O$_3$/10	$q_1''(t)$	87.08	86.55	0.53
6	Al$_2$O$_3$/10	$q_2''(t)$	604.90	599.59	5.31
7	Al$_2$O$_3$/1	$q_1''(t)$	86.72	86.55	0.17
8	Al$_2$O$_3$/1	$q_2''(t)$	601.28	600.90	0.38

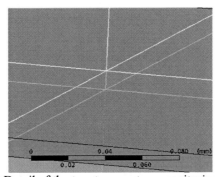

(a) Detail of the two temperature monitoring points.

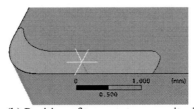

(b) Position of temperature monitoring.

Figure 3. Temperature monitoring points located on and under the 10 μm TiN coating layer.

Table 1 shows temperature values obtained after cutting for 63.14 s on the chip-tool (CT) and the coating-substrate (CS) interfaces. This was determined using the ANSYS CFX® Academic Research software, v.12. For the 10 μm coating with flux $q_2''(t)$, case 4 (TiN coated K10 substrate) had the highest calculated temperature difference. For 1 μm coating thickness with flux $q_1''(t)$, case 7 (Al$_2$O$_3$ coated K10 substrate) had the lowest calculated temperature difference.

Figure 3 shows the two points where temperature was monitored during the simulation. For the coordinates on the tool substrate-coating interface: x=1.5 mm, y=0.25 mm and z=10 µm and on the coating: x=1.5 mm, y=0.25 mm and z=0 mm. Case 04 (Table 1), with a thickness of 10 µm, exhibited the greatest temperature decrease, dropping 8.20 °C (from 607.37 to 599.17 °C). Figures 4a and 4b show, respectively, heat rate q and heat flux q_1'' W m^{-2} against cutting time.

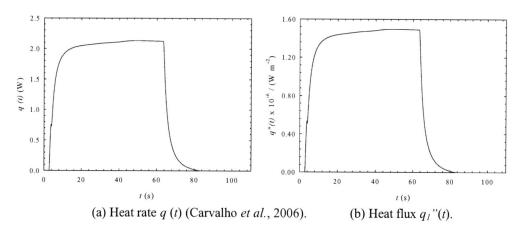

(a) Heat rate q (t) (Carvalho et al., 2006). (b) Heat flux $q_1''(t)$.

Figure 4. Heat rate and heat flux utilized in the present work.

Figures 5 through 8 show the simulation results for coated cutting tools. The influence of heat flux and coating thickness on the temperature fields on the chip-tool coating-substrate interfaces can thus be assessed. It can be seen that owing to the fact that the coating did not affect temperature reduction, it possesses negligible influence on thermal insulation.

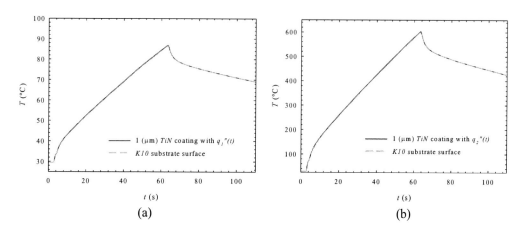

(a) (b)

Figure 5. Effect of heat flux variation on temperature (TiN coating with 1 µm thick).

Figures 9a, 9b and 9c show the temperature fields at instant 63.14 s on the top and bottom of the insert and the heat flux surface, respectively, for case 4 (TiN coating with 10 µm).

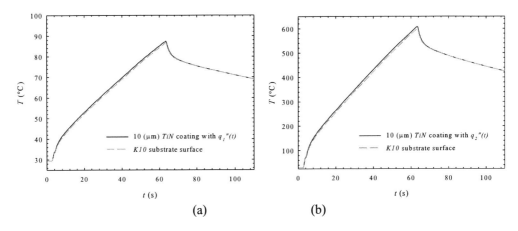

Figure 6. Effect of heat flux variation on temperature (TiN coating with 10 μm thick).

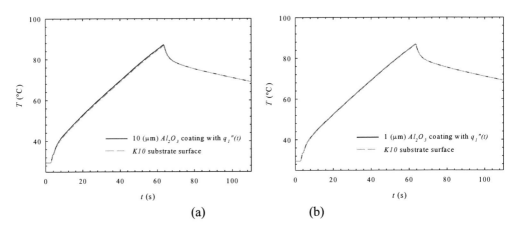

Figure 7. Effect of Al$_2$O$_3$ coating thickness variation on temperature with $q_1''(t)$ heat flux: (a) 10 μm thick and (b) 1 μm thick.

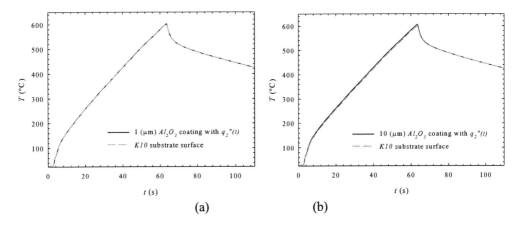

Figure 8. Effect of Al$_2$O$_3$ coating thickness variation on temperature with $q_2''(t)$ heat flux: (a) 1 μm thick and (b) 10 μm.

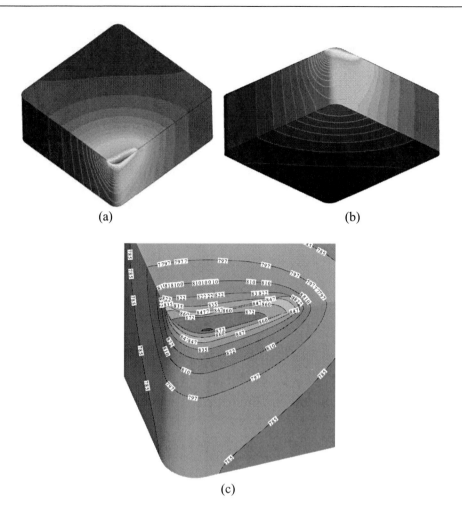

Figure 9. Temperature fields on: (a) top and (b) bottom and (c) heat flux surface (measured in K) on the TiN coated tool (case 4) at instant $t=63.14$ s.

5.1. Design of Experiment (DoE)

The DoE was configured with 3 factors (thickness, heat flux and coating material) at two levels (-1 and +1) aiming to determine their influence on the temperature field. The coatings were TiN and Al_2O_3, the thickness values were 1 μm and 10 μm and the heat flux values were 1 and 10 times (Montgomery, 2000). Table 2 shows the parameters specified for the design of experiment. The experiments were planned and analyzed with the Minitab® Inc. v.14 software.

Figure 10 shows the factors main effects on the temperature gradient (ΔT). The best levels defined for the parameters are a TiN coating, coating thickness of 10 μm and heat flux of $q_2''(t)$. This combination presented the highest temperature gradient.

Table 2. DoE matrix

Trial no.	Parameter			
	Coating	Thickness h μm	Heat Flux	Difference of temperature ΔT °C
1	Al$_2$O$_3$	10	10	5.31
2	TiN	10	1	0.82
3	TiN	1	1	0.18
4	TiN	10	10	8.20
5	Al$_2$O$_3$	10	1	0.53
6	Al$_2$O$_3$	1	10	0.38
7	TiN	10	1	0.82
8	TiN	1	10	1.86

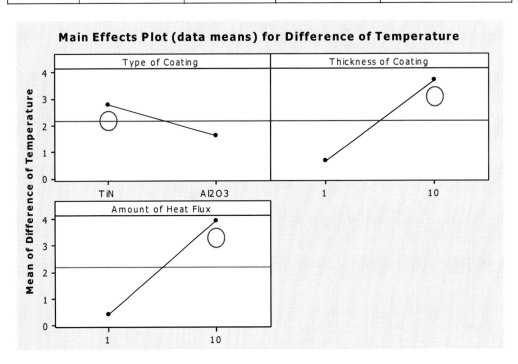

Figure 10. Influence of the main effects on the temperature gradient.

CONCLUSION

One of the contributions of this work is its numerical approach. This methodology permits the simulations of complex geometrical forms. It also includes, relative to experimental cases, a more realistic heat flux (Carvalho et al., 2006). In addition to that, the following conclusions can be drawn regarding the numerical results obtained for the thermal model of heat transfer in coated cutting tools:

- For a uniform heat source varying in time with constant surface contact between chip and tool, the coatings may slightly influence the temperature on the tool. This is true, however, when the thermal properties of the coating are rather different from those of the substrate, even for a thin coating (1 μm thick).

- During continuous cutting, the coating on the cemented carbide tool presented unsatisfactory results. The calculated temperature gradient was lower than for those found in the literature.
- None of the films analyzed showed a significant change in the temperature gradient across the tool coating.
- The DoE methodology proved to be an excellent tool for test planning. The best levels defined for the highest temperature gradient in the coated cutting tool were obtained with a coating of TiN with a thickness of 10 μm at the highest heat flux.
- Future studies should consider the influence of temperature variation on the thermal conductivity k and specific heat capacity C_p.

ACKNOWLEDGMENTS

The authors would like to thank Scientific Community Support Fund (CAPES), Research Support Foundation of the State of Minas Gerais (FAPEMIG), and The National Council for Research and Development (CNPq) for the financial support granted for the present work.

REFERENCES

ANSYS, Inc., (2009), *"ANSYS CFX® Academic Research Software, Release* v.12, Help System, Coupled Field Analysis Guide".

Barth, T. and Ohlberger, M., (2004), "Finite Volume Methods: Foundation and Analysis, *Encyclopedia of Computational Mechanics"*, Ed. Wiley, 57 p.

Carvalho, S.R., Lima e Silva, S.M.M., Machado, A.R. and Guimarães, G., (2006), "Temperature Determination at the Chip-Tool Interface using an Inverse Thermal Model considering the Tool and Tool Holder", *Journal of Materials Processing Technology*, 179, 97-104.

Coelho, R.T., Ng, E.G. and Elbestawi, M.A., (2007), "Tool Wear when Turning Hardened AISI 4340 with Coated PCBN Tools using Finishing Cutting Conditions", *International Journal of Machine Tools and Manufacture*, 47, 263-272.

Engqvist, H., Hogberg, H., Botton, G.A. Ederyd, S. and Axén, N., (2000), "Tribofilm Formation on Cemented Carbides in Dry Sliding Conformal Contact", *Wear*, 239, 219-228.

Grzesik, W., (2003), "A Computational Approach to Evaluate Temperature and Heat Partition in Machining with Multiplayer Coated Tools", *International Journal of Machine Tools and Manufacture*, 43, 1311-1317.

Kusiak, A., Battaglia, J.L. and Rech, J., (2005), "Tool Coatings Influence on the Heat Transfer in the Tool during Machining", *Surface and Coatings Technology*, 195, 29-40.

Löhner, R., (2008), *"Applied Computational Fluid Dynamics Techniques: an Introduction based on Finite Element Methods"*, Second ed., Ed. Wiley, 544 p.

Marusich, T.D., Brand, C.J. and Thiele, J.D., (2002), "A Methodology for Simulation of Chip Breakage in Turning Processes using an Orthogonal Finite Element Model", Proceedings of the Fifth CIRP International Workshop on *Modeling of Machining Operation*, West Lafayette, USA, 139-148.

Montgomery, D.C., (2000), *"Designs and Analysis of Experiments"*, 2nd ed., John Wiley & Sons, 538 p.

Rech, J., Kusiak, A. and Battaglia, J.L., (2004), "Tribological and Thermal Functions of Cutting Tool Coatings", *Surface and Coatings Technology*, 186, 364-371.

Rech, J., Battaglia, J.L. and Moisan, A., (2005), "Thermal Influence of Cutting Tool Coatings", *Journal of Materials Processing Technology*, 159, 119-124.

Ross, P.J., (1995), *"Taguchi Techniques for Quality Engineering"*, 2nd ed. New York: McGraw-Hill Professional, 329 p.

Sahoo, P., (2009), "Wear Behaviour of Electroless Ni–P Coatings and Optimization of Process", Materials and Design, 30, 1341-1349, Technical Report.

Shaw, C.T., (1992), *"Using Computational Fluid Dynamics - An Introduction to the Practical Aspects of using CFD"*, Ed. Prentice Hall Publications, 300 p.

Taguchi, G., (1986), "Introduction to Quality Engineering: Designing Quality into Products and Processes*", Quality Resources*; illustrated edition, 191 p.

Versteeg, H. and Malalasekra, W., (2007), *"An Introduction to Computational Fluid Dynamics: the Finite Volume Method"*, 2nd ed., Ed. Prentice Hall, 520 p.

Yen, Y.C., Jain, A., Chigurupati, P., Wu, W.T. and Altan, T., (2003), *"Computer Simulation of Orthogonal Cutting using a Tool with Multiple Coatings"*, Proceedings of the Sixth CIRP International Workshop on Modeling of Machining Operation, McMaster University, Canada, 119-130.

Yen, Y.C., Jain, A., Chigurupati, P., Wu, W.T. and Altan, T., (2004), "Computer Simulation of Orthogonal Cutting using a Tool with Multiple Coatings", *Machining Science and Technology*, 8, 305-326.

In: Machining and Forming Technologies, Volume 3
Editor: J. Paulo Davim

ISBN: 978-1-61324-787-7
© 2012 Nova Science Publishers, Inc.

Chapter 6

EFFECT OF THE LOADING SEQUENCE ON THE WORK-HARDENING OF LOW CARBON STEEL AND CUZN34 BRASS SHEETS

Wellington Lopes[1], Elaine Carballo Siqueira Corrêa[2], Haroldo Béria Campos[3], Maria Teresa Paulino Aguilar[3] and Paulo Roberto Cetlin[3]*

[1] CEFET-MG, Campus VII, Av. Amazonas, 1193 – Vale Verde,
Timóteo MG, Brazil
[2] CEFET-MG, Av. Amazonas, 5253 – Nova Suiça,
Belo Horizonte MG, Brazil
[3] Escola de Engenharia, Universidade Federal de Minas Gerais,
Av. Antônio Carlos, 6627, Pampulha,
Belo Horizonte MG, Brazil

ABSTRACT

Metal forming processes apply a variety of stresses on metallic parts in order to obtain the final shape and dimensions. The adequate description of the stress and strain states developed in these operations requires information on the processes (e.g. temperature and lubrication) and on the material properties (e.g. work hardening characteristics). These data are necessary due the unusual hardening behavior of metals caused by the complex strain path changes present in forming operations. Considering that the mechanical behavior of a material during and after a forming operation is affected by the material properties (including its structural characteristics) and by the successive deformation modes, this paper presents experimental results for the work hardening of low carbon steel and for CuZn34 brass, covering a processing route including rolling/tension/rolling/tension/shear. All tests were conducted at 0° rolling direction and both metals were used in the as received condition. The results suggest that the dislocation substructure evolution of the two materials was different for the same loading sequence. The low carbon steel presented early strain localization at the

---

* E-mail: pcetlin@demet.ufmg.br

beginning of the second tension step, whereas the brass displayed a delayed onset of the plastic instability.

Keywords: shear test, strain path, low carbon steel and brass

1. INTRODUCTION

Sheet metal forming employs relatively simple tooling and can produce parts with a variety of shapes and dimensions under economically attractive conditions. The success of this process hinges, however, on the formability of the material being formed, involving variables such as the material initial conditions, ductility, strength and anisotropy. In addition to that, the process parameters (strain rate, temperature, friction between the material and the tools, etc.) are also of importance (Taylan *et al.*, 1998).

A commonly studied feature of the materials subjected to sheet metal forming is their work hardening characteristics under monotonic deformation (usually pure tension) and considering various strain rates (Wagoner, 1981). These characteristics, however, can be deeply affected by the strain path followed by the material up to the various final effective strain values reached at each point of the formed part. The material formability limit, for instance, is significantly reduced when tension follows previous biaxial stretching (Rauch, 2000). Such phenomena are associated with flow stress and work hardening transients observed upon strain path changes, which may lead to premature or delayed plastic instabilities in the material. The evaluation and eventual prediction of such effects can be of great importance in the analysis of complex sheet metal forming operations, especially those involving multiple forming steps.

The magnitude of the stress and work hardening transients seems to depend on the degree of change in the straining direction of two successive loading modes, according to an α parameter (Schmitt *et al.*, 1985), given by the scalar product of the strain tensors before and after the strain path change, as indicated in Eq. (1). Parameter α is given by the cosine of the angle between two vectors representing the initial and the final deformation modes, which varies from 1 (monotonic loading) to -1 (reversed: Bauschinger loading).

$$\alpha = \frac{(\varepsilon_p . \varepsilon)}{|\varepsilon_p| \, |\varepsilon|} \tag{1}$$

In Equation (1), ε_p and ε are the deformation tensors for pre-straining and final straining, respectively. Lopes *et al.* (2001) showed that for strain path changes involving rolling followed by tension, the anisotropy of the material has to be measured in order to calculate the parameter α, as indicated in Eq. (2):

$$\alpha_{rolling\text{-}tension} = \frac{\cos^2(\Theta) - \dfrac{R_\Theta}{1+R_\Theta} \, sen^2(\Theta) + \dfrac{1}{1+R_\Theta}}{\sqrt{2} . \sqrt{1 + \left(\dfrac{R_\Theta}{1+R_\Theta}\right)^2 + \left(\dfrac{1}{1+R_\Theta}\right)^2}} \tag{2}$$

In Equation (2), Θ is the angle between the rolling and the tension loading and R_Θ is the anisotropy factor measured at this angle.

The stress and work hardening transients responsible for the formability changes of the material caused by strain path changes have been ascribed to the reorganization of dislocation substructures and/or to crystallographic texture effects. According to Barlat et al. (2003), in the case of a succession of loading modes involving tension, compression and shearing for Bauschinger type loadings (with full reversion of the straining direction), the evolution of the dislocation substructures would be responsible for the stress and work hardening transients observed during strain path changes.

Rauch and Schmitt (1989) subjected samples of low carbon steel sheets (0.06%C in weight) to loading sequences involving an initial tension followed by shearing at angles of 45°, 90° or 135° in relation to the previous tension direction. Their results are presented in Figure 1, where it can be seen that the final shearing at 45° to the tension direction, for two levels of initial strain (the tensile strain was transformed in shearing strains of 0.08 and 0.35), led to negligible changes in the hardening material, in relation to the behavior under previous tension. On the other hand, shearing at 90° or at 135° to the previous tension direction displayed increased or decreased initial flow stresses, in relation to the tension flow stress, respectively. In addition, work hardening transients were observed in both cases.

According to these authors, crystallographic texture effects were of no importance in the reported results, which were attributed to polarization and dissolution of the dislocation substructure introduced by tension, caused by the subsequent shearing.

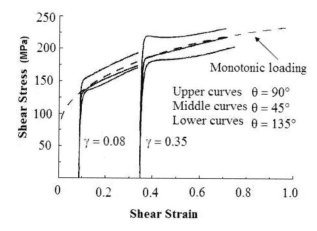

Figure 1. Shear stress (τ) – shear strain (γ) curves for low carbon steel sheets pre-strained in tension (full lines) or without pre-strain (dashed line), (Rauch and Schmitt, 1989).

The crystallographic texture of metal sheets before their final forming is controlled by many variables in their previous processing; in the case of steel sheets, these variables range from those in the steel making practice to those in their final heat treatments (Davenport and Higginson, 2000). Gracio et al., (2000) showed that the aluminum alloy AA1050-O presented the same dislocation substructure after various processing paths involving pre-strain by cold rolling followed by shearing, but the corresponding mechanical behavior, described by their shear stress –shear strain curves displayed different characteristics. These authors concluded that such results, (which included the observed stress and work hardening transients), should be ascribed to crystallographic texture effects.

The identification of the contributions of dislocation substructures evolution and/or of crystallographic texture effects to the stress and work hardening transients typically associated with strain path changes represents a significant challenge, made even more complex when one considers further aspects of the problem, such as the stress stagnation and localized softening of the material upon the strain path change. These phenomena are affected by the amount of pre-strain, by the structural characteristics of the material, by the amplitude of straining in the case of cyclic loading, by the initial condition of the material (initially annealed or work hardened), and by the loading sequence itself (Barlat *et al.,* 2003 and Gracio *et al,.* 2004). This paper presents experimental results for the processing of two materials (low carbon steel and CuZn 34 brass) subjected to a sequence of loadings involving rolling, tension and shearing. The results also cover the conditions under which the formability limit of the material, connected to the onset of plastic instability (necking) would occur. The choice of these materials represents situations where extensive cross-slip and dynamic recovery (DRV) predominates during deformation (low carbon steel) and where cross slip and the resulting DRV is limited (CuZn34).

Table 1. Chemical composition (weight %) of the steel and brass sheets

Low carbon steel		CuZn34 brass	
C	0.052	Cu	65.75
Mn	0.316	Zn	34.19
P	0.015	Pb	0.010
S	0.015	Fe	0.025
Si	< 0.05		

Table 2. Mechanical properties of the steel and brass sheets

Material	Mechanical properties			
	Hardness [HV]	$YS^{(1)}$ (MPa)	$UTS^{(2)}$ (MPa)	$\varepsilon u^{(3)}$
Low carbon steel	106	160	290	0.250
CuZn34 brass	130	260	415	0.317

[1]: Yield Strength.

[2]: Ultimate Tensile Strength.

[3]: Uniform Elongation.

2. MATERIALS AND METHODS

2.1. Materials

The materials utilized in the present research were sheets of low carbon steel and of CuZn34 brass 0.60mm and 0.51mm thick, respectively. The materials were in their "as-received" condition. Table 1 shows the chemical composition of both materials and Table 2 displays their initial mechanical properties. The high initial hardness of the brass sheet in comparison with that of the low carbon sheet suggests it was received in an initially work hardened state.

2.2. Experimental Procedure

Both materials were subjected to a deformation sequence composed of rolling$_{15\%}$/tension$_{8\%}$/rolling$_{8\%}$/tension$_{8\%}$/shearing. All loadings were applied along the sheet rolling direction (RD). The percentage figure after each deformation step indicates the level of true effective plastic strain applied. Rolling was performed in a mill with 200mm diameter rolls, no lubrication and at speed of 6.25m/min. The effective strain in rolling was calculated according to Eq. (3), see Hundy and Singer (1954).

$$\varepsilon_{efet(rolling)} = 2/\sqrt{3} \cdot \ln(t_0/t_f) \tag{3}$$

where:
t_0: initial thickness of the sheet [mm] and t_f: final thickness of the sheet [mm].

An INSTRON model 5582 machine was used for the tensile and the shearing tests, which were performed at an initial strain rate of $0.002s^{-1}$. Rectangular specimens were prepared in according to the ISO 50 standard for the tensile tests.

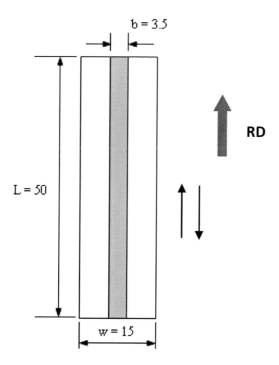

Figure 2. Dimensions of the shearing samples.

The shearing specimens were cut with a shearing machine after the last tensile deformation step. The shearing tests required a special fixture mounted on the INSTRON machine, as described by Lopes *et al.* (2007). Figure 2 illustrates the shearing samples, and Eq. (4) and (5) were used in order to convert the shearing strains (γ), into effective strains (ε_{efet}) and the shearing stresses (τ), into effective stresses (σ_{efet}), respectively.

These equations employ the conversion factor of 1.84, according to recommendations of Rauch (1992).

$$\varepsilon_{efet} = \gamma / 1.84 \tag{4}$$

$$\sigma_{efet} = \tau \cdot 1.84 \tag{5}$$

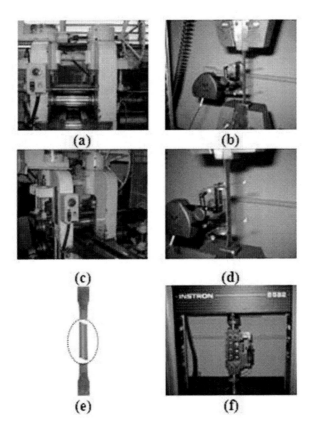

Figure 3. Scheme of the loading sequence: (a) rolling$_{15\%}$, (b) tension$_{8\%}$, (c) rolling$_{8\%}$, (d) tension (e) cutting of specimen for shear testing and (f) final shearing.

Figure 3 illustrates the deformation steps employed in this research, and Figure 4 shows the α parameters for each strain path change, where the angle between successive loadings was always zero.

The values of parameter α for the loading sequence exhibited in Figure 3 are shown in Figure 4. The parameters α were calculated according to Eq. (1) and (2). The lower value of the parameter α for the low carbon steel ($\alpha = 0.78$), for the rolling – tension sequence indicates that this material is more anisotropic that brass ($\alpha = 0.89$).

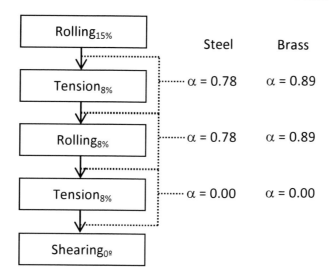

Figure 4. Values of the α parameter for the sequence: rolling15%/tension8%/rolling8%/tension8%/shearing0° used in this research.

3. RESULTS AND DISCUSSION

Figure (5a) presents the effective stress – effective strain curves for the low carbon steel and brass in their "as received condition". According to Zandrahimi *et al.*, (1989), strain localization under tension occurs when the product of the work hardening rate (θ) and the inverse of the effective stress ($1/\sigma$) becomes less than unity. Figure (5b) indicates that the effective strain value where this instability condition is satisfied is approximately the same for both materials. Initially, the work hardening rates for brass are lower than those for the low carbon steel, but for strains above 0.18, both materials work harden at approximately the same rate. It should be remembered, however, that the tensile testing of brass, in the present context, already represents some type of strain path change, since the material was in an initially work hardened state, probably obtained through cold rolling. The curves for brass given in Figures 5a and 5b indicates this pre-strain and the strain path change.

Figures 6a and 6b show the effective stress – effective strain curve for steel and brass, respectively, following the deformation sequence rolling$_{15\%}$/tension$_{8\%}$/rolling$_{8\%}$/ tension$_{8\%}$/ shearing. The materials presented a tensile flow stress after the initial rolling below the corresponding value for monotonic tension. Additionally, Figure 6 shows also that the low carbon steel presented only 0.035 effective strain (3.5%) before the onset of the plastic instability, during the quarter deformation step (tension), as exhibited in the legend of this figure and of the Figure 7.

The low carbon steel, however, presented a pronounced decrease in its work hardening rate during this tensile deformation, which was not observed for brass; the phenomenon being clarified in Figure 7, where the normalized work hardening rate $\theta.(1/\sigma)$ versus effective strain is plotted for both materials. The second tensile deformation for both materials (following the second rolling step) led to flow stresses initially above that for monotonic tension. This increase in the initial flow stress was higher for brass than for low carbon steel.

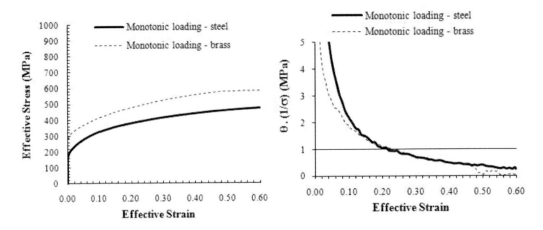

Figure 5. (a) Effective stress – effective strain curves and (b) curves of θ 1/σ - effective strain for the low carbon steel and for the CuZn34 brass.

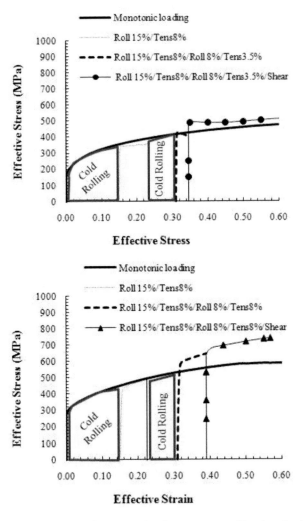

Figure 6. Effective stress – effective strain curves for the sequence rolling/tension/rolling/tension/ shearing for: (a) steel and (b) brass.

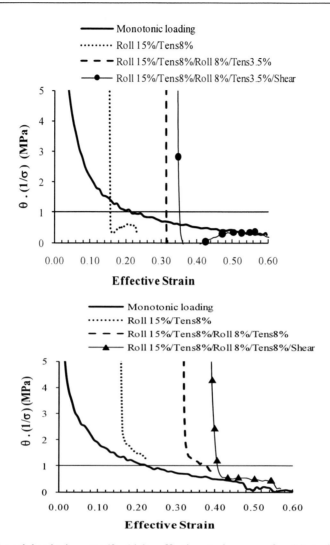

Figure 7. Normalized work hardening rate ($\theta \cdot 1/\sigma$) - effective strain curves for: (a) steel and (b) brass.

In addition to that, Figure 7 indicates that this tensile deformation led to a pronounced work hardening rate fall in the case of steel, which was not observed for brass. In both cases, it is considered that the dislocation substructure generated in the previous deformation steps represented an initial barrier to further movement during the second tensile step. (Wilson and Bate, 1994; Fernandes and Schmitt, 1983 and Barlat *et al.*, 2003), but, for the case of steel, this substructure was unstable during the additional tensile straining and underwent some dissolution, whereas for brass the substructure was stable and led to further work hardening.

Final shearing after the second tensile deformation step led to a behavior quite similar to that observed in the second tension after the second rolling step: The initial flow stresses were higher than that for monotonic tension and the final flow stress in the previous tension, but shearing in the low carbon steel led to a precipitous fall in the work hardening rate of this material, which reached negative values, (see Figure 7a), whereas brass displayed falling but positive work hardening rates during shearing. This confirms the suggested evolution of the dislocation substructure of both materials along the second tensile step. It is interesting to note that for the case of steel, Figures 6a and 7a indicate that an effective strain of only 0.035 (3.5%) in the second tensile step was sufficient to bring the material to a plastic instability condition. In the same step, brass allowed the imposition of an effective strain of almost 0.08 without reaching a condition for plastic instability. This is a consequence of the instability of the dislocation substructures developed in steel, where profuse cross-slip and cell formation seem to lead to a higher substructural instability than in brass, where cross-slip is hindered and cell formation is difficult. This is confirmed by the fact that brass reached a value below unity of the $\theta.(1/\sigma)$ parameter only at the end of the second tensile step, whereas steel had already reached this situation in the first tensile step.

It is also interesting to note that low carbon steel exhibited increasing work hardening rates following the initial precipitous fall of this parameter, observed after the strain path changes. This is agreement with the observations of Zandrahimi *et al.,* (1989), who observed that the recovery of the work hardening rate was more pronounced in metals displaying cross-slip ease, in comparison with materials where cross-slip is difficult; they reached this conclusion based on experiments with low carbon steel (easy cross-slip) and AISI 304 stainless steel (difficult cross-slip). In the present experiments, brass displayed a stabilization of the work hardening rate, after its initial fall, only in the last shearing deformation step.

Rauch and Schmitt (1989) showed that the strain amplitude in the transient range under shearing following tension, for low carbon steel, was approximately ¾ of the value of the tensile pre-strain. Figure 7a shows that the same behavior was observed for the present results: the hardening transient in the last shearing step for low carbon steel (see detail in Figure 7a) involved a strain of about 0.062, which is $\cong 0.772$ of the tensile pre-strain of 0.08. For brass, Figure 7b indicates that the same transient involved an effective strain of about 0.052, which is $\cong 0.651$ of the previous tensile pre-strain of 0.08, but with a work hardening rate always above the corresponding values under monotonic tension.

CONCLUSION

Low carbon steel and CuZn34 brass sheets presented the following behaviors during their deformation according to the strain path of rolling$_{15\%}$/tension$_{08\%}$/rolling$_{8\%}$/ tension/shearing, all steps taken at 0° to the material rolling direction:

- A different sensitivity of their effective stress − effective strain curves and work hardening − effective strain curves was observed to the various strain path changes.
- The low carbon steel presented a pronounced instability in its mechanical behavior, caused by the strain path changes, which caused steep changes in its work hardening rate.
- The CuZn34 brass presented higher ability to withstand plastic instabilities than the low carbon steel and it was able to maintain positive work hardening rates throughout all the imposed strain path changes.

The above results are considered to be connected to the lower stability of the dislocation substructures developed in low carbon steel, in comparison with those found in brass. This would be caused by the ease of cross-slip in low carbon steel (with the attending formation of dislocation cells), as opposed to the difficulty for cross slip in brass.

ACKNOWLEDGMENTS

The authors would like to thank CAPES, CNPq and FAPEMIG for supporting the present research work.

REFERENCES

Barlat, F., Ferreira Duarte, J. M., Gracio, J. J., Lopes, A. B. and Rauch, E. F. (2003), "Plastic flow for non-monotonic loading conditions of an aluminum alloy sheet sample", *International Journal of Plasticity*, 19, 1215-1244.

Davenport, S. B. and Higginson, R. L. (2000), "Strain path effects under hot working: an introduction", *Journal of Materials Processing Technology,* 98, 267-291.

Fernandes, J. V. and Schmitt, J. H. (1983), "Dislocation microstructures in steel during deep drawing", *Philosophical Magazine*, 48 (6), 841-870.

Gracio, J. J., Lopes, A. B. and Rauch, E. F. (2000), "Analysis of plastic instability in commercially pure Al alloys", *Materials Processing Technology,* 103, 160-164.

Gracio, J. J., Barlat, F., Rauch, E. F., Jones, P. T., Neto, V. F. and Lopes, A. B. (2004), "Artificial aging and shear deformation behavior of 6022 aluminium alloy", *International Journal of Plasticity,* 20, 427-445.

Hundy, B. B. and Singer, A. R. E. (1954), "The distribution of strains in the rolling process", *J. Inst. Metals*, 83, 401-407.

Lopes, Augusto Luís Barros. (2001), "Análise microestrutural das instabilidades plásticas em materiais metálicos", *Ph. D. Thesis,* Universidade de Aveiro, Departamento de Engenharia Cerâmica e do Vidro, Porto/Portugal, 264p.

Lopes, W., Corrêa, E. C. S., Campos, H. B., Aguilar, M. T. P. and Cetlin, P. R. (April 15-18, 2007), "Uso da técnica de cisalhamento planar simples para alterar a trajetória de deformação do aço AISI 430E", Anais do 4° Congresso Brasileiro de Engenharia de Fabricação, 4, 1-10, Estância de São Pedro/SP, Brazil.

Rauch, E. F. and Schmitt, J. H. (1989), "Dislocation substructures in mild steel deformed in simple shear*", Materials Science and Engineering*, 113A, 441-448.

Rauch, E. F. (1992), "The flow law of mild steel under monotonic or complex strain path", *Solid State Phenomena*, 23, 317-334.

Rauch, E. F. (2000), "Plasticity of metals during cold working", Multiscale phenomena in plasticity: from experiments to phenomenology, *Springer,* Paris, France, 529 p.

Schmitt, J. H., Aernoud, E. and Baudelet, B. (1985), "Yield loci for polycrystalline metals without texture", *Materials Science and Engineering,* 75A, 13-20.

Taylan, V., Wagoner, R. H.and Lee, J. K. (1998), "Formability of stainless steel", *Metallurgical and Materials Transactions,* 29A, 2161-2172.

Wagoner, R. H. (1981), "A technique for measuring strain-rate sensitivity", *Metallurgical and Materials Transactions,* 12A, 71-75.

Wilson, D. V., Bate, P. S. (1994), "Influences of cell walls and grain boundaries on transient responses of an IF steel to changes in strain path", *Acta Metall. Mater.,* 42 (4), 1099-1111.

Zandrahimi, M., Platias, S., Price, D., Barrett, D., Bate, P. S., Roberts, W. T. and Wilson, D. V. (1989), "Effects of changes in strain path on work hardening in cubic metals", *Metallurgical and Materials Transactions,* 20A, 2471-2482.

In: Machining and Forming Technologies, Volume 3
Editor: J. Paulo Davim

ISBN: 978-1-61324-787-7
© 2012 Nova Science Publishers, Inc.

Chapter 7

NUMERICAL AND EXPERIMENTAL ANALYSIS OF CRIMPING TUBULAR STEEL COMPONENTS BY MEANS OF A SPHERICAL PUNCH

João Emanuel Soffiati and Sérgio Tonini Button[*]
School of Mechanical Engineering, University of Campinas –
UNICAMP - C.P. 6122 Campinas – SP - CEP: 13.083-970 – BRAZIL

ABSTRACT

Crimping is a cold metal forming process which presents low cost when applied to assemble tubular components. The process of crimping analyzed in this work is used to manufacture components that transmit rotating motion, like in the joint system of brake pedals. In this study if the joint clearance is less than 0.02mm the brake movement could be restrained and could not close the vacuum valve of the break system causing reduction of efficiency. Otherwise, if the clearance is greater than 0.15mm, it could cause excessively noise. Another problem is the presence of cracks or fissures generated by the wrong choice of crimping parameters. Usually these parameters are not specified in the product design leading to high process variability. The main objective of this work was to analyze crimping and define the best process conditions to avoid defects commonly observed. Six factors were chosen to start the analysis: crimping punch geometry, punch surface roughness, diameter of the rod end, thickness of the tubular component, indentation depth, and position of indentation force. Crimping was analyzed experimentally with an apparatus specially designed and built for this work and compared to results obtained with the software MSC Superforge 2005.

Keywords: Metal forming, Statistical analysis, Failures, Numerical Analysis

---

[*] Email: sergio1@fem.unicamp.br

1. INTRODUCTION

The increasing necessity to supply products with low cost and high level of quality has been forcing Manufacturing Engineering to develop and search alternatives in the fabrication and assembling elements of machines and auto-parts.

Mechanical elements for fixtures such as screws, rings and rivets have been extensively used. However depending on the application of these elements they tend to become expensive and not trustworthy throughout the time.

In this work an alternative for the design of such mechanical elements is presented in a case that requires the transmission of movements of rotation in one given joint, as shown in Figure 1, which presents a joint of connecting rod of automobile brake pedal.

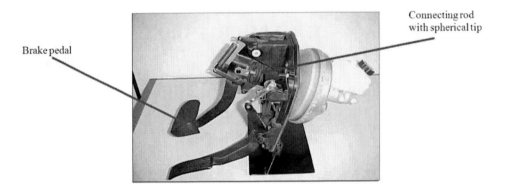

Figure 1. Components of automotive brakes.

In this specific case of automotive brake pedal, the clearance of the joint is of extreme importance, therefore if it is less than 0.02mm, a practical value in the usual manufacturing process, it will not be able to hold back the movement of the brakes pedal and then would affect the valve of vacuum control system in the brakeage amplifier.

Otherwise, if this clearance is greater than 0.15mm, an undesirable noise will be generated what is not accepted for the consuming market.

2. OBJECTIVES

The main objective of this project is to study some variables involved in the cold crimping of mechanical elements in connecting rods with spherical tip, as presented in Figure 2, in order to achieve a manufacturing process more robust which could provide the requirements of quality and security.

A complementary objective is the design, manufacturing and assembling of an experimental set for the physical simulation of crimping, and the development of a numerical model based in the method of the finite volumes.

The numerical results will supply information to verify the most important variables, the possible presence of cracks in the indentations, the influence of the clearance on the crimping, and finally the determination of the curve of indentation force versus depth and of the sphere extraction force, as a function of the selected process variables.

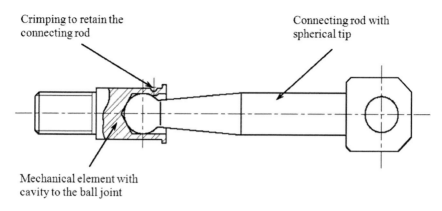

Figure 2. Mechanical elements assembled in the spherical tip by crimping.

3. THEORETICAL BACKGROUND

The influence of the geometry of the punch used in crimping can be verified in the model of Energy of Indentation for Fracture defined by Murty, *et al.*, (1998) that assumes conditions in which fracture can happen if the stresses generated for a spherical indenter are to exceed the critical stress of fracture for cleavage *(σf)*. From this criterion, the stress in the contact point is known as a function of the depth of indentation.

Ciavarella, *et al.*, (1998) studied the influence of rounded edges on the indentation with flat punches usually applied in fatigue tests. They concluded that if a perfect flat contact is established, the pressure at interface is very sensible to small variations in the geometry of the contact surface, particularly in the presence of small particles near to the punch edges.

If, however, punch edges are rounded off, many of these problems disappear, and therefore small particles will have a lesser influence on punch sliding, and the contact pressure will fall continuously until zero on the contact edges.

Also using the method of Muskhelishvili (1977), to determine the field of stresses, Ciavarella, *et al.*, (1998), developed a model to evaluate contact problems in the penetration of a punch with rounded ends, and concluded that the stresses present in the contact represent about 60% of the values found for a Hertzian equivalent case. This means that indenters with small bending radius influence significantly the stress distribution in the contact region.

Dini and Nowell (2003), presented a model based on the Hertzian formulation to determine the friction coefficient in the sliding contact area of a flat surface with an curved surface, i.e., a contact geometry similar to that found in the crimping with spherical punches.

Lee, et al., (2002) modeled the friction in sheet forming as a function of the lubrication system and the surface roughness of tools and metal sheet.

Shirgaokar, *et al.*, (2003), studied crimping with six indenters to deform a cartridge and fill a groove machined in the projectile. One of the objectives of the authors was to determine a good position between crimping punches and the machined groove. They demonstrated that the variation of this position has great influence in the extraction force of the projectile.

4. MATERIALS, METHODS AND EQUIPMENTS

To better understand the process of crimping studied in this work and to show the process variables, a simplified physical model was designed (Figure 3).

The indentation force (Find) is a function of the following variables (typed in italic): the *friction force*, in this case evaluated by the *surface roughness of the punch*; the *diameter of the sphere*, the *thickness of the wall of the tube*, the *geometry of the punch* (spherical or square), the *depth of indentation* and the *distance between the center of the sphere and the punch*, that also can be understood as the *gap or clearance for indentation*.

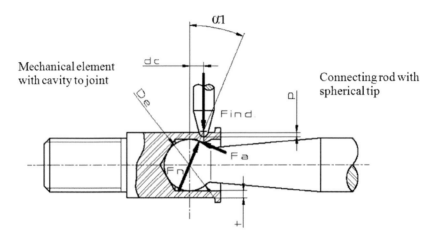

Figure 3. Crimping variables and composition of crimping forces.

For the analysis of crimping and the definition of the main influent variables on the process (Figure 4), it was designed a numerical model based on the method of finite volumes solved by the software MSC.Superforge 2005.

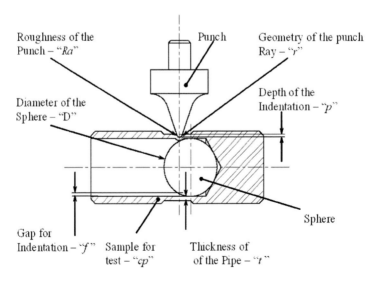

Figure 4. Representation of the main variables that affect crimping.

The press adopted in the numerical simulation represents the hydraulic press used in the experimental tests with an average speed of 0.25 mm/s.

To characterize the conditions of friction at the interfaces indenter-workpiece and sphere-workpiece it was assumed a Coulomb friction coefficient of 0.2 since the variable roughness of the punch cannot be simulated with the software. However, it is known that there is a relation between roughness and friction coefficient according to Lee et al., (2002). The room temperature was assumed to be 20°C considering crimping as a cold forming process.

In all models the size of the side of the volume element was assumed to be 0.3mm. The numerical models were designed considering five of the six variables shown in Figure 4: thickness of the tube; depth of indentation; gap for indentation; roughness of the punch; geometry of the punch and diameter of the sphere.

The influence variables of the experimental model were selected based on the theoretical revision, on the physical model, from the results of the numerical method, and from previous experience of the authors.

Besides these five influence variables, the variable hardness of the workpiece material was considered with a variation of 25 Brinell observed in lots of commercial materials used to manufacture the actual joint.

Considering all these variables, a first set of experiments was designed with a fractional factorial 2^{6-1} (Montgomery, 1991), which has six influence variables with two levels for each variable, to analyze their interaction and influence on the process.

The response variables selected were: indentation force, force to extract the sphere after crimping and gap between the sphere and the tubular workpiece.

All the experimental results of the crimping tests were evaluated through an analysis of variance - ANOVA, available in the commercial software Minitab version 2000 to calculate the significance of the effect of each variable studied as well as its order of influence on the response variable indentation force.

The analysis of fracture in the indentations was done after measuring the gap of indentation and consisted in the verification of cracks by scanning electron microscopy.

To verify the presence of passing cracks, a leaking test was done to evaluate the loss of vacuum during a specific period of time.

After these analyses tests were carried out to analyze the force necessary to extract the crimped sphere. The two variables of higher influence defined from the results of the first experiments were selected to plan the second set of experiments to validate the results of the preliminary factorial design, to replicate some tests, and to make a better conclusion on the influence of these more significant process variables.

After having reduced the study to two influence variables, new tests of crimping, measurement of the gap for indentation, measurement of the gap between the sphere and the tubular component, and leaking tests were carried out. The results of these tests were also evaluated through an analysis of variance – ANOVA.

The crimping experimental set (Figure 5) was designed to simulate the actual process of crimping and to carry out crimping tests with a steel tubular workpiece and spheres with diameter 7 or 12mm. With this device it was also possible to carry out the test of extraction ("pull out") of the sphere after crimping.

This device consists of three punches with adjustable movement to form indentations with variable depth on the external surface of the tubular workpieces.

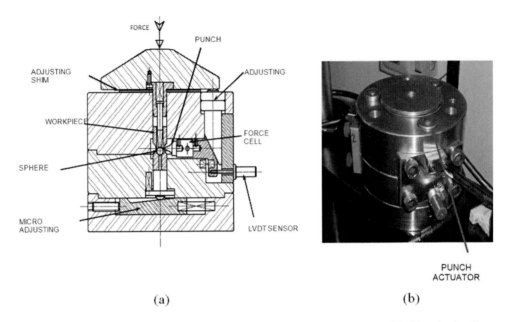

Figure 5. Experimental set used in crimping tests: (a) drawing details, (b) assembled in a hydraulic press.

These three punches were defined to make possible a condition of three points to support the sphere, with stability to measure the gap of the sphere and the extraction force of the sphere after crimping. The set also allows the adjustment of the position of the crimping punches in relation to the center of the sphere, the adjustment of the depth of indentation, and the equalization of the three indentations.

A load cell was connected in one of the punches to evaluate the indentation force necessary in each tested crimping condition.

To adjust the position of the punches in relation to the center of the sphere and to equalize the stroke of each punch it was used a 150mm caliper Vernier with a resolution of 0.05mm. A linear transducer assembled directly in the punch measured the depth of indentation as function of punch displacement pre-defined in 0.50 and 0.85 mm.

Workpieces of the crimping tests were made of steel AISI 12L14 equivalent to 9SMnPb28 - DIN1651 – as hot rolled, recommended for machining in automatic high speed machines, for the manufacturing of small and medium size parts.

The range of hardness measured in this material was from 114 to 140 HB. As this variation of the hardness is normally found in a lot of workpieces, it was assumed that this variation is an inherent condition in the process, and it was considered that to control a range smaller than 25 Brinell would become the process very expensive and impracticable because it would be necessary the inspection of hardness in practically 100% of the parts.

The surface roughness in the workpieces was achieved by machining, with average roughness Ra around 3.00 μm, which represents an economic condition and allows the easy slide of the punches that will perform the crimping.

Each variable was evaluated in two levels in a fractional factorial design 2^{6-1}, having as response variable the indentation force, and with only one replicate for each one of the factorial combination, representing a fractional factorial experiment 2^{6-1} - ½.

The value considered as result of each test was the maximum indentation force measured in each combination of variables levels, in such way that was possible to evaluate which combination of levels demanded the greater amount of energy to crimping.

With regard to the variable geometry of the punch the surface of indentation and the volume of material displaced in function of the depth of indentation had been taken in consideration to the dimensions of the square and spherical punches.

The force of extraction of the sphere is an important response variable and can be associated with the force of extraction of the connecting rod of the brake that is one of the requirements of the product. To evaluate this force it was used the same experimental method already presented, with a punch to push the sphere from the hole with the smallest diameter of the workpiece. The speed of extraction adopted was 0.40mm/s.

Considering that the results obtained in the crimping tests for the variable geometry of the punch had presented a critical condition of deformation for the square shaped form, it was assumed a maximum indentation force and the variable geometry of the punch was fixed in the level spherical shape with radius 1mm.

Therefore, for these tests five influence variables had been considered: thickness of the tube; indentation depth; gap for indentation; roughness and diameter of the sphere, and each one of the variables was evaluated in two levels in a fractional factorial design 2^{5-1}.

After the crimping tests, the same workpiece was used to measure the gap of the sphere, i.e., the gap between the sphere and the internal diameter of the tube, being also considered the six influence variables previously defined.

To verify the presence of passing cracks after crimping, tests of leaking were carried out with all workpieces tested in the fractional factorial experiments and in the experiments randomized by levels. In the leaking tests a line of industrial vacuum was used and the stabilization and fall of pressure had been monitored with a vacuum manometer.

5. RESULTS AND DISCUSSION

Table 1 shows the process conditions applied in the numerical simulation of crimping, which represent the experimental tests.

Table 1. Numerical simulation parameters

	Test number	Wall thickness (mm)	Gap for Indentation (mm)	Geometry of the Punch - radius (mm)	Sphere Diameter (mm)
Squared Punch	3	0.85	0.25	0	12
	4	1.25	0.25	0	7
	7	0.85	0.75	0	7
	8	1.25	0.75	0	12
Spherical Punch	19	0.85	0.25	1	7
	20	1.25	0.25	1	12
	23	0.85	0.75	1	12
	24	1.25	0.75	1	7

Figure 6 presents the results of indentation force as a function of the depth of indentation for a spherical punch and with a gap for indentation of 0.25mm (tests 19 and 20). In these tests the inclination of the curves is uniform until the depth of 0.85mm, and there is not an inflexion point.

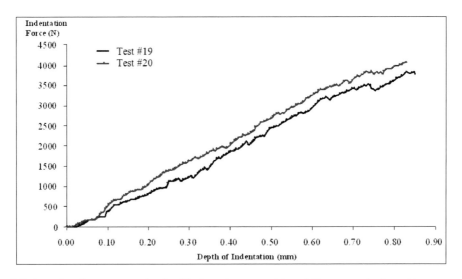

Figure 6. Indentation force versus depth of indentation, tests 19 and 20 – numerical results.

Table 2. Results of the numerical simulations

Test	Maximum Indentation Force Fl_{max} (N)	Results of Numeric Simulation			
		Maximum Effective Tension σeffec max (MPa)	Ratio σeffec max / σ_o	Maximum Effective Plastic Deformation εp effect max	Intensity Factor (Fi) x 10^3 N
3	3242	875.20	3.72	0.904	10.91
4	3116	852.40	3.63	0.722	8.16
7	1929	780.80	3.32	0.441	2.83
8	2146	758.20	3.23	0.367	2.54
19	3838	878.70	3.74	0.938	13.46
20	4073	820.50	3.49	0.598	8.50
23	2047	750.30	3.19	0.343	2.24
24	3028	783.80	3.34	0.448	4.52

Table 2 shows the results obtained in the numerical simulation for the maximum indentation force, maximum effective stress, relation maximum effective stress/flow stress and the effective plastic deformation.

All the conditions shown in Table 2 present the same effect on the deformation energy, i.e., the values of each parameter increases as the depth of indentation increases, and allows the calculation of a factor that relates these parameters, called intensity factor (Fi), which follows Equation (5.1):

$$F_i = \frac{FI_{max} \; \sigma_{efet-max} \; \varepsilon_{p-efet-max}}{\sigma_0}$$

(5.1)

Considering the intensity factor, it can be observed that tests 19, 03, 20 and 04 had presented the biggest forces in the numerical simulation. The common variable in these tests is the gap of indentation of 0.25mm that promotes the largest deformation of the workpiece and can cause the formation of passing cracks.

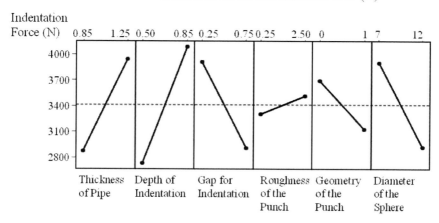

Figure 7. Main effects on the indentation force.

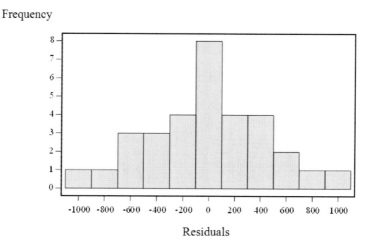

Figure 8. Histogram of residuals of the indentation force.

Figure 7 shows the variables that most influence the indentation force with a significance level of 10%: thickness of tube, depth of indentation, gap for indentation, geometry of the punch and diameter of the sphere. Despite the total of samples is 32, smaller than the standardized value of 50 samples to evaluate normality, a high degree of normality can be observed in the histogram of Figure 8.

This high normality confirms the good selection of process variables and that the range of hardness used in the analysis (114 to 135 HB) does not represent a variable of significant influence.

Figure 9 shows the results of indentation force as a function of the indentation depth and gap of indentation.

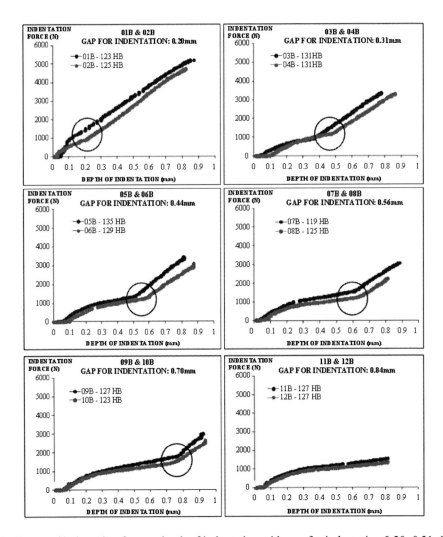

Figure 9. Curves of indentation force x depth of indentation with gap for indentation 0.20; 0.31; 0.44; 0.56; 0.70 and 0.84mm. Thickness of the tube: 0.85mm.

The inflexion point observed in Figure 9 is probably related with the depth where occurs the contact of the workpiece with the surface of the sphere, what characterizes a large energy of deformation able to cause the nucleation of cracks, and therefore, must be prevented. The inflexion point can be also verified in the results obtained in tests with wall thickness of 1.25mm.

In the second set of tests, for the extraction of the sphere and measurement of the extraction force it was considered the results of the first set of crimping tests which showed that the higher resistance to indentation was achieved with the punch of square end. Therefore, the spherical punch was adopted for the tests as the only condition for the punch geometry.

Figure 10 shows that the most influent variables on the extraction force with a significance level of 10% were: thickness of wall of the tube, depth of indentation, and diameter of the sphere.

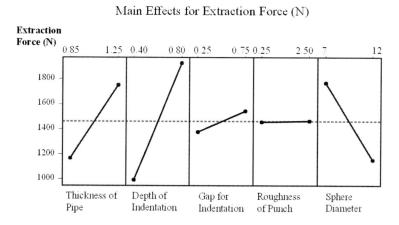

Figure 10. Main effects on the extraction force.

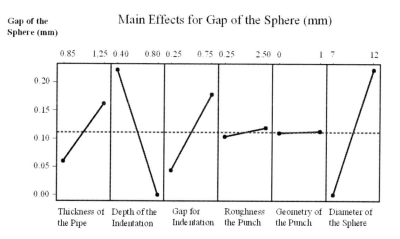

Figure 11. Main effects on the gap of the sphere.

In the third set of experiments (response variable gap of the sphere and significance level of 10%), influence variables depth of indentation, diameter of the sphere and gap for indentation were evaluated as causes of variations in the gap of the sphere after crimping, as shown in Figure 11.

After the crimping tests all samples were tested to evaluate leaking and it was not detected any presence of passing cracks. Table 3 summarizes all the analysis of the experimental results and Table 4 shows a comparison between simulation and experimental results.

Table 3. Summary of the analysis of the experimental results

Objective					
Variables	Level	Indentation Force	Force of the Extraction	Gap of Sphere	Conclusion
Diameter of the sphere (mm)	7.00		↑	↓	Variable adjusted according to necessity of the product
	12.00	↓			
Depth of the indentation (mm)	0.40	↓			Variable of control in the manufacturing proces
	0.80		↑	↓	
Thickness of the pipe (mm)	0.85	↓		Not significant	Variable adjusted according to necessity of the product
	1.25		↑		
Gap for indentation (mm)	0.25		Not significant	↓	Variable of control in the manufacturing proces
	0.75	↓			
Geometry of the punch	Square		Not significant	Not significant	The best form: spherical
	Spheric al	↓			
Roughness of the punch - Ra (μm)	0.25	Not significant	Not significant	Not significant	The best condition of the punch surface: polished surface
	2.5				

Table 4. Comparison of numerical and experimental results for indentation force, considering the depth of indentation equal to 0.85mm

	Test	Thickness of the Pipe (mm)	Gap for Indentation (mm)	Geometry of the Punch - Raio (mm)	Diameter of the Sphere (mm)	Indentation Force Numeric Method (N)	Indentation Force Experimental Method (N)	Difference in the Result (N)	Diference in the Result (%)
	3	0.85	0.25	0	12	3190	4605	1415	31%
Square Punch	4	1.25	0.25	0	7	3115	5763	2648	46%
	7	0.85	0.75	0	7	1914	3680	1766	48%
	8	1.25	0.75	0	12	2142	3521	1379	39%
Spherical Punch	19	0.85	0.25	1	7	3838	4237	399	9%
	20	1.25	0.25	1	12	4073	4164	91	2%
	23	0.85	0.75	1	12	2040	1695	-345	20%
	24	1.25	0.75	1	7	3037	4427	1390	31%

It is observed for the square end punch end that the numerical results are 31 to 48% smaller than the experimental results indicating that the numerical method is not adequate to determine the indentation force. Otherwise, the spherical punch presented numerical and experimental results that are very close especially with a gap for indentation of 0.25mm.

CONCLUSION

Considering that the quality and the confidence of results are implicit in the high indices of statistical significance achieved in the experiments, it can be concluded that:

- The variable diameter of the sphere has different effects on the response variables studied in this work. Considering that the primordial requirement for the functionality of the project is the absence of passing cracks, the use of a sphere with diameter of 12mm is recommendable to crimping if compared to the sphere of 7mm, because the first presents smaller indentation forces. However, if it is desired an increase of the force of extraction of the sphere and reduction of the gap of the sphere, the sphere of 7mm will present better results.
- The variables thickness of the tube and gap for indentation must have a mandatory control in crimping, since they present a significant influence on the response variable indentation force.
- The variable geometry of the punch has a significant influence in the indentation force; the spherical shape offers a smaller indentation force and therefore it is recommendable that crimping punches shall be designed with spherical ends.
- The variable depth of indentation presented the higher influence on the indentation force. Thus, the maximum permissible crimping depth must be evaluated to avoid the formation of cracks.
- The variables depth of indentation, diameter of the sphere and gap for indentation are significant causes of variation in the gap of the sphere after crimping.
- In order to define the diameter of the sphere some aspects must be considered: the position of the sphere and depth of indentation must be defined to achieve the gap for indentation necessary to deform the tubular workpiece without generating undesirable stresses and avoiding the generation of cracks; the force of extraction of the sphere shall be maximized by increasing the thickness of the tube and allowing an adequate gap (0.05mm to 0.15mm) between the tube and the sphere by the control of the depth of indentation.
- As much the indentation force as the effective stress and the effective plastic deformation had been maximized in the tests with the minimum value for the variable gap for indentation. Using the index intensity factor it was concluded that the variable gap for indentation is the most influent on the indentation force and effective strain when compared to the variables thickness of the wall of the tube and geometric form of the punch.
- In the comparative analysis between the numerical method and the experimental method, a significant difference was verified between the results with the square and

spherical punches, showing that the numerical method is more suitable to the analysis of crimping with spherical punches.

ACKNOWLEDGMENTS

The authors would like to thank "Fundação de Amparo à Pesquisa do Estado de São Paulo (FAPESP)", for the financial support to this work.

REFERENCES

Ciavarella, M., Hills, D.A. and Monno, G., 1997, "The influence of rounded edges on indentation by a flat punch", Proceedings of the Institution of Mechanical Engineers, Part C: *Journal of Mechanical Engineering Science*, 212 (4), 319-327.

Dini, D. and Nowell D., 2003, "Prediction of the slip zone friction coefficient in flat and rounded contact", *Wear,* 254 (3-4), 364–369.

Lee, B.H., Keum, Y.T., Wagoner, R.H., 2002, "Modeling of the friction caused by lubrication and surface roughness in sheet metal forming", *Journal of Materials Processing Technology,* 130-131, 60-63.

Murty, K.L., Mathew, M.D., Miraglia, P. Q., Shah, V.N. and Haggag, F.M, 1998, "Non-destructive evaluation and fracture properties of materials using stress-strain microprobe", *Non-Destructive Characterization of Materials in Aging Systems,* 503, 327–337.

Muskhelishvili, N.I., 1977. *Singular Integral Equation,* 1st ed. Springer: Noordhoff International Publishing.

Montgomery, D.C., 1991. *Design and Analysis of Experiments*, 3rd ed., John Wiley and Sons.

Shirgaokar, M., Cho H., Ngaile G., Altan, T., Yu, J.-H., Balconi J., Rentfrow R. and Worrell, W.J., 2004, "Optimization of mechanical crimping to assemble tubular components", *Journal of Materials Processing Technology,* 146 (1), 35-43.

In: Machining and Forming Technologies, Volume 3
Editor: J. Paulo Davim

ISBN: 978-1-61324-787-7
© 2012 Nova Science Publishers, Inc.

Chapter 8

PREDICTION OF LOCAL NECKING LIMIT CURVE IN SHEET METAL FORMING

José Divo Bressan
Department of Mechanical Engineering - UDESC Joinville,
Joinville SC – Brazil

ABSTRACT

The present work examines the mathematical models for the industrial processes involving biaxial stretching and drawing of sheet metal such as deep drawing and stamping operations, aiming at obtaining the theoretical forming limit strain curve or the local necking limit curve for the final shaped product. Historically, sheet metal formability has been assessed by simple laboratory testing such as the Erichsen test. Lately, the concept of experimental Forming Limit Curve (FLC) was developed to evaluate formability. The Forming Limit Diagram shows the FLC which is the plot of principal true strains on the sheet metal surface, ε_1 and ε_2, occuring at crital points obtained in the laboratory formability tests or in the real fabrication process. Two types of curves can be plotted: the local necking limit curve FLC-N and the shear fracture limit curve FLC-S. However, formability is a complex attribute that involves different variables such us the process parameters and the material plastic properties. In addition, material defects or heterogeneities such us thickness variations, roughness, porosities and local variations in the plastic properties affects the limit strains of sheet metals. The approaches of Marciniak-Kuczinski and D-Bressan for the theoretical prediction of forming limit curve owing to the onset of local necking, FLC-N, in sheet metal forming are utilized to investigate the influence of thickness and the mechanical plastic properties such us the plastic anisotropy, pre-strain, work hardening coefficient, strain rate sensitivity coefficient in the steel formability. The M-K model utilizes the concept of evolution of the initial defect in the sheet thickness f, while the D-Bressan's model uses the concept of local strain gradient evolution λ from the initial thickness gradient μ, i.e., the initial waviness in the sheet thickness. Some experimental results of forming limit curve obtained from literature for steel sheets are compared with the theoretical predicted curves. A new model for the theoretical prediction of the forming limit curve, based on the D-Bressan model for the development of the initial thickness defects and material properties which generates a strain gradient and ends in the local necking and sheet metal rupture, is analised. Limit strains are obtained from a software developed by the author.

Keywords: Forming limit curve, mathematical model, plasticity, M-K model, strain gradient model

1. INTRODUCTION

Sheet metal blanking and forming technologies are the principal fabrication processes in the automotive, aeronautic, kitchen utensils, metal packaging and other metal-mechanic sectors. The usual sheet metals utilized in these industries are steel, aluminum, brass and titanium alloys. The success of these fabrication processes are due to various factors which make them attractive and competitive as for example: good surface finishing, low weight, flexibility to change dies, production of parts with complex shapes, near net shape forming process, and sometimes is of low cost owing to its high mass production. However, a set of dies and stamping presses costs in the automotive industry is in the order of million dollars.

Globalization of economy has increased the industrial competition, forcing the reengineering of all activities and processes in the industries in order to decrease the production costs. Nowadays, in the engineering practice in industries, there is the formation of working group to study and advance the fabrication processes technologies aiming at improving the product quality and equipment efficiency as well as to reduce production and maintenance costs.

Formability of sheet metals is an important and complex issue related to the optimization and quality control of the final product. New developments and research have been carried out, aiming at improving the materials and these shearing, blanking, stamping, bending and other sheet metal forming processes and also the equipments in order to increase its productivity, quality and to lower costs through the finished product shape more simple, with zero defects and obtained by less number of operations. In addition, researches have been carried out to increase equipment productivity by faster presses and dies of high resistance to wear and fatigue.

Recently, a new research area was established in sheet metal forming: the incremental sheet metal forming process (ISFP) (Park and Kim, 2003; Jesweit et al., 2005). It is a flexible sheet metal forming process under the action of a rigid punch which moves around such that any peace with a tridimensional shape can be fabricated without the necessity to use a stamping die. Hence, the process is also named "dieless sheet metal forming" and has received the attention of various research groups in Europe. The forming operation consists of a small punch that deforms plastically a sheet metal by following a pre-determined contour path or a spiral path, while the sheet is clamped at its border only. However, many aspects related to the process mechanics, material formability and parameters which affects the process are not sufficiently understood. This fabrication process is recognized as a new sustainable technology owing to its potential to provide a fabrication process for small batches of unique under measure parts manufactured from a sheet metal, as well as to rebuild a damaged or obsolete product. The potential applications of incremental sheet metal forming are for example: fabrication of prototypes from formed sheet, fabricate or fix an automotive or airplane part (for ex. nose of airplane jet), orthopedic components, electro-domestic devices, casing for electronics and others.

The conventional sheet metal forming of fine sheets implies in utilizing a flat blank, a die, a blank hold and a punch that press the blank into the die to produce the appropriate shape of the final desired product. It is a metal forming method used to fabricate parts of complex or axisymmetric shape. Consequently, the strains and stresses that occurs during the plastic flow process are equally complex and of difficult evaluation and prediction.

The goal of present work is to examine the mathematical models by Marciniak-Kuczinski and D-Bressan for the theoretical prediction of the local necking limit curves, FLC-N, in sheet biaxial stretching. The D-Bressan's model uses the concept of local strain gradient evolution λ from the initial thickness defect μ, i.e., the initial waviness in the sheet thickness. The influences of the material parameters on the limit strains are studied. Experimental results of FLC-N obtained from literature for steel sheets are compared with the predicted curves for three thickness gauge.

2. Formability of Sheet Metals

Historically, in the early period of forties and fifties, formability of sheet metals was evaluated by simple tests of deep drawing or biaxial stretching as the Erichsen and Olsen tests. However, formability of sheet metals is a complex attribute which involves many variables as the forming process parameters and the material properties. In addition, material defects and heterogeneities as thickness variations, porosity, roughness and the plastic properties variations also influence the forming limits in sheet metal forming. The usual laboratory tests performed to obtain the formability characteristics or parameters of sheet metals presents the mechanical properties related to a particular strain path, generally linear strain path. However, the industrial forming process to produce an automobile panel is complex with non-linear strain path at the critical points, thus, the evaluation in general must involve more than one test. In addition, the material characterization by laboratory tests or the numerical simulation, do not avoid testing in the industrial scale or the "try out" phase, since the practical forming conditions will be correctly evaluated only by real experiments in the press shop which results should be compared with the laboratory tests and numerical simulations for better formability evaluation. This can be performed with the aid of the forming limit curve, FLC, which will be described below. The FLC curves are obtained by printing circles in the sheet surface and measuring the diameters evolution of circles or ellipses after the plastic deformation process, see Figure 1.

The concept of formability is based firstly on *rupture* or *local necking* in the sheet metal. This means that a material with good formability characteristics should not fracture or shows a visible local necking during the forming operations, but these are not unique factors. Secondly, there are the concepts of rigidity of shape (occurrence of spring back effect or elastic recovery, looseness), surface roughness or texture and the occurrence of wrinkles or sheet wrinkling. Therefore, evaluation of sheet metal formability in the press operation should be carried out considering not only the material mechanical properties, shape, roughness, etc., but also the operation conditions in the press or the process parameters and the tooling conditions in the real industrial scale production process.

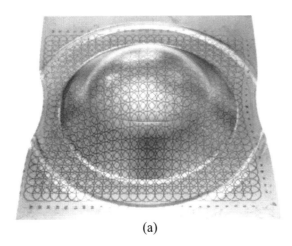 

(a)   (b)

Figure 1. Stamped sheet steel with printed circle marks for obtaining the Forming Limit Curve – FLC. a) shows local necking in the region of punch contact. b) shear fracture without necking (Koronen et al., 2006).

The main important characteristics of sheet metal forming identified in the practical press shop and laboratory experiences are the followings:

- stamping operation is the combination of sheet stretching and deep drawing processes,
- the quantity of sheet material displacement and rotation is high,
- the relative displacement between the sheet and the blank hold should be controlled by pressure, friction coefficient and by an appropriate geometric profile of blank hold so to avoid wrinkling or sheet rupture,
- small strains occurs only in a small portions of deformed surface area, less than 10%. Hence, the major part of the sheet surface area undergoes large deformations,
- elastic springback occurs after unloading the punch or pressure,
- in deep drawing, the higher deformations points occur in the region of approximately constant thickness, i.e., in the lateral cup wall and near the punch edge, but depend on material and friction,
- the limit strains of stamped parts in the press shop, usually are coincident to the experimental curves obtained from laboratory forming tests, despite the non linearity of strain paths in the manufactured parts.
- the strain path in the deformed sheet during the forming of pieces are largely non linear.
- sheet metal rupture is due to local necking or shear fracture mechanisms.

Therefore, the industrial process of sheet metal forming consists in changing the shape from a simple plane blank to a complex shape part with no rupture, local necking, wrinkling or finished rough surface. This practical observations leads to the conclusion that it is essential to describe and control the sheet surface and thickness strains evolution and roughness developments during sheet metal forming processes (Keeler, 1965).

The factors that help to attain uniform thickness distribution and high strain levels in the formed sheet part, avoiding local necking formation by distributing the strains and increasing the resistance to sheet thinning are:

- higher thickness gauge,
- higher material hardening coefficient *n*,
- higher material strain rate sensitivity coefficient *M*,
- high normal anisotropy coefficient *R*,
- high forming temperature,
- superplastic forming,
- very small initial thickness defect size f_o , very small grain size, very low microstructure defects as porosity and microcracks, high homogeneous microstructure,
- low friction.

Sheet wrinkling is controlled by sufficient pressure and appropriate profile of blank hold, low friction and low planar anisotropy coefficient of sheet material. Final surface roughness is controlled by the initial roughness parameters, material grain size, onset of surface shear bands or Lüders lines and local necking, friction against die and punch.

3. FORMING LIMIT CURVES: FLC-N AND FLC-S

As stated above, evaluation of sheet metal formability was initially performed by simple deep drawing laboratory tests as the Erichsen and Olsen tests. However, lately the concept of Forming Limit Diagram, FLD, or the Principal Strain Diagram to assess formability of sheet metals was developed by industry (Keeler, 1965). The FLD shows the principal true strains, ε_1 and ε_2 , or the major and minor engineering strains, e_1 and e_2 , measured at critical points near the rupture site in a printed mesh in the sheet surface in laboratory formability tests or in the press shop of sheet metal forming processes, i.e., presents the forming limit curve, FLC. Two types of curves can the plotted: local necking limit curve FLC-N and the shear fracture limit curve, FLC-S, which can be seen in Figure 2. Experimental curves and theoretic prediction of local necking and shear fracture limit have been intensively investigated by academic researchers as well as by industry professionals.

Since 1952, various mathematical models have been proposed to predict the forming limit curves of sheet metals in processes such us stamping, deep drawing, stretching, bending and other operations with constant or variable strain path based on the plasticity theory and the mechanics of sheet metal deformation (Hill, 1952; Swift, 1952; Marciniak and Kuczynski, 1967; Stören and Rice, 1975; Needleman and Triantafyllidis, 1978; Bressan and Williams, 1983).

One first work published to treat theoretically the mechanics of triggering local necking in sheet metal forming was the paper by Hill in 1952. The author approach was limited to analyze the negative quadrant (or the deep drawing region) of the forming limit diagram where directions of zero extension in the plane of the sheet exist, see Figure 2. In the positive quadrant, or region of biaxial stretching, the author postulated that there were no mathematical conditions in the plasticity theory for the onset of local necking, i.e., direction of zero extension in the sheet plane which allows velocity discontinuity.

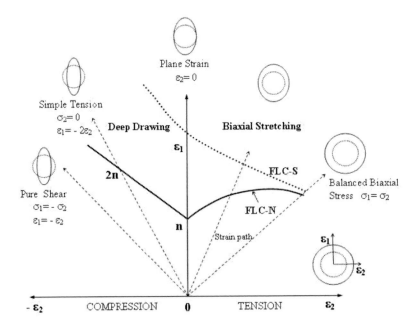

Figure 2. FLD or the Diagram of Principal True Strains in the sheet plane, $\varepsilon_1 \times \varepsilon_2$, showing the Local Necking Forming Limit Curve, FLC-N, and the Shear Fracture Forming Limit Curve, FLC-S, of sheet metals.

This theoretical paradox was solved later on in 1967 by the mathematical model presented by Marciniack and Kuczynski (1967). The authors proposed that the onset of local necking under the conditions of biaxial stretching deformations, in the positive quadrant of FLD, was due to the occurrence of an initial thickness defect $f_o = h/h_o$ in the sheet thickness h which evolves during sheet straining to plane strain condition and, therefore, local necking was mathematically admissible. Thus, forming limit strains in the sheet plane were attained only when the strain state inside the initial defect region moved to plane strain state, i.e., $d\varepsilon_2 = 0$. While this condition did not occur, the sheet would continue straining without the onset of local necking. Despite of the rigorous mathematical treatment of the M-K model, the theoretical results for the forming limit curve are very sensitive to the initial defect size f_o and presents discrepancies with experimental results which are considered non negligible in modelling the mechanics of sheet metal forming. Another disadvantage of the model is the defect parameter f_o to be one-dimensional or a through-thickness defect, without considering the width or the length of the initial defect grove. However, this model has given a new momentum to the theoretic researches on sheet metal forming limits, mainly on the mechanics of sheet metal deformation and on the influence of microstructure on fracture strains which goes on till now. Nowadays, it is without doubt the model most used in the theoretical investigations on the forming limit curves of sheet metal forming.

Another theoretical model was proposed in 1975 by Stören and Rice (1975) to explain the triggering of local necking in the positive quadrant of sheet biaxial stretching. The authors, based on the theory of strain bifurcation, considered the possibility of onset of plane strain condition in the sheet plane owing to the development of a vertex in the subsequent yield stress locus during the plastic flow or the yielding curve expansion in the continuous sheet metal deformation process. This strain bifurcation corresponds to the onset of local necking. The vertex would allow the change in the local strain path to plane strain and would be associated with discrete crystallographic nature of a polycrystalline material exhibiting various sliding systems, appearing during the expansion of the yield stress locus without necessity of an initial geometric defects existence in the sheet as postulated by Marciniack et al. However, the theoretical results were fairly good in the positive quadrant but very poor in the negative quadrant of the FLD, as they present large discrepancies in relation to the Hill's model and to the experimental values of various metallic alloys.

Alternatively, various other researchers (Ghosh, 1976; Needleman and Triantafyllidis, 1978) proposed criterions for rupture limit in sheet metals, based on the microstructure evolution of internal damage such as the nucleation, grow and coalescence of voids or micro-porosities at the failure site. Hence, these analyses aimed at investigating the superior forming limit curve or the shear fracture limit curve FLC-S in the forming limit diagram which is located after the local necking limit curve FLC-N as can be seen in Figure 2. The FLC-S is obtained from the critical strain values at the failure site or the rupture points in the sheet thickness.

In 1983, Bressan and Williams (1983) proposed a phenomenological model based in the rupture mechanism due to shear stress in the through thickness direction, based on the practical experience observation that the sheet rupture could occurs before as well as after the onset of local necking. Limit strains are calculated from a shear stress criterion: local necking or fracture occurs when the pure shear stress in the through thickness direction of zero extension attains a critical value which is a material property. This critical shear stress value can be obtained from the sheet material properties in the tensile test. The theoretical results for the limit strains obtained from this model for the positive quadrant of the FLD is very close to the results calculated by the Rice's model mentioned above. In general, the theoretical curve is situated below the experimental points, being the greater discrepancy between the predicted and experimental points for the aluminum killed steel which presents a very clear local necking before rupture. Its major advantage is the simplicity of the equation that predicts the FLC-N and the possibility to plot a region for minimum and maximum limit strain curve values.

In 1985, another model to predict the forming limit strain was presented by the present author (Bressan and Williams, 1985). A new approach for the mechanics of plastic deformations localization was proposed: the continuous grows of the local plastic strain gradient from an initial thickness imperfection, leading to the onset of local necking and fracture of sheet metals in the forming processes. This model is described in details below.

4. THEORETICAL ANALYSES OF LIMIT STRAIN FOR LOCAL NECKING: FLC-N

As mentioned above, sheet metal forming operations by biaxial stretching, stamping and deep drawing are considered failed when the formed part shows a visible local necking or a shear fracture. Within the biaxial stretching region of the FLD, the experimental investigations have shown that rupture is generally preceded by local necking or is due to a shear fracture mechanism (Bressan and Williams, 1983; Koronen et al., 2006).

Following, the M-K and D-Bressan mathematical models for predicting local necking in sheet metal for biaxial stretching operations are presented.

4.1. D-Bressan Mathematical Modelling

The deformation process for generating, development and localization of local necking in sheet metals have been investigated by the present author, using the concept of local strain gradient evolution approach from initial thickness imperfections (Bressan and Williams, 1985). The visible local necking arises from the development of an initial defect in the sheet thickness which is considered to present waviness variations in the sheet roughness: waviness of roughness. This initial defect is defined as thickness variation which is characterized by the normalized parameter of the initial gradient in the transversal area $\mu = (1/A_o)(dA_o/dx)$, where A_o is the initial area of the transversal section in the through-thickness direction and x is the coordinate axis in the direction perpendicular to the local necking and is situated in the sheet plane. It is also defined the local strain gradient as $\lambda = \partial \bar{\varepsilon}/\partial x$, where $\bar{\varepsilon}$ is the equivalent strain.

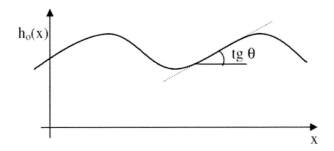

Figure 3. Initial profile of the waviness of roughness and the thickness $h_o(x)$ of sheet metal in the D-Bressan model.

In the present analysis, it is assumed that the sheet thickness varies according to the profile of the waviness of surface roughness, i.e., $h_o = h_o(x)$ is the thickness profile in the x-axis direction. Therefore,

$$\mu = \frac{1}{A_o}\frac{dA_o}{dx} = \frac{1}{h_o}\frac{dh_o}{dx} = \frac{1}{h_o} tg\,\theta \qquad (1)$$

where tg θ is the *inclination profile angle of the initial waviness of roughness in the sheet surface*, see Figure 3, x is the axis of coordinate system perpendicular to the local necking.

A visible local necking in the sheet plane occurs when the strain gradient λ attains a critical value which is assumed to be 20 on base of experimental results of biaxial stretching of various metallic alloys. The theoretical limit strains are in good agreement with the experimental results of FLC-N obtained from laboratory tests.

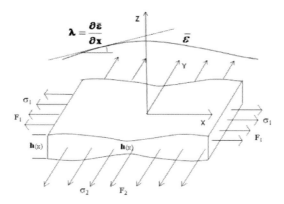

Figure 4. Element of sheet metal under biaxial stretching, showing the local necking and the definition of local strain gradient λ.

The theoretical analyses identifies a critical point of FLC-N, point of minimum limit strain value which correspond to the plane strain condition similar to the sheet bending at the punch corner or die edge. Hence, this is the most probable site of onset of local necking or sheet fracture occurrence. This model was lately extended (Bressan, 1997), aiming at considering the existence of initial imperfections of thickness μ and heterogeneities in the material mechanical proprieties such as the coefficient of hardening n, coefficient of strain rate sensitivity M, coefficient of strength k and pre-strain ε_o which are parameters of plastic behavior of sheet material constitutive flow equation presented in the following.

Present analyses consider the straining of a fine sheet metal which material exhibits both strain and strain rate hardening plastic behavior. Hence, the constitutive equation of plastic flow is,

$$\bar{\sigma} = k\left(\varepsilon_o + \bar{\varepsilon}\right)^n \dot{\bar{\varepsilon}}^M \qquad (2)$$

where $\bar{\sigma}$ is the equivalent flow stress, k is the strength coefficient, ε_o is the pre-strain, $\bar{\varepsilon}$ is the equivalent strain, n is the hardening coefficient, $\dot{\bar{\varepsilon}}$ is the strain rate and M is the coefficient of strain rate sensitivity.

The yield stress criteria proposed by Hill (1979), which accommodates the coefficient of normal anisotropy R less than 1, is used in the present analysis,

$$\bar{\sigma}^m = \frac{1}{2(1+R)}[(1+2R)|\sigma_1 - \sigma_2|^m + |\sigma_1 + \sigma_2|^m] \qquad (3)$$

where m is also an intrinsic parameter of normal anisotropy, $m = 1{,}14 + 0{,}86R$ (Bressan and Williams, 1983), σ_1 and σ_2 are the principal stresses in the sheet plane as can be seen in Figure 4. The equation that governs the formation and growth of local necking from the initial thickness imperfections μ in sheet metal forming processes is (Bressan and Williams, 1985),

$$\frac{\partial \lambda}{\partial \overline{\varepsilon}} = \frac{\mu}{M} + \frac{1}{M}\left\{\frac{\alpha}{(1+\alpha)z} - \frac{n}{(\varepsilon_o + \overline{\varepsilon})}\right\}\lambda \qquad (4)$$

where $\lambda = \partial \overline{\varepsilon}/\partial x$ is the local necking strain gradient, $\alpha = \partial \varepsilon_1 / \partial \varepsilon_2$ is the strain path and z is the subtangent,

$$z = \frac{[2(1+R)]^{1/m}}{2(1+\alpha)}\left\{\frac{|\alpha-1|^{m/(m-1)}}{(1+2R)^{1/(m-1)}} + |\alpha+1|^{m/(m-1)}\right\}^{\frac{m-1}{m}} \qquad (5)$$

Equation (4) can be solved either analytically or numerically to obtain the detailed description of the strain gradient evolution during the sheet straining process. Varying the strain path α, the local necking limit curve or FLC-N can be calculated for the positive quadrant of FLD when the strain gradient λ attain a critical value λ_{crit} = 20 or when $\lambda/\mu = (\lambda/\mu)^*$ = constant.

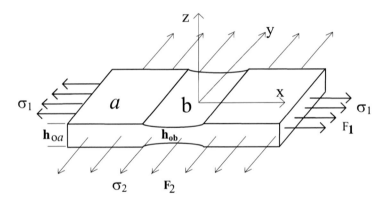

Figure 5. Sheet metal element under biaxial stretching, showing the narrow band of initial thickness defect for the M-K model.

4.2. Marciniack-Kuczynski Mathematical Modelling: The M-K Model

Marciniack and Kuczynski (1967) presented a model for the onset of local necking in sheet metal stretching processes based on the assumption that local necking develops from an initial local defect or thickness heterogeneity in the sheet metal. This initial imperfection is assumed to be a narrow band or groove in the sheet surface which has reduced thickness and is characterized by the defect parameter f_o, $f_o = \frac{h_{ob}}{h_{oa}}$, where h_{ob} is the initial thickness inside the defect and h_{oa} is the initial homogeneous thickness outside the region of defect as indicated in Figure 5.

Considering the instantaneous equilibrium of forces between region a and b in the sheet metal element seen in Figure 5, and for a given instant during the sheet straining process, the equilibrium is $F_{1a} = F_{1b}$, hence,

$$\sigma_{1a}.h_a = \sigma_{1b} h_b \tag{6}$$

Defining the stress ratio $\beta = \sigma_2/\sigma_1$, the equivalent flow stresses in region a and b of sheet element, for an isotropic material, are calculated respectively by,

$$\overline{\sigma}_a = \sqrt{1-\beta_a + \beta_a^2}\ \sigma_{1a} \quad \text{and} \quad \overline{\sigma}_b = \sqrt{1-\beta_b + \beta_b^2}\ \sigma_{1b} \tag{7}$$

Additionally, considering equal strain increments in the direction along the neck , $d\varepsilon_{2a} = d\varepsilon_{2b}$, defining the current thickness defect parameter as $f = h_b/h_a$, the strain path in region a $\alpha = d\varepsilon_1/d\varepsilon_2$, the true thickness strain $\varepsilon_3 = ln(h/h_o)$ and assuming isotropic material with strain hardening plastic behavior given by Equation (2), the governing equation to the local necking evolution for the M-K model, from the initial Equation (6), is given by,

$$\frac{\sqrt{1-A}\left(\dfrac{d\overline{\varepsilon}_a}{d\overline{\varepsilon}_b}\right)^M}{\sqrt{1-A\,(\dfrac{d\overline{\varepsilon}_a}{d\overline{\varepsilon}_b})^2}} = f\,\frac{(\varepsilon_o + \overline{\varepsilon}_b)^n}{(\varepsilon_o + \overline{\varepsilon}_a)^n} \tag{8}$$

where,

$$f = f_o\,\exp\left(\frac{\overline{\varepsilon}_a}{Z} - \varepsilon_{3b}\right)\ ;\quad \varepsilon_{3b} = \int_0^{\overline{\varepsilon}_b}\left[\frac{\sqrt{3}}{2}\sqrt{1-A\,(\frac{d\overline{\varepsilon}_a}{d\overline{\varepsilon}_b})^2} + B\,\frac{d\overline{\varepsilon}_a}{d\overline{\varepsilon}_b}\right]d\overline{\varepsilon}_b \tag{9}$$

The equivalent plastic strain increment is $d\overline{\varepsilon} = \sqrt{\dfrac{2}{3}}\sqrt{d\varepsilon_1^2 + d\varepsilon_2^2 + d\varepsilon_3^2}$ and the coefficients are,

$$A = \frac{3}{4\,(1+\alpha+\alpha^2)}\quad B = \frac{\sqrt{3}}{4}\frac{1}{\sqrt{1+\alpha+\alpha^2}}\quad Z = \frac{2\sqrt{1+\alpha+\alpha^2}}{\sqrt{3}\,(1+\alpha)} \tag{10}$$

Equations (8) and (9) must be solved by the iterative method for each strain increment $d\overline{\varepsilon}_b$ inside the neck or in region b. The maximum limit strain outside the neck $\varepsilon_{1a}{}^*$ is attained when the total strain ε_{1b} or $\overline{\varepsilon}_b$ tends to a very large value while the strain ε_{1a} in region a

remains constant. The forming limit strain $\varepsilon_{1a}{}^*$ is very sensitive to the initial defect parameter f_o, as shown in Figure 8.

The initial sheet inhomogeneities or imperfections characterized by the conceptual parameter f_o might include thickness variations, non-uniform distribution of impurities, varying texture, different size and orientation of grains, porosity and others. Although this concept provides a powerful and fruitful model, there are discrepancies with experimental results obtained from laboratory testing not yet fully explained.

5. RESULTS AND DISCUSSION

Considering that the maximum inclination angle of waviness of roughness on the sheet metal surface are less than $\theta = 10^o$, from experimental observations, the normalized critical strain gradient $(\lambda/\mu)^*$ in the local necking for sheet metals of thickness 1 mm is calculated approximately as,

$$ (\lambda/\mu)^* = \frac{h_o}{tg\,\theta}\,\lambda_{crit} = 113.4 $$

$$(11)$$

Equation (11) shows that the local necking limit strain curve depend upon the initial thickness gauge and on the waviness profile of roughness of the sheet metal.

Figure 6 presents the theoretical results for the local necking forming limit curve, FLC-N, predicted from the present D-Bressan model of the local necking strain gradient evolution, $\lambda = \partial\bar{\varepsilon}/\partial x$. It shows the influence of plasticity parameters ε_o and M on the major limit true strain $\varepsilon_1{}^*$ for material which hardening law is $\bar{\sigma} = k\left(\varepsilon_o + \bar{\varepsilon}\right)^{0.22}\dot{\bar{\varepsilon}}^{0.012}$.

The increase of pre-strain from $\varepsilon_o = 0.05$ to $\varepsilon_o = 0.20$ causes a decreasing on the limit strain $\varepsilon_1{}^*$ from 0.31 to 0.21, i.e., a reduction of about 50 %. Hence, eliminating the initial material work hardening by heat treatment of full annealing can increase up to 50 % the formability of fine steel sheets. Usually, the final thickness of fine sheet metal is accomplished by cold rolling which increases the material pre-strain and, consequently, decreases the FLC-N. It is also noticed that the increase in M-value from 0.012 to 0.018 provides an increase of 10% in the FLC-N.

Alternatively, Figure 7 shows the theoretical prediction of present model for the local necking forming limit curve for two hardening coefficients, $n = 0,22$ and $0,30$ and two pre-strain values of $\varepsilon_o = 0,05$ and $0,15$, i.e., the graph shows the influence of hardening coefficient n and pre-strain on the limit strain $\varepsilon_1{}^*$. Despite of the pre-strain increase by 3 times, 200%, the increase of 36% in the work hardening coefficient was enough to compensate the curve lowering and to produce an increase in the forming limit strain $\varepsilon_1{}^*$ by approximately 26 % from 0.30 to 0.38.

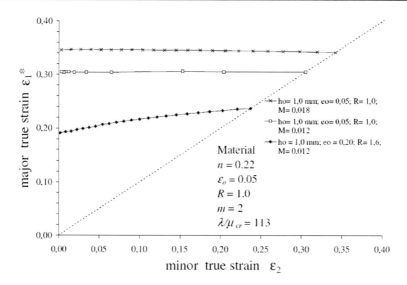

Figure 6. Theoretical prediction of Local Necking Forming Limit Curve, FLC-N, according to the D-Bressan model of local necking strain gradient evolution. Influence of parameters ε_o and M on the limit strain ε_1^*.

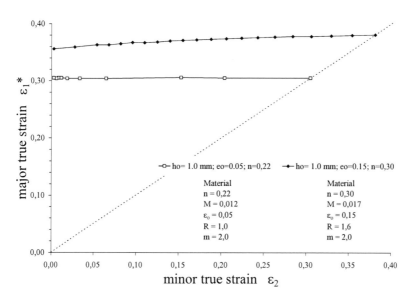

Figure 7. Theoretical prediction of Local Necking Limit Strain according to the present model of local necking strain gradient evolution. Influence of work hardening coefficient n on the limit strain ε_1^*.

Figure 8 shows the theoretic prediction of FLC-N for various values of the initial defect f_o according to the M-K model (Mattiasson and Sigvant, 2006). It is observed that the limit strains curve is very sensitive to small variations in the defect parameter f_o. Hence, this is the main disadvantage of M-K model, so this do not correspond to real experimental results verified in the laboratory sheet metal testing studies, although it is the model most used in the investigation of sheet metal formability in biaxial stretching.

In Figure 9, comparison between the experimental FLC-N and the present D-Bressan model prediction for sheet steels of thickness gauge 1, 2 and 3 mm (Müschenborn and Sonne,

1977) is presented. The increase in thickness gauge yields an increase in the limit strain ε_1^* which is anticipated by the present model: the theoretical prediction is in good agreement with the experimental results. The greater difference is in the plane strain state condition, where possibly the adopted steel parameters are different, mainly the work hardening coefficient should be lower than that for balanced biaxial stress.

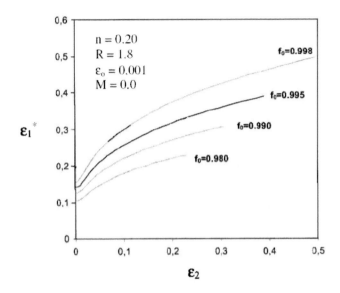

Figure 8. Prediction of FLC-N for various values of thickness defect f_o, according to the M-K model (Mattiasson and Sigvant, 2006).

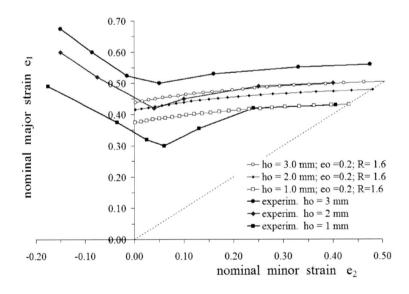

Figure 9. Comparisons between the experimental FLC-N and present model theoretical prediction for steel sheet of 1, 2 and 3 mm. (Müschenborn and Sonne, 1977).

CONCLUSION

In the light of the experimental results from literature and the predictions of the forming limit curves for the onset of local necking, FLC-N, in sheet metals under biaxial stretching the following conclusions can be drawn:

a) the present model for the theoretical prediction of the FLC-N is in fairly good agreement with the experimental values obtained from laboratory tests of steel sheets,

b) the two main material parameters that affects the FLC-N in the present model are the coefficients of strain hardening n and the strain rate sensitivity M,

c) the initial imperfections in sheet metals are due to thickness variations or the waviness of roughness in the sheet metal surface. Local necking occurs when the local strain gradient attains critical value $(\lambda/\mu\ _{crit} = \dfrac{h_o}{tg\ \theta}\ \lambda\ _{crit} = 113.4\ h_o$, where $(\lambda/\mu$ is the normalized strain gradient, μ is the initial thickness defect, h_o is the initial thickness and θ is the inclination angle of the waviness profile.

d) the theoretical prediction of the FLC-N curves by present model increases directly with the sheet thickness gauge h_o . The M-K model does not predict the influence of the thickness gauge on the limit strains.

ACKNOWLEDGMENTS

The author would like to thank the research scholarship received from CNPq as financial support and the University of Santa Catarina State, UDESC.

REFERENCES

Bressan, J. D. and Williams, J. A. (1983), "The Use of a Shear Instability Criterion to Predict Local Necking in Sheet Metal Deformation", Int. J. Mech. Sci., 25, 155.

Bressan, J. D., Williams, J. A. (1985), "Limit Strains in the Sheet Forming of Strain and Strain-Rate Sensitive Materials", *J. Mech. Working Tech.,* 11, 291.

Bressan, J. D. (1997), "The Influence of Material Defects on the Forming Ability of Sheet Metals". *J. Mater. Process. Technol.,* 72, 11-14.

Ghosh, A. K. (1976), "A criterion for ductile fracture in sheets under biaxial loading". *Metall. Trans. A,* 7A, 523-541.

Hill, R., 1952, "On Discontinuous Plastic States with Special Reference to Localised Necking in thin Sheets", *J. Mech. Phys. Solids,* vol. 1, pp. 19-30.

Hill, R. (1979), "Theoretical Plasticity of Textured Aggregates". *Math. Proc. Cam. Phil.* Soc., 85, 179-191.

Jesweit, J., Micari, F., Hirt, G., Bramley, A., Duflou, J. and Allwood, J. (2005), "Asymmetric single point incremental forming of sheet metal*", Ann. CIRP* 54 (2).

Keeler, S. P. (1965), "Determination of Forming Limits in Automotive Stampings". Sheet Metal Industries, 683-691.

Koronen, A. S., Manninen, T. and Kanervo, K. (2006), "On Necking, Fracture and Localization of Plastic Flow in Austenitic Stainless Steel Sheets". In: ESAFORM 2006, Glasgow/UK, 93-98.

Marciniak, Z. and Kuczynski, K. (1967), "Limit Strains in the Process of Stretch-Forming Sheet Metal", *Int. J. Mech. Sci.*, 9, 609-620.

Mattiasson, K. and Sigvant, M. (2006), "On necking prediction in ductile metal sheets". In: ESAFORM 2006, Glasgow/UK, 335-338.

Müschenborn, W. and Sonne, H. M. (1977*), "Forming limits of steel sheet determined by cine film technique"*. In: Communication to the IDDRG Conference, Warsaw, Poland, October 1977.

Needleman, A. and Triantafyllidis, N. (1978), "Void Growth and Local Necking in Biaxially Streched Sheets". Journal of Eng. Materials and Tech., Trans. ASME, 100, 164-169.

Park, J. J. and Kim, Y. H. (2003), "Fundamental studies on the incremental sheet metal forming technique", *J. Mater. Process. Technol.*, 140, 447–453.

Stören, S. and Rice, J. R. (1975), "Localized Necking in Thin Sheets*", J. Mech. Phys. Solids,* 23, 421-441.

Swift, H. W. (1952), "Plastic instability under plane stress". *J. Mech. Phys. Solid,* 1(1), 1.

In: Machining and Forming Technologies, Volume 3
Editor: J. Paulo Davim

ISBN: 978-1-61324-787-7
© 2012 Nova Science Publishers, Inc.

Chapter 9

STUDY OF AN AUTOMOTIVE COMPONENT PRODUCTION THROUGH THE UTILIZATION OF THE FORMING LIMIT STRAIN AND STRESS DIAGRAMS

S. M. R. Cravo[1], B. P. P. A. Gouveia[2] and J. M. C. Rodrigues[2]*
[1]Instituto Superior Técnico, Av. Rovisco Pais, 1049-001 Lisboa, Portugal
[2]Instituto Superior Técnico, Departamento de Engenharia Mecânica,
Av. Rovisco Pais, Lisboa, Portugal

ABSTRACT

Strain based forming limit diagram is well known and has been widely applied in sheet metal forming operations, during the design and project of the stamping tools, in order to prevent the occurrence of localized necking and fractures. However, the strain based forming limit diagram is very sensitive to strain path changes. This fact limits the use of the conventional forming limit diagrams for assessing forming severity of complex operations, such are the multi-step forming processes. Recently, forming limit stress diagrams have been intensively studied and have been established that the stress based forming limit curves exhibit no significant dependence on strain path for a wide range of materials.

This paper presents the utilization of forming limit strain and stress diagrams to eliminate formability problems that occur during the fabrication of an industrial automotive part. This automotive part is presently produced in a sequence of two stamping operations that originate excessive thickness reduction and fractures due to localized necking on the component. The aim of this work was the elimination of these defects through the suggestion of a new geometry for the first operation tool. The study was conducted using an existing explicit finite element analysis program. Forming limit curves based on strain and stress were determined using the theoretical model proposed by Stören and Rice and by Hill. The work was supported by experimental observations and data measurements.

---

* Fax: +351-21-8419058 E-mail: bgouveia@ist.utl.pt

Keywords: sheet metal forming, finite element analysis, forming limit strain and stress diagrams, forming limit curve

1. INTRODUCTION

During design of sheet metal forming tools, the knowledge on sheet metal formability limits plays an important role, since the amount of plastic deformation in the sheet metal part is limited by the occurrence of undesirable phenomena such as localized necking and fractures. To study and prevent these formability problems during stamping operations, strain-based forming limit curves have been widely used (Keeler and Backofen, 1963). However, its utilization is restricted to linear or near linear loading conditions, leading to erroneous assessments on the analysis of sheet metal formability under complex strain paths such as multi-step forming. Recently, stress based forming limit diagrams and the respective curves (Arrieux, 1982) have been successfully applied under these circumstances (Wu, 2005; Stoughton, 2004 and Chen, 2007).

This paper presents the utilization of forming limit strain and stress diagrams to study an automotive part fabrication done through a sequence of two stamping operations. The produced parts exhibit formability problems related with the occurrence of fractures induced by localized necking onset and high values of thickness reduction. The aim of this work was the achievement of a new geometry for the tool used in the first operation in order to eliminate these defects. The analyses of both stamping operations were made by means of the numerical simulation using an existing explicit finite element analysis program. Major and minor principal strains and stresses in the plane of the sheet metal component were plotted in the respective forming limit strain and stress diagrams at the end of both stamping operations. The forming limit curves based on strain and stress were determined using the theoretical model proposed by Stören and Rice (1975). According to some studies, Stören and Rice model underestimates the limit strains for localized necking when the ratio of the principal strains in sheet metal plane $\beta = \varepsilon_2 / \varepsilon_1$, is negative, therefore, Hill's theoretical model (1952) was considered in this region instead. Under these circumstances, considering a power law stress-strain relation ($\sigma = K\varepsilon^n$) to reproduce the mechanical behaviour of the sheet material, the critical strain condition considered for the localized necking onset, on the left side of the forming diagram (according to Hill's model, 1952) is:

$$\varepsilon_1 = \frac{n}{1+\beta} \quad \text{and} \quad \varepsilon_2 = \beta\left(\frac{n}{1+\beta}\right) \qquad , -1 < \beta \le 0 \tag{1}$$

and,

$$\varepsilon_1 = \frac{3\beta^2 + n(2+\beta)^2}{2(2+\beta)(1+\beta+\beta^2)} \quad \text{and} \quad \varepsilon_2 = \beta\left(\frac{3\beta^2 + n(2+\beta)^2}{2(2+\beta)(1+\beta+\beta^2)}\right) \quad , 0 \le \beta \le 1 \tag{2}$$

on the right side of the forming diagram, according to Stören and Rice model (1975).

On the other hand, the stress based forming limit curves were obtained following the method proposed by Stoughton (2004), resulting in the following locus of stress points for the Hill model (1952):

$$\sigma_1 = \frac{K(2+\beta)}{\sqrt{3(1+\beta^2)}}\left(\frac{2n\left(3(1+\beta^2)\right)^{1/2}}{3(1+\beta)}\right)^n \quad \text{and} \quad \sigma_2 = \frac{K(1+2\beta)}{\sqrt{3(1+\beta^2)}}\left(\frac{2n\left(3(1+\beta^2)\right)^{1/2}}{3(1+\beta)}\right)^n , -1 < \beta \le 0 \tag{3}$$

and the following locus of stress points for Stören and Rice model (1975):

$$\sigma_1 = \frac{K(2+\beta)}{\sqrt{1+\beta+\beta^2}}\left(\frac{3\beta^2 + n(2+\beta)^2}{(2+\beta)\sqrt{1+\beta+\beta^2}}\right)^n \quad \text{and} \quad \sigma_2 = \frac{K(1+2\beta)}{\sqrt{1+\beta+\beta^2}}\left(\frac{3\beta^2 + n(2+\beta)^2}{(2+\beta)\sqrt{1+\beta+\beta^2}}\right)^n , 0 < \beta \le 1 \tag{4}$$

where K and n are the material constants of the Hollomon's power law stress-strain relation, $\sigma = K\varepsilon^n$ (please refer to section 3.2.1).

2. CASE STUDY DESCRIPTION

The case study presented in the present work reports to the fabrication of an automotive part produced at a Portuguese company. The part is a suspension support produced in a sequence of two stamping operations followed by other finishing operations, using a 2 mm thick high strength low alloy steel S420MC (EN 10149-2, 1995) sheet. At the end of both stamping operations the produced parts exhibit formability problems related with the occurrence of fractures induced by localized necking and values of thickness reduction above the limits usually accepted for this kind of components (10~15%). Figure 1 shows the component exhibiting fractures after the first (a) and after the second (b) stamping operations. It is also possible to identify a localized necking in the component after the first stamping operation, showed in detail at Figure 1a. Even thought Figure 1a shows a component exhibiting a fracture after the first stamping operation it is worth mentioning that fractures do not always occur during the first operation.

The approach adopted in this study to propose a new geometry for the first forming operation tool, comprised three phases: (i) characterization of the sheet material mechanical behaviour by means of uniaxial tensile tests; (ii) reproduction and analysis of the present fabrication conditions through numerical modelling based on a dynamic explicit finite element computer program. This phase allowed to indirectly determine the unknown process parameters, i.e., the blank holder force and the friction coefficient. These values were obtained, for both operations, through the comparison between numerical results and experimental measurements of thickness along selected cross sections, geometries and contours of the stamped parts. Another important objective of this phase was to evaluate the capability of the commercial finite element program, utilised in the present work, to reproduce the present stamping operations; (iii) study and analysis of different tool geometries in order to find the most suitable geometry to solve/reduced the formability problems previously described. In this stage, the results of the analysis of the present fabrication conditions made in the previous phase were used. Several tool geometries were tested. The analysis of the numerical results, for both operations, was carried out based on the forming diagrams of strain and stress considering the limits given by the Equations (1) and (2) and Eq. (3) and (4), respectively.

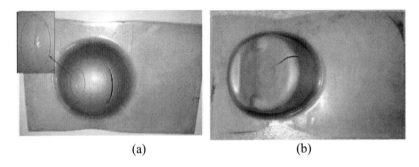

(a)   (b)

Figure 1. Sheet metal parts showing fractures at the end of: (a) the first stamping operation with a detail of the localized necking, (b) the second stamping operations.

3. METHOD

3.1. Modelling Conditions

Sheet metal forming analyses of the stamping operations were performed using the commercial finite element program, MSC.Dytran®, based on an explicit dynamic formulation. Hypermesh® and Patran® computer softwares were utilized for pre and post processing operations of the finite element analysis, respectively.

The numerical analyses were carried out through the discretization of the sheet blank by means of four-noded shell elements using three integration points across the thickness. Although it was not done an exhaustive mesh sensitive study, the elements dimension was established considering a balance between the overall CPU time and the quality of the finite element predictions. Tooling was assumed to be perfectly rigid (non-deformable). Figures 2 and 3 show, respectively, the discretized forms for the first and second stamping operations of the automotive part. In the first operation the blank was discretized with 2100 four-noded shell elements, the tool components were discretized with rigid quadrilateral elements, 1440 for the punch, 4307 for the die and 2329 elements for the blank holder. In the second stamping operation the mesh of the stamped part was the same of the first operation, the punch, the die and the blank holder were discretized with 4858, 12817 e 11728 elements, respectively.

The sheet was modelled as an elasto-plastic material following von Mises's yield criterion and its associated flow rule. The plastic material anisotropy was not considered.

The friction interface material/tooling was modelled by means of Coulomb's friction law and the friction coefficient for the contact surfaces was set to 0,15 according to the procedure described in Section 3.2.2.

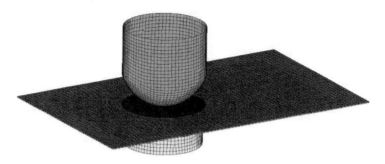

Figure 2. First stamping operation: discretized mesh of the initial blank sheet and tooling (punch, die and blank holder).

Figure 3. Second stamping operation: discretized mesh of the initial blank sheet and tooling (punch, die and blank holder).

Punch speed was set at 500 mm/s for both operations in order to satisfy the stability condition inherent to a proper utilization of the explicit methods (Chung, 1998). Since the punch displacement is 75 mm in the first operation and 86 mm in the second operation, the total time for both operations was approximately 0,324 s. The size of the incremental step was automatically established by the program. The blank holder forces utilized in both operations were set to the values determined as described in Section 3.2.2.

Finally, it is worth mentioning that the values of the principal stress components utilized to construct the forming limit diagrams were determined from the strain numerical results since the correspondent stress results, with the exception of the effective value, have revealed a poor correlation with the respective modes of deformation. These inaccurate stress predictions within the sheet metal formed parts are related with the convergence characteristics of the explicit schemes (Yang, 1995 and Chung, 1998). Details of the procedure adopted to determine the principal stress components can be found in Cravo (2009).

3.2. Experimental Background

3.2.1. Material behaviour

The mechanical properties of the high strength low alloy steel S420MC sheet where obtained by means of experimental tensile tests (NP EN 10002-1). The experimental data was utilized to obtain the material rigid-plastic relations between the effective stress, σ, and the effective strain, ε, according to Hollomon's power law (Equation 5) and to Krieg's law (Equation 6). Table 1 summarizes data obtained in the experimental tensile tests and the material constants obtained by fitting the experimental data to the above cited material laws.

$$\sigma = K(\varepsilon)^n \tag{5}$$

$$\sigma = a + b(\varepsilon + c)^n (1 + k(\dot{\varepsilon})^m) \tag{6}$$

The Holloman's law was utilized to determine the forming limit curves based on stress, as mentioned in Section 1, while the Krieg's law, without the viscoplastic response, was utilized in the finite elements analysis using the MSC.Dytran® software.

Table 1. HSLA S420MC experimental tensile test results. Holloman and Krieg material constants (referred to Equations 5 and 6)

Tensile test		Hollomon's law			Krieg's law		
E (MPa)	$\sigma_{0.2}$ (MPa)	R (MPa)	K (MPa)	n	b (MPa)	c	n
198534	498.81	539.39	795.68	0.127	797.15	0.01	0.1269

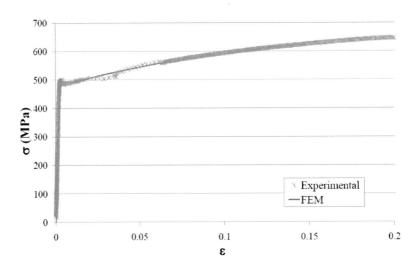

Figure 4. Comparison between numerical and experimental results of true stress-strain curve in the uniaxial tensile test of HLSA S420MC steel sheet.

The accuracy of the Krieg's material law, obtained by fitting the experimental data, was also tested through the finite element simulation of the uniaxial tensile test. Figure 4 illustrates the comparison between the numerical results and the experimental data for the uniaxial tensile test of the HSLA S420MC. The agreement of these results allowed to validate the material curve fitting and have permitted to test the effective stress results of the numerical explicit dynamic formulation.

3.2.2. Process parameters determination

The blank holder force and the tribological conditions for both stamping operations were determined using an inverse procedure through the comparison between numerical results and experimental measurements of thickness along selected cross sections, of geometries and of contours of the stamped parts.

Figures 5 and 6 show, respectively, the predicted and measured values of sheet metal thickness along the longitudinal and transversal section at the end of the first stamping operation. The overall agreement between numerical and experimental distribution of thickness across the two cross sections is good. Major reductions in thickness part can be well identified in the numerical analysis in close accordance with the experimental results, along both sections.

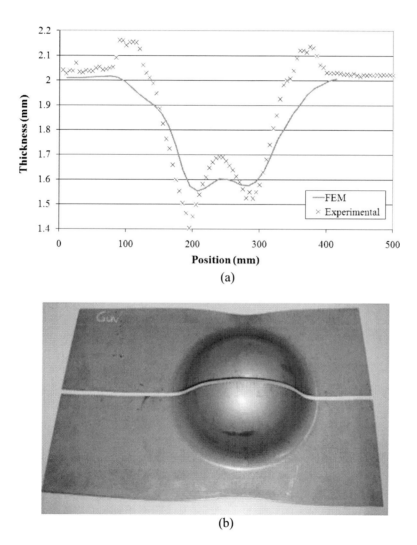

Figure 5. (a) Predicted and measured values of thickness along the longitudinal section of the stamped part at the end of the first stamping operation; (b) illustration of the longitudinal section of the stamped part.

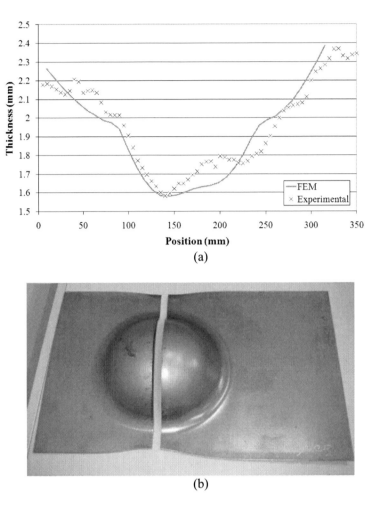

Figure 6. (a) Predicted and measured values of thickness along the transversal section of the stamped part at the end of the first stamping operation; (b) illustration of the transversal section of the stamped part.

It should be noticed, however, that along the longitudinal section the numerical predictions smooth the experimental results, not only near the die corner, but also in the areas where localized necking is likely to occur. These differences may be related with the element type characteristics coupled with the mesh size. Besides these effects it is important to retain the accuracy of the numerical results in predicting the location of the major thickness reductions along the longitudinal section in close agreement with the onset of the localized necking and/or fracture of the real part. The lowers thickness values occur along the longitudinal section due to the fact that the material flow into the die has a bigger constrain along this direction leading to the occurrence of localized necking and/or fracture in the transversal direction as observed in the real component.

The comparison between numerical predictions and the experimental results of stamped part contour is illustrated in Figure 7. The agreement between numerical predictions and experimental results is very good.

These numerical results, corresponding to the replication of the fabrication conditions, were obtained with a blank holder force of 50 kN and a friction coefficient of 0.15 along the interfaces between the sheet and the tooling.

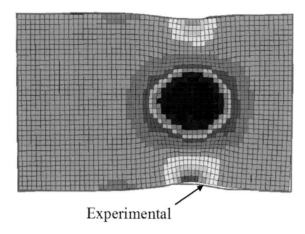

Figure 7. Sheet metal part contour at the end of the first stamping operation. Comparison of the finite element prediction (with the vertical displacement distribution) and the experimental results (exterior line).

A similar procedure was adopted to determine the process parameters to be used in the numerical simulation of the second stamping operation. The process parameters evaluation was done through the comparison between the numerical and the experimental values of thickness along selected cross sections, since this variable revealed to be the most sensitive to the process parameters variation. Again the friction coefficient considered was 0.15. Relatively to the blank holder force, the numerical results indicated that the specific pressure of the previous operation should be kept. Details of this methodology can be found in Cravo (2009).

4. RESULTS AND DISCUSSION

4.1. Present fabrication conditions

In order to suggest changes to tool geometry for the first operation, with the aim of enlarging the formability limits and minimizing the thickness thinning in the part at the end of this operation, it was necessary to carefully study and analyze the deformation modes in present fabrication conditions. This investigation was supported by the numerical results of the principal strains and stresses in the plane of the sheet metal component plotted in the respective forming limit strain and stress diagrams, at the end of both stamping operations. The forming limit curves based on the localized necking criteria were determined considering the theoretical models proposed by Stören and Rice, for positive values of $\beta = \varepsilon_2/\varepsilon_1$ and by Hill, for negative values of β, as described in Section 1. The results at the end of the first stamping operation are shown in Figure 8. It can be easily observed that this operation is mainly characterized to introduce critical deformations in the part varying between plane strain and balanced biaxial stretch. The deformation magnitude in this operation originates excessive thickness reduction (above 20%) and, according to the fact that the numerical results reach or overstep the forming limit curves in both strain and stress diagrams, there are conditions of necking with a subsequent failure at the location of the elements concerned.

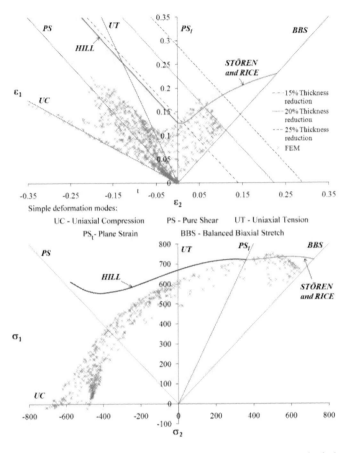

Figure 8. Strain and stress based forming limit diagrams obtained in the numerical simulation at the end of the first stamping operation.

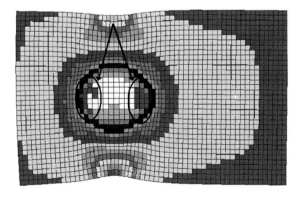

Figure 9. Numerical prediction of the vertical displacement distribution at the end of the first stamping operation. The elements that overstepped the forming limit curves on both diagrams are highlighted.

Figure 9 shows the mesh elements (marked with arrows), that overstepped the forming limit curves on both diagrams at the end of the first operation. It is worth noting that these elements are approximately located along the longitudinal cross section, symmetrically disposed to the pole. These results are in accordance with the existence of friction between hemispherical punch and the blank and in good agreement with the occurrence of local necking and fracture in the real part (please refer to Figure 1a), i.e., the limit strains and stresses predicted for the present operation are in accordance with the experimental results.

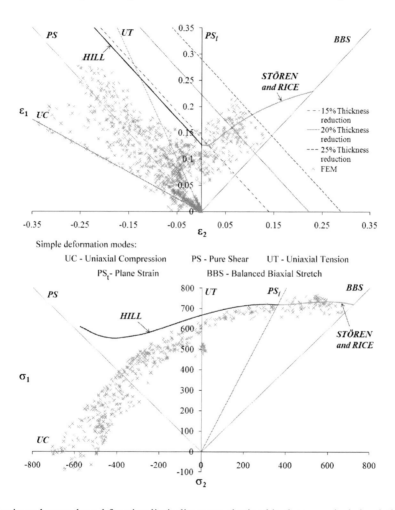

Figure 10. Strain and stress based forming limit diagrams obtained in the numerical simulation at the end of the second stamping operation.

Figure 10 shows the major and minor strains and stress in the sheet metal plane, obtained at the end of the numerical analysis of the second stamping operation, plotted in the respective forming limit diagrams. Again, this operation is mainly characterized to introduce critical deformations in the part varying between plane strain and balanced biaxial stretch. Consequently, this second operation is an extension of the first one, contributing to an additional thickness reduction and to a higher risk of localized necking. It is also important to observe that the numerical results in terms of critical stress points are in agreement with those based on the critical strains.

Although the utilization of the strain based forming limit diagrams should be limited to the first stamping operation, their application was also considered in this study to help the analysis of the second stamping operation. According to the above mentioned results, in the actual fabrication conditions, the most critical sheet metal areas, in terms of formability, have deformation modes varying between plane strain and balanced biaxial stretch. Studies conducted by Graf e Hosford (1993), revealed that the strain based forming limit curves of steels with pre-deformations from plane strain to balanced biaxial stretch suffer translations to the right side on the forming limit diagram. Under these circumstances the strain based forming limit curve considered in this study is more conservative for the above cited deformation modes. As a consequence, due to the impossibility of determining all the strain based forming limit curves for second stamping operations in this study, and to the fact that their determination is almost unrealistic considering that each element in the finite element analysis will have a different forming limit curve in a subsequent stamping operation (Stoughton, 2000), the authors have considered the same strain based limit curve in the analysis of both stamping operation bearing in mind the restrictions of this procedure.

The mesh elements situated above the forming limit curves at the end of the second stamping operation are marked in Figure 11a. The critical region, exceeding the forming limits in terms of local necking, is situated in the deepest part of the automotive component. The comparison between the location of the two elements most distanced from the forming limit curves (see Figure 11b) and the experimental location of the fractures in the real component (see Figure 1b) is in close agreement. These results confirm the performance of the finite element program, utilised in the present work, to replicate the present stamping operations and the ability of Stören and Rice's model to estimate the onset of local necking on the automotive component in study.

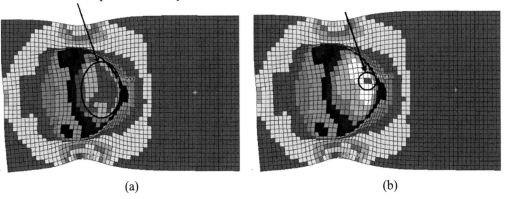

(a) (b)

Figure 11. Numerical prediction of the vertical displacement distribution at the end of the first stamping operation. (a) Elements that overstepped both forming limit curves at the end of the second stamping operation are highlighted; (b) the two elements most distanced from the limit curves are highlighted.

Globally, in terms of formability, the analysis of the actual fabrication conditions reveals that the most sacrificed sheet metal area in the second operation it is the same as in the first operation with the same type of deformation modes, between plane strain and balanced biaxial stretch. It was also possible to analyse that at the end of the first operation, according to the numerical results and to the experimental observations, the sheet metal part is at the forming limit borderline in terms of localized necking and thickness. Therefore, small variations in the material or in the process parameters are enough to originate a local necking or fracture. As a result of this analysis, the strategy adopted to improve the formability limits of the automotive component consisted in changing the tool geometry for the first stamping operation in order to reduce the magnitude of the biaxial stretch to attain a safe margin for the subsequent operation (allowing to enlarge the safe margin for the subsequent operation).

4.2. New Geometry for the First Stamping Operation Tool

A systematic study, with different geometries, was conducted in order to achieve a new tool geometry for the first stamping operation. Accordingly it was possible to conclude that the punch diameter and the die corner radius should be increased to 200 mm (in the present conditions it is 182 mm) and to 25 mm, respectively. The former is suitable for the subsequent forming operation and the latter benefits the sheet metal flow into the die with no risk of wrinkling. On the other hand, the deformation magnitude in biaxial stretching was controlled through the redrawing of the punch geometry according to Figure 12(a).

The appropriate punch corner radius, R_p, and punch bottom radius, R_{db}, dimensions where found regarding the following observations: when the punch corner radius increases the plane strain and biaxial stretch deformations increase while the principal strains on the left side of the strain based forming limit diagram decrease; with respect to the punch bottom radius, its increase originates the biaxial stretch deformation mode reduction, while there is an increasing of the principal strains on the left side of the strain based forming limit diagram. The latter is due to an increase in friction force along the sliding interface between the sheet and the punch as consequence of the sliding interface area increase. Accordingly the expansion in the bottom of the sheet metal component is restricted and the punch travel mainly induces the plastic deformation of the walls under plane strain conditions. As a result of this study the values adopted for the corner radius and the bottom radius of the punch geometry for the first stamping operation were 125 mm and 50 mm (see Figure 12b), respectively.

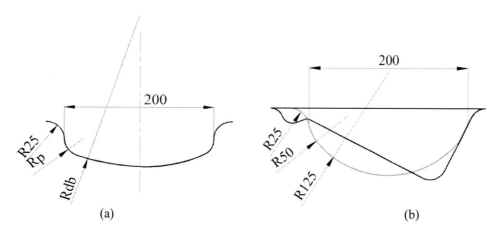

Figure 12. (a) Tool geometry proposed for the first stamping operation; (b) superposition of the tool geometry proposed (in grey) with the second stamping operations geometry (in black).

The procedure adopted for the numerical simulation of the first stamping operation with the new tool was similar to that used in the present fabrication conditions, described in section 3.1. Only the holder force had to be increased in order to avoid the wrinkling phenomenon.

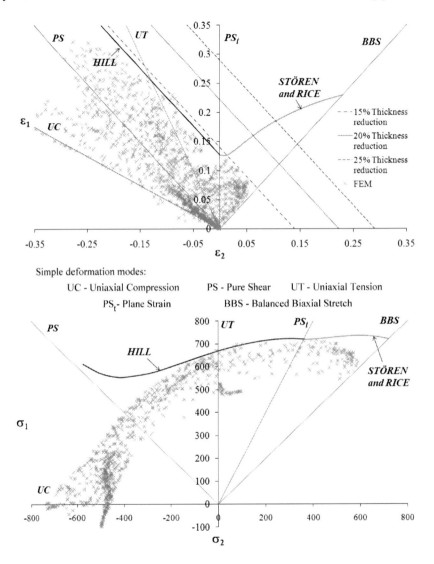

Figure 13. Strain and stress based forming limit diagrams obtained in the numerical simulation at the end of the first stamping operation with the new tool geometry proposed for this operation corresponding to a punch travel of 80 mm.

The numerical analysis of the first stamping operation with the new tool (see Figure 13) showed that from the very beginning of the operation the bottom of the sheet metal part deforms under biaxial stretching. Although, at the mid of the operation this biaxial stretch deformation ends and punch travel essentially originates the plastic deformation of the walls with strains localized and the left side of the forming limit diagram. In consequence, thickness reduction is considerable reduced, reaching acceptable values for this kind of automotive component, and the number of mesh elements close to the localized necking conditions has decreased. Thus, regarding the previous benefits, and in order to approximate the sheet metal part geometry at the end of this operation to the tool geometry of the subsequent operation, the punch displacement was increased to 80 mm in the this operation. Even thought, under these circumstances, the maximum thickness thinning predicted reduction is 13.6%, quite inferior to the value reached at the end of the first operation in the present fabrication conditions, 21.2%, for a punch travel of 75 mm.

Figure 14 shows the forming limit diagrams obtained with the numerical values of the principal strains and stress in sheet metal plane of the component at the end of the second stamping operation. This numerical analysis was performed after the first stamping operation with the tool geometry changes above described. As expected, the second stamping operation introduces biaxial stretching deformations on the bottom of the sheet metal part. Although, in face of the biaxial expansion during the first operation has been limited, due to the modifications introduced in the tool geometry and also by the decrease of the punch displacement during this operation, the thickness reduction was considerably reduced from 25.1% (in the present fabrication conditions) to 16.9%. Finally is worth noting that despite the existence of mesh elements situated above the forming limit curves these were significantly reduced indicating that the critical area in the sheet metal part in terms of local necking onset decreases.

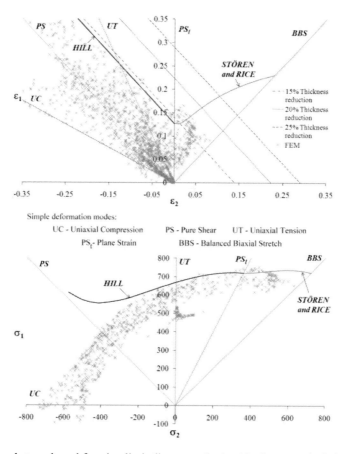

Figure 14. Strain and stress based forming limit diagrams obtained in the numerical simulation at the end of the second stamping operation subsequent to the utilization of the new tool geometry proposed for the first operation.

4.3. Raw Material

With the new geometry proposed for the first operation tool, according to the numerical results, the overall formability after both operations was considerably improved. However, the fabrication success cannot be assured since there are still some mesh elements with strain and stress values situated above the forming limit curves. Thus, in order to attain a safe margin for successfully form the automotive part, independently of any variation in the raw material composition and in the process parameters, a different raw material, the HSLA S355, was considered.

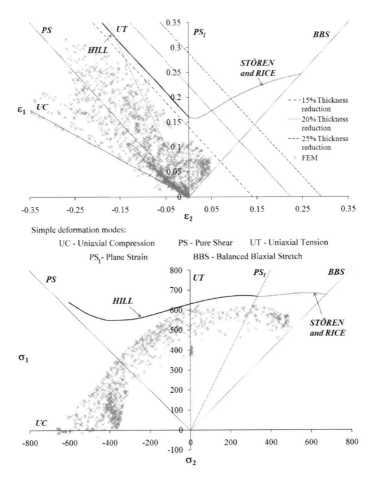

Figure 15. Strain and stress based forming limit diagrams obtained in the numerical simulation at the end of the first stamping operation with the new tool geometry proposed. Raw material: HSLA S355.

Even thought this steel has lower yield strength, its work-hardening exponent, n, is higher so that the strain distribution will be relatively homogeneous, avoiding the development of sharp strain gradients resulting in the improvement of formability. However, it was verified that, this material change was not enough to achieve the expected goals in terms of formability, keeping the present fabrication conditions. Therefore it was necessary to make further studies including the new raw material and tool geometry, proposed for the first stamping operation (see section 4.2). Figure 15 shows the forming limit diagrams obtained at the end of the numerical simulation with the previous cited conditions. The results indicate that the use of HSLA S355 steel is advantageous for the part fabrication, since, at the end of the first stamping operation, there are no elements with strain and stress values above the respective forming limit curves and simultaneously the biaxial stretching at the component's bottom was reduced with the inherent reduction of thickness thinning. The results indicate that the use of HSLA S355 steel is advantageous for the part fabrication, since, at the end of the first stamping operation, there are no elements with strain and stress values above the respective forming limit curves and simultaneously the biaxial stretching at the component's bottom was reduced with the inherent reduction of thickness thinning.

In what concerns to the second stamping operation the numerical results (see Figure 16) also exhibit a significant improvement in formability (in terms of localized necking) and in thickness thinning. Even though, as expected, there is an increase of deformation under biaxial stretching, being the safe margin of this operation effectively higher than in the present fabrication conditions.

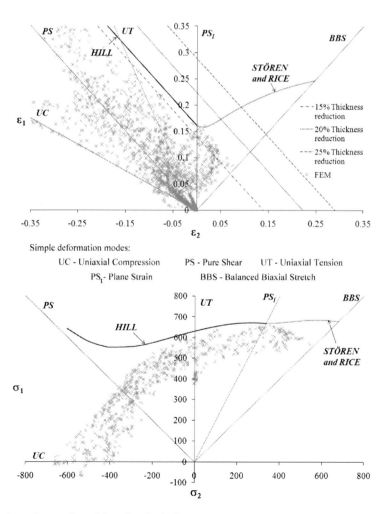

Figure 16. Strain and stress based forming limit diagrams obtained in the numerical simulation at the end of the second stamping operation subsequent to the utilization of the new tool geometry proposed for the first operation. Raw material: HSLA S355.

CONCLUSION

The fabrication of an automotive component in to two-step stamping operations was studied. The objective was the achievement of a suitable geometry for the first stamping operation in order to eliminate existing formability problems, related with the occurrence localized necking, fractures and excessive thickness thinning. The theoretical models proposed by Stören and Rice and by Hill correctly identified the localized necking onset in the present fabrication conditions. The geometry chances suggested for the first operation tool considerably reduce the thickness thinning in the sheet metal part. In what concerns to localized necking occurrence, due to the proximity of the numerical results to the forming limit curves in the forming limit diagrams, only the realization of experimental test can confirm the effectiveness of the proposed solution. Thus, in order to attain a safe margin for successfully form the automotive part, independently of any variation in the raw material and in the process parameters, a raw material change is proposed.

ACKNOWLEDGMENTS

The authors wish to express thanks to the company MCG, Lda. and to CEIIA (Centre for Excellence and Innovation in Automotive Industry) for all the support during this work.

REFERENCES

Arrieux, R., Bedrin, C., Boivin, M. (1982), "Determination of an intrinsic forming limit stress diagram for isotropic metal sheets", In: *Proceedings of the 12th Biennial Congress IDDRG*, 61–71.

Chen M.H., Gao L., Zuo D.W., Wang M. (2007), "Application of the forming limit stress diagram to forming limit prediction for the multi-step forming of auto panels", *Journal of Materials Processing Technology*, 187-188, 173-177.

Chung, W.J., Cho J.W., Belytschko T. (1998), "On the dynamic effects of explicit FEM in sheet metal forming analysis", *Computational Mechanics.*, 15, 750–776.

Cravo, S. (2009), "Estudo do fabrico de um componente para a indústria automóvel com base na utilização de curvas limite de estampagem", *Master Thesis*, Instituto Superior Técnico, Lisboa (in Portuguese).

Graf, A.F., Hosford, W. (1993), "Calculation of the forming limit diagrams for changing strain paths", *Metall. Trans. A*, 24, 2497–2501.

Hill, R. (1952), "On Discontinuous Plastic States, With Special Reference to Localized Necking in Thin Sheets", *Journal of the Mechanics and Physics of Solids*, 1, 19-30.

Keeler, S. P., Backofen, W. A. (1963), "Plastic Instability and Fracture in Sheets Stretched Over Rigid Punches", *Trans ASM*, 56, 25–48.

Krieg, R.D. (1975), "A practical two surface plasticity theory", *Journal of Applied Mechanics*, 42, 641-646.

NP EN 10002-1, (2006), *"Materiais metálicos. Ensaio de tracção. Parte 1: Método de ensaio à temperatura ambiente"*, Instituto Português da Qualidade (in Portuguese).

MSC.Dytran Reference Manual, (2005), *MSC.Software Corporation.*

Stören, S., Rice, J. R. (1975), "Localized Necking in thin sheets", *J. Mech. Phys. Solids,* 23, 421-441.

Stoughton, T.B. (2000), "A general forming limit criterion for sheet metal forming", *International Journal of Mechanical Sciences,* 42, 1–42.

Stoughton, T.B., Zhu, X. (2004), "Review of theoretical models of the strain-based FLD and their relevance to the stress-based FLD*", International Journal of Plasticity,* 20, 1463–1486.

Wu P.D., Graf A., MacEwen S.R., Lloyd D.J., Jain M., Neale K.W. (2005), "On forming limit stress diagram analysis", *International Journal of Solids Structures,* 42, 2225–2241

Yang, D.Y., Jung, D.W., Song, I.S., Yoo, D.J.;Lee, J.H. (1995), "Comparative investigation into implicit, explicit, and iterative implicit/explicit schemes for the simulation of sheet-metal forming processes", *Journal of Materials Processing Technology,* 50, 39-53.

In: Machining and Forming Technologies, Volume 3
Editor: J. Paulo Davim

ISBN: 978-1-61324-787-7
© 2012 Nova Science Publishers, Inc.

Chapter 10

THE ASSESSMENT OF HOT FORGING BATCHES THROUGH COOLING ANALYSIS

Edilson Guimarães de Souza, Wyser José Yamakami, Alessandro Roger Rodrigues, Miguel Ângelo Menezes, Juno Gallego, Vicente Afonso Ventrella and Hidekasu Matsumoto

UNESP - Univ Estadual Paulista, Engineering Faculty of Ilha Solteira,
Ilha Solteira - SP, BRAZIL

ABSTRACT

Using the available thermal energy in hot forging through controlled cooling as soon as the part is forged allows cost reduction by avoiding post-forging thermal treatments. This paper assesses, through transient analysis of the cooling process, the batch size of cylindrical hot forged parts that must be subjected to controlled cooling in the furnace in order to provide a microstructure similar to that obtained by normalizing. Thus, the first forged part of the batch must present a minimum surface temperature that keeps it in austenitic condition when controlled cooling started. Therefore, by assessing the time that the first forged part takes to reach the minimum temperature and measuring the productive flux, it is possible to calculate batch size for the subsequent controlled cooling in the furnace. The size of the batches obtained for the analyzed parts was small, indicating the need for continuous forging and controlled cooling process. It was demonstrated that forged parts that have larger volume/area ratio require larger batches since the forging time is the same for all parts.

Keywords: batch size, controlled cooling, hot forging, cooling analysis

INTRODUCTION

Several research works concerned with hot forging followed by controlled cooling have been performed (Rasouli et al., 2008; Kazeminezhad and Karimi Taheri, 2003; Rong Zhou et al., 2001; Bakkaloglu, 2002; Liu et al, 1998; Kaspar et al, 1997). Other works correlated to this subject, such as developments metallurgic of alloys, that permit the direct cooling (Lin, 1994), and the application of the thermomechanical treatment (Ghosh et al, 2003) have allowed to obtain more adequate microstructures and mechanical properties required in forging applications, thus enabling the reduction of production cycle and energy costs (Luo et al., 2009).

In conventional hot forging processes, the batch of forged parts is cooled to room temperature and then heated in a continuous furnace for additional normalizing, commonly used after hot forging. Thus, it is observed that the conventional procedure used in hot forging wastes the thermal energy available in the hot forged parts, hence requiring their reheating for subsequent normalizing.

On the other hand, obtaining a normalized microstructure by cooling in furnace the forged parts immediately after the hot forging for a given temperature and time lower than that used in the normalizing can simultaneously represent lower thermal energy and time costs, when compared to the conventional process.

In hot forging tests followed by furnace cooling, the forged parts are immediately put in the furnace to avoid excessive loss of thermal energy to the environment, hence avoiding any undesired microstructural transformations. Each part is individually monitored for a more precise control, thus avoiding significant parameter variations as time, temperature and cooling rates defined for each testing condition.

Lou et al. (2009) showed the formation of different microstructures for a same cooling condition, according to the chemical composition of the hot forged steel. In the same way, obtaining a normalized microstructure for a particular steel requires determining the temperature and cooling rates from the moment the forged parts are obtained up to their insertion in the furnace in order to avoid that microstructural transformations take place out of the furnace. However, during forging the movement of parts in the production steps generally occurs such as batches and not individually. Thus, depending on the productive flux and the mass of the forged parts, the air cooling times that the first and last forged parts from a given batch were submitted will lead to quite different cooling rates and final temperatures, which can prevent obtaining a totally normalized microstructure through the furnace cooling of this batch. Analyzing the thermal energy transfer phenomena is important for determining the best conditions for the controlled cooling process to occur and to define the amount of time that the first forged part can stay out of the furnace without any microstructural transformation. In practical situations in which the complexity does not allow an analytical approach, as for instance, in the analysis of forged parts of complex geometries, numerical simulations can assist in solving these problems.

The aim of this work was to apply the concepts of heat energy transfer to determine the size of hot forged part batches in order to allow the formation of a normalized microstructure similar to the real parts in terms of microstructure and hardness using the cooling furnace.

MATERIALS AND PROCEDURES

AISI 1008, 1042 and 5132 steels were selected for air cooling simulations due to the similarities of their thermal properties with steels employed by the automotive industry. The cylindrical geometry of forged parts named after FOR 1 and FOR 2 present length and diameter defined so that their masses are equal to billets used in actual forged gears production. The former hot forged has 144 mm length, 88.9 mm diameter, 6.99 kg mass and the latter has 101 mm length, 63.5 mm diameter and 2.51 kg mass. The chemical composition range of steels is shown in Table 1.

Controlled Cooling after Hot Forging

The hot forging followed by controlled cooling proposed in this research was idealized according to the Figure 1. The billets were heated between 1473 and 1523 K (T_b) in a continuous furnace, forged in eccentric-shaft press with 1000, 2000 and 3000 ton (depending on forged part dimensions) and stored in a container near presses. After forging the first part ("1") under temperature $T_1 = T_i$, the initial time $t = 0$ is defined. As forged parts are put in container, they start to cool due to thermal energy loss by radiation and convection.

In order to allow the formation of pearlite and ferrite similar to that obtained by normalizing in industrial process, it is relevant to determine the cooling time "ct" that forged part "1" spent to attain the critical temperature "T_c" so that the material has an austenitic microstructure when put into furnace of controlled cooling.

Table 1. Chemical composition range of steels employed in this work

AISI Grade	Element (% weight)					
	C	Mn	P	S	Si	Cr
1008	0.05 0.10	0.30 0.50	0.04 (max)	0.05 (max)	0.01 0.25	-
1042	0.40 0.47	0.60 0.90	0.04 (max)	0.05 (max)	0.15 0.35	-
5132	0.30 0.35	0.60 0.80	0.035 (max)	0.04 (max)	0.15 0.30	0.75 1.00

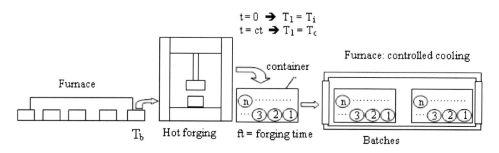

Figure 1. Hot forging followed by controlled cooling in furnace.

On the other hand, the determination of temperatures, rates and cooling times of first forged parts exposed to the environment is necessary in order to avoid some austenitic transformation in other microconstituents before parts being put into furnace of controlled cooling.

The transformation diagrams for continuous cooling extracted from Atkins (1980) predict the formation of pearlite/ferrite for air cooled bars with 75 mm medium diameter. In addition, the initiation of austenitic-ferritic transformation for these diameters takes place at 1108 K, 997 K and 1005 K, for AISI 1008, 1042 and 5132 steels, respectively.

The normalizing of steels employed in industry, which air cooling simulations are the aim of the present work, is performed in a conveyor type furnace at 1123 K for 2 hours, after hot forged parts are air cooled up to room temperature. This process aims to promote a microstructure of pearlite and ferrite evenly distributed along forged parts with a hardness ranging from 163 to 187 Brinell.

The pearlitic-ferritic microstructure predicted by the transformation diagram in continuous cooling for the air cooling condition does not contain the same characteristics and hardness through normalizing. The controlled cooling of forged parts in furnace at 873 K for 20 minutes carried out in laboratory immediately after their forming allowed obtaining a similar microstructure to that from industrial normalizing with hardness levels specified for this thermal treatment.

The temperature Ti adopted in simulations was 1473 K owing to it is the minimal value of billet after it is removed from the furnace and before forging. From this temperature the steel is cooled to the room temperature. The critical temperature $T_c = 1200$ K, assumed in simulations, is above the temperature required for ferrite formation initiation for assuring the existence of an unique austenitic phase in materials. Besides, this temperature allows time to introduce a container with a batch of "n" forged parts into the cooling furnace, since this movement would not be performed in a instantaneous way after cooling time "ct" is reached.

The number of forged parts of a batch (n) can be determined through mean time "ft" to forge the parts (time/forged part) and cooling time "ct" (Figure 1) that the first forged part takes to reach the critical temperature T_c, see Equation (1).

$$n = \left(\frac{ct}{ft} \right) + 1 \tag{1}$$

Estimate of the Cooling Time

To estimate cooling time, the spatial distribution of the temperature as a function of time must be known. Thus, the heat diffusion equation must be solved. Therefore, the equation for a cylinder with initial and boundary conditions independent of the angular coordinate is given by Equation (2):

$$\frac{1}{r} \frac{\partial}{\partial r} \left(kr \frac{\partial T}{\partial r} \right) + \frac{\partial}{\partial z} \left(k \frac{\partial T}{\partial z} \right) = \rho \, c \, \frac{\partial T}{\partial t} \tag{2}$$

The boundary conditions are as follows:

$$- k \left. \frac{\partial T}{\partial r} \right|_{r=r_0} = h \left[T \left(r_0, z, t \right) - T_\infty \right] \tag{3}$$

$$-k\left.\frac{\partial T}{\partial z}\right|_{z=2L} = h\left[T(r,2L,t)-T_\infty\right] \qquad (4)$$

$$-k\left.\frac{\partial T}{\partial z}\right|_{z=0} = 0 \qquad (5)$$

and its initial condition is:

$$T(r,z,0) = T_i \qquad (6)$$

where T is the temperature; t is the time; ρ, c, k are the density (kg/m^3), specific heat (J/kg·K) and the thermal conductivity (W/m·K), respectively; T_i is the initial temperature (K); h is the generalized heat transfer coefficient (W/m^2·K) and 2L (m) is the total length of billet.

The generalized heat transfer coefficient is found by adding the convective heat transfer coefficient, h_c (W/m^2·K), with the radiation heat transfer coefficient, h_r (W/m^2·K):

heat transfer coefficient, h_r (W/m^2·K):

$$h = h_r + h_c \qquad (7)$$

The coefficient h_r is obtained as followed:

$$h_r = \varepsilon\sigma\left(T + T_{sur}\right)\left(T^2 + T_{sur}^2\right) \qquad (8)$$

where ε is the surface emissivity, adopted as 0.8 since it is a more adequate value for oxidized surfaces (Kaviany, 2002); σ is the Stefan-Boltzmann constant (5.67×10^{-8} W/m^2·K^4); T is the surface temperature and T_{sur} is the surrounding temperature.

The coefficient h_c can be estimated through correlations, although, in this paper the typical maximum value of 25 W/m^2·K for free convection in gaseous environment was adopted, redrawn from Incropera et al (2007).

Equations 2, 3, 4 and 5 are solved analytically by the multidimensional effects method. For this purpose, Equations 7 and 8 are used. The solution is given by:

$$\left\{ \begin{array}{l} \dfrac{T(r,z,t)-T_\infty}{T_i - T_\infty} = \dfrac{T(z,t)-T_\infty}{T_i - T_\infty}\cdot\dfrac{T(r,t)-T_\infty}{T_i - T_\infty} \\[2ex] 0 \leq z \leq 2L \\[1ex] 0 \leq r \leq r_0 \\[1ex] t \geq 0 \end{array} \right. \qquad (9)$$

where T_∞ is the room temperature (K). $T(z,t)$ and $T(r,t)$ are solutions for one-dimensional problems, plane wall and infinite cylinder, respectively.

The solution for plane wall is given by Equation (10):

$$T(z,t) = T_\infty + (T_i - T_\infty).\sum_{n=1}^{\infty} \frac{4.sen\left(\zeta_n\right)}{2.\zeta_n + sen\left(2.\zeta_n\right)}\exp\left(-\zeta_n^2.\frac{\alpha t}{(2L)^2}\right).\cos\left(\zeta_n.\frac{z}{2L}\right) \qquad (10)$$

The values of ζ_n are the positive roots of:

$$\zeta_n.\tan\left(\zeta_n\right) = \frac{h2L}{k} \qquad (11)$$

and the solution for infinite cylinder is given by Equation (12):

$$T(r,t) = T_\infty + (T_i - T_\infty).\sum_{n=1}^{\infty} \frac{2}{\xi_n} \frac{J_1.\left(\xi_n\right)}{[J_0\left(\xi_n\right)]^2 + [J_1\left(\xi_n\right)]^2}.\exp\left(-\xi_n^2.\frac{\alpha t}{r^2}\right).J_0\left(\xi_n.\frac{r}{r_0}\right) \qquad (12)$$

where J_0 and J_1 are zero-order and one-order Bessel function of the first kind, respectively. The values of ξ are positive roots of:

$$\zeta_n.\frac{J_1\left(\xi_n\right)}{J_0\left(\xi_n\right)} = \frac{hr_0}{k} \qquad (13)$$

An algorithm in the Python programming language was developed to speed up the calculations owing to the fact that Equations (11) and (13) are transcendental.

The cooling time was obtained numerically by considering Equation (9) equal to the critical temperature T_c, r equal to r_0 and z equal to $2L$, due to the heat transfer rates are more intense in these points (r_0, $2L$) at the border.

RESULTS AND DISCUSSION

The average forging time (ft) for forged parts FOR1 and FOR2 was 15.9 and 10.1 s, respectively. It is noted that heavier parts are more difficult to be handle and consequently waste longer time to be obtained from the billet.

Figure 2 shows the temperature distribution in a cross section of AISI 1008 forged part with 88.9 mm diameter at 46 seconds after forging. The critical temperature, T_c=1200 K, was reached at the borders, which confirms that these points (r_0, $2L$) are those that cool more rapidly. The critical temperature was selected for a value at which the materials still remain in the austenitic condition.

The cooling rates on surface and closed regions are larger than that provided to forged parts inner and give the microstructure formation with morphology and hardness distinct along material even they contain qualitatively only pearlite and ferrite.

This difference in cooling rates between the surface and interior of forged part, as well as in microstructure and hardness, was observed by Luo et al. (2009) for different steels employed for mold applications used in plastic injection.

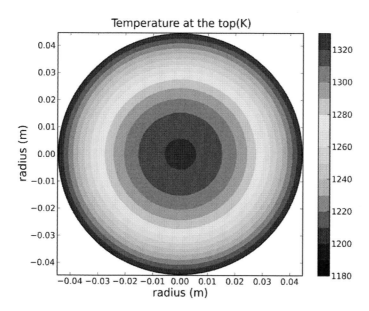

Figure 2. Temperature distribution in the AISI 1008 forged part at 46 seconds after forging and with initial temperature of 1473 K.

The chemical composition and the thermomechanical processing of the steels are important for austenite transformation, which have usually shown to affect microstructure-mechanical properties relationships of the hot forged steel parts. Considering the chemical composition and carbon content in AISI 1008, 1042 and 5132 steel grades, in addition to the chrome content in this latter, this is the critical element for the A3 and eutectoid transformation temperatures.

For a medium diameter of 75 mm, Atkins (1980) found for AISI 1008, 1042 and 5132 steels cooling rates of 45, 41 and 39 K/min, respectively. It has been noticed that for these cooling rates, which are normally applied in industrial forging, there is a ferrite-pearlite microstructure formation in all these steels. The increase in the amount of carbon has contributed to reduce the temperature of austenite decomposition, effect that was reinforced by the presence of chrome in the AISI 5132 steel - an element considered a very strong ferrite stabilizer (Bhadeshia and Honeycombe, 2009).

The morphology of ferrite and pearlite could be affected for the cooling conditions which are imposed by the industrial process. Higher cooling rates have not contributed to the solute redistribution by diffusion and beyond increase the dislocation density in the new microconstituents.

As consequence, ferrite grain size after transformation has showed to decrease, as well as the interlamellar spacing of pearlite (Zhao et al., 2003). Both effects contribute to a significant increase in strength and toughness of the forged steels.

Figure 3 and 4 show the evolution of the temperature as a function of time. The curves were obtained from a point at the border of the forged part with 88.9 and 63.5 mm diameter, respectively. It is known that thermal diffusivity measures the ability of a material to conduct thermal energy relative to its ability to store thermal energy (Incropera et al., 2007).

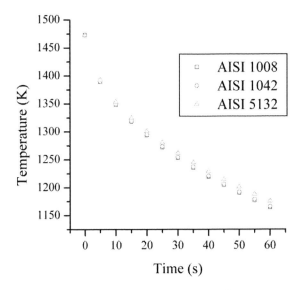

Figure 3. Reduction of temperature for forged part with 88.9 mm diameter.

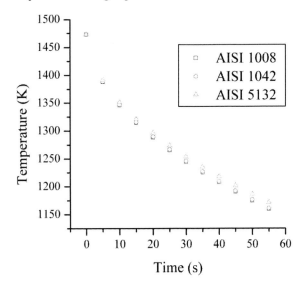

Figure 4. Reduction of temperature for forged part with 63.5 mm diameter.

Materials of large thermal diffusivity respond quickly to changes in their thermal environment, while materials of small thermal diffusivity respond more slowly, taking longer to reach a new equilibrium condition (Incropera et al., 2007).

Figure 5 shows the thermal diffusivity (ratio between the thermal conductivity and the product of the specific heat by density) for the three materials analyzed in this work as a function of temperature. The data was obtained from thermal conductivity and specific heat values collected from ASM International (ASM, 1995). This figure shows that the values of diffusivity have the same magnitude for all materials. The average values between 900 and

1300 K are 4.3 x 10^{-6}, 4.4 x 10^{-6} and 3.9 x 10^{-6} m^2/s for the steels AISI 1008, 1042 and 5132, respectively.

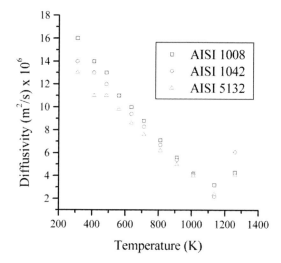

Figure 5. Thermal diffusivity variation with temperature.

Owing to the fact that the difference between the thermal diffusivity values for the three materials was negligible, the curves shown in Figure 3 and 4 remained practically superposed.

Table 2 shows the cooling time for both forged parts. Despite the fact that the cooling time for forged part with 63.5 mm diameter was shorter than that with 88.9 mm diameter, the batch for the former is larger due to the shorter forging time.

The size of the batches obtained were quite small as observed in Table 2 and that occurred because the forged part was left to cool freely, i.e., there was no concern in keeping its internal energy by placing it in a thermal box that could reduce the convection, radiation and conduction heat transfer rates. Like the analyzed forged parts had a small batch size, a forging process and continuous controlled cooling seemed to be a better option for the process.

Table 2. Batches sizes of forged parts FOR1 and FOR2 for all materials

	Cooling time [s]		ct/ft [-]		Batches [-]	
AISI GRADE	FOR		FOR		FOR	
	1	2	1	2	1	2
1008	46	41	2.89	4.05	4	6
1042	45	40	2.83	3.96	4	5
5132	47	43	2.95	4.25	4	6

FOR1 is the forged part with 88.9 mm diameter; FOR2 is the forged part with 63.5 mm diameter. ct is the cooling time and ft is the forging time.

Figure 6 shows the batch size behaviour as a function of the forging time for cylindrical forged parts with diameter equal to height. The material is the AISI 5132 steel and it can be observed that the forged with larger ratio between volume and area present larger batches.

The reason for this behaviour is that the larger this ratio, the smaller the heat changing area with the environment and, therefore, longer is the time that the first forged part can stay outside before being put into the controlled cooling furnace. This tendency is well pronounced for small forging times. Thus, the reduction of production time leads to an increase in the batch size, especially for large parts.

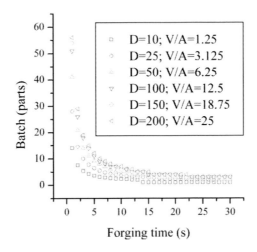

Figure 6. Size of the batches as a function of the forging time for AISI 5132 steel. D is the diameter and V/A is the ratio between volume and area.

CONCLUSION

- The three steels employed in air cooling simulations (AISI 1008, 1042 and 5132) presented similar properties and thermal behavior between 1200 and 1473 K.
- The number of forged parts from each batch submitted to the controlled cooling was: 4 parts with 6.99 kg for the three steels, 5 parts with 2.51 kg for AISI 1042 steel and 6 parts with 2.51 kg for AISI 1008 and 5132 steels.
- Forged parts with greater or smaller masses presented similar batch sizes due to the fact that larger forged parts, even possessing slower cooling, required longer forging time whilst smaller forged parts (which cool faster) are forged faster.
- The size of the batches obtained for the analyzed parts was small, indicating the need for continuous forging and controlled cooling process.
- The controlled cooling may minimize the consumption of electrical energy and processing time for forging and normalizing commonly used in the automotive industry.
- For the same forging time, the forged parts of larger volume/area ratio presented greater batch size compared to the smaller ratio. The higher the forging time, the lower the difference among batches sizes. For small forging times, this difference is large.

ACKNOWLEDGMENTS

The authors would like to thank FUNDUNESP (Foundation for the Development of UNESP) for the financial support.

REFERENCES

ASM International: The Materials Information Society, (1995), "ASM Handbook Vol. 1: Properties and Selection: Irons, Steels, and High – Performance Alloys", 10ª Ed., *Carbon and Low-Alloy Steels,* pp. 149, 179, 197-198.

Atkins, M., (1980), "Atlas of Continuous Cooling Transformation Diagrams for Engineering Steels". American Society for Metals. Metals Park, Ohio. British Steel Corporation. Sheffield, England, 260 p.

Bakkaloglu, A., (2002), "Effect of processing parameters on the microstructure and properties of an Nb microalloyed steel", *Materials Letters,* 56, 263-272.

Bhadeshia, H. K. D. H., Honeycombe, R., (2009*), "Steels: Microstructure and Properties",* 3rd edition. Elsevier Ltd., Oxford. 71-93.

Ghosh, A., Das, S., Chatterjee, S., Mishra, B. and Rao, P. R., (2003), "Influence of thermo-mechanical processing and different post-cooling techniques on structure and properties of an ultra low carbon Cu bearing HSLA forging", *Materials Science and Engineering A,* 348, 299-308.

Incropera, F., Dewitt, D. P., Bergman, T. L. and Lavine, A. S. (2007), *"Fundaments of heat and mass transfer",* 6th Ed., John Wiley and Sons, Inc, New York - N.Y., USA, pp. 8, 68.

Kaspar, R. Gonzalez-Baquet, I., Schreiber, N., Richter, J., Nussbaum, G. and Köthe, A., (1997), "Application of thermomechanical treatment on medium-carbon microalloyed steels continuously cooled from forging from forging temperature", *Materials Technology,* Steel Research*,* 68(1), 27-31.

Kaviany, M., (2002), "Principles of Heat Tansfer", John Wiley and Sons, Inc, New York - N.Y., USA, *Appendix C,* pp. 974.

Kazeminezhad, M. and Karimi Taheri, A. (2003), "The effect of controlled cooling after hot rolling on the mechanical properties of a commercial high carbon steel wire rod", *Materials and Design*, 24, 415-421.

Lin, H. R. and Chen, Y. K. (1994), "Development of new alloy steel grade which facilitates elimination of process annealing*", Ironmaking and Steelmaking*, 21(1), 27-31.

Liu, C., Liu, Y., Ji, C., Zhu, Q. and Zhang, J. (1998), "Cooling process and mechanical properties design of high carbon hot rolled high strength (HRHS) steels", *Materials and Design, 19*, 175-177.

Luo, Y., Wu, X., Min, Y., Zhu, Z. and Wang, H. (2009), "Development of Non-Quenched Prehardenend Steel for Large Section Plastic Mould", *Journal of Iron and Steel Research, International,* 16, 61-67.

Rasouli, D., Asl, S. K., Akbarzadeh, A. and Daneshi, G. H. (2008), "Effect of cooling rate on the microstructure and mechanical properties of microalloyed forging steel", *Journal of Materials Processing Technology,* 206, 92-98.

Zhao, M. C., Yang, K., Xiao, F. R. and Shan, Y. Y. (2003), "Continuous cooling transformation of undeformed and deformed low carbon pipeline steels". *Materials Science and Engineering A,* 355, 126-136.

Zhou, R., Jiang, Y., Lu, D., Zhou, R. and Li, Z. (2001), "Development and characterization of a wear resistant bainite/martensite ductile iron by combination of alloying and a controlled cooling heat-treatment", *Wear,* 250, 529-534.

In: Machining and Forming Technologies, Volume 3
Editor: J. Paulo Davim

ISBN: 978-1-61324-787-7
© 2012 Nova Science Publishers, Inc.

Chapter 11

3D MODELING OF PRECISION HIGH SPEED TURNING AND MILLING FOR THE PREDICTION OF CHIP MORPHOLOGY USING FEM

A. P. Markopoulos[*], K. Kantzavelos, N. Galanis and D. E. Manolakos

Laboratory of Manufacturing Technology, National Technical University of Athens, Athens, Greece

ABSTRACT

In this paper high speed turning and milling models are considered and chip formation and cutting forces are investigated. For the simulation the Finite Element Modeling software Third Wave Systems AdvantEdge is employed and the cutting forces results as well as the morphology of the chip are compared with the results of experiments conducted with CNC machine tools. For the experimental work the principles for design of experiment were used in order to minimize the required amount of experiments and obtain useful results at the same time. All parameters and cutting conditions were chosen to be close to the common industrial ones, with the material of the workpiece being C45 steel and the tools being coated carbides. Although published work on FEM simulations on high speed machining exist, this paper provides validated 3D models that cover the cases of precision high speed turning and high speed milling. The validation of the numerical data with the experimental ones show a good agreement and the proposed models can be used for the prediction of various machining parameters.

Keywords: High Speed Machining, Turning, Milling, Simulation, Finite Elements Method

---

[*] Email: amark@mail.ntua.gr

1. INTRODUCTION

High Speed Machining (HSM) refers to processes with cutting speed or spindle rotational speed substantially higher than some years before or also than the still common and general practice, according to a definition by Tlusty (1993) in a keynote paper of CIRP. In this definition, a spectrum of speeds or a lower value above which a machining process is characterized as HSM is avoided, because the speed involved in high speed machining depends upon the actual cutting process, the cutting tool and mainly on the workpiece material; for instance aluminum alloys can be machined at significantly higher speeds than stainless steels or titanium alloys (Erdel, 2003).

The original research on high speed machining combined three objectives: the technological breakthrough in machining specific kinds of materials such as aluminum and titanium, the improvement of the final product quality and mainly its surface characteristics and the achievement of higher productivity (Schulz and Moriwaki, 1992). The first of the objectives is related to the areas of application of HSM and especially the aerospace and automotive industry, where advanced materials such as aluminum alloys, titanium alloys, steels and superalloys are incorporated into the final products. The second objective is a constant pursue in machining and its combination with increased Material Removal Rates (MRR) is ideal. The third objective accomplishes exactly that, namely higher productivity with high quality at the same time. This advantageous combination is realized through the use of suitable cutting tools, machine tools' stiffness and damping capacity, toolholders and workholders, special spindles, fast feed drives and the level of automation implemented in the machine tools. The aforementioned advantages have placed HSM in a high position in modern production engineering with constantly growing industrial interest.

High speed machining concentrates the interest of researchers also because it exhibits some special characteristics in relation to cutting forces and chip morphology. Experimental work has indicated that cutting forces reduce by 10-15% as the speed is increased to high values (Trent and Wright, 2000; Erdel, 2003). Force reduction is attributed to the reduced strength of the workpiece material due to the elevated temperatures of the process. High temperatures are produced due to the lack of use of cutting fluids; cutting fluids in high speed machining do not offer any benefits and considering their cost, the environmental pollution and their inability to cool and lubricate in time-limited contact, it may be stated that dry machining is one-way in high speed metal cutting (Mamalis, Kundrák, Markopoulos and Manolakos, 2008; Nouari and Ginting, 2004). Besides the benefit of lowering the process cost by fluid omission, lower loads simplify part fixture design and allow for the machining of thin-walled sections, a common geometry of workpieces for the aerospace industry.

However, limitations also come into view. According to Ernst's classification of chip types, three possibilities exist, namely continuous, built-up edge and discontinuous chip (Kalpakjian and Schmid, 2003). However, at high speeds a fourth possibility appears; the serrated or segmented type (Trent and Wright, 2000; Balaji, Ghosh, Fang, Stevenson and Jawahir, 2006). This saw-toothed chip also referred as "shear localized" results in increased chip velocity, chip-tool friction and temperatures at the rake face of the tool that consequently provoke significant wear and tool life reduction (Tang, Wang, Hu and Song, 2009). Although it is generally accepted that chip segmentation is energetically favorable, tool wear issues exist (Trent and Wright, 2000; Bäker, 2004). Additionally, other investigators observe increase in the friction force and tool wear, at high speeds, after the transition from continuous chip to saw toothed chip, especially for hard ferrous materials (Lin, Liao and Wei, 2008; Thamizhmanii and Hasan, 2009; Yang and Li, 2010). However, tool life can be prolonged by optimizing cutting parameters, cutting conditions and machining strategy (Robinson and Jackson, 2005).

A lot of research on HSM and especially on high speed turning and milling but as well in grinding, boring and drilling is carried out. The results pertain to chip morphology, cutting forces, temperatures, surface integrity and tool wear among other characteristics. As in traditional machining, numerical modeling and especially the finite element method (FEM) is used for the analysis and the prediction of the cutting performance in high speed machining.

Numerical modeling and especially the finite element method is widely used for the analysis and the prediction of the cutting performance in machining operations. Simulations of orthogonal machining using the finite element method have a background of about three decades; in Refs (Mackerle, 1999; Mackerle 2003; Davim, 2008; Dixit and Dixit, 2008) a wide collection of papers can be found.

High speed machining has already been investigated with FEM by a few researchers (Marusich and Ortiz, 1995; Özel and Altan, 2000; Koshy, Dewes and Aspinwall, 2002; Robinson and Jackson, 2005; Bäker, 2006; Hortig and Svendsen, 2007; Davim, Maranhão, Jackson, Cabral and Grácio, 2008; Tang, Wang, Hu and Song, 2009). The proposed models deal mainly with features such as chip morphology, cutting forces, temperatures, surface integrity and tool wear. Most of the relative work examines turning but milling is considered as well. However, 3D simulation is scarcer.

In the present paper 3D models using the Finite Elements Method are proposed. High speed turning and milling are simulated and the chip morphology and the cutting forces are predicted. For the validation of the models experiments are carried out. In order for the results to be more efficient without having to perform a large number of experiments, the procedure is designed based on the theory of orthogonal arrays. Simulations and experiments refer to the same cutting conditions and compared as discussed in the following sections.

2. EXPERIMENTAL DESIGN AND MODELING

Turning and milling experiments were carried out first in order to determine the cutting conditions to be simulated with FEM. For turning four cutting parameters were taken into account. i.e. tool type, depth of cut, feed and cutting speed, while for milling the four cutting parameters were tool type, depth of cut, feed and spindle speed.

For the most efficient design for the experimental procedure, the orthogonal arrays were employed. This way a full factorial analysis, which is rather laborious, is avoided; instead a fractional factorial analysis is used. The value of fractional factorial experiments in general lies in the fact that higher order interactions are usually negligible. This leads to a considerable reduction in the number of parameters that need to be considered in the analysis of the data from such experiments. This, in turn, also leads to a reduction in the number of treatment combinations to be used in an experiment and hence to a reduction in the number of observations to be taken. Orthogonal arrays are the foundation for design of experiments in

Taguchi methodology and are capable of providing useful data for a small amount of experiments (Peace, 1992; Hinkelman and Kempthorne; 2008).

For both turning and milling, four control factors, i.e. the four cutting parameters, with three levels each were used. A three level design parameter counts for two degrees of freedom while the interaction between the cutting parameters is neglected. Therefore, there are eight degrees of freedom owing to the four cutting parameters selected. The L_9 orthogonal array with four columns and nine rows can be used in both turning and milling, as it can handle adequately the on hand problem (Yang and Tarng, 1998; Kirby, Zhang, Chen and Chen, 2006; Nalbant, Gökkaya and Sur, 2007). With L_9 orthogonal array only 9 experiments need to be carried out instead of the $3^4=81$ that the full factorial analysis would require. The orthogonal array used can be seen in Table 1, while factors and levels for turning and milling are tabulated in Table 2 and Table 3, respectively.

Table 1. Experimental layout using an L_9 orthogonal array

$L_9 (3^4)$				
	A	B	C	D
1	1	1	1	1
2	1	2	2	2
3	1	3	3	3
4	2	1	2	3
5	2	2	3	1
6	2	3	1	2
7	3	1	3	2
8	3	2	1	3
9	3	3	2	1

Table 2. Turning parameters and their levels

Symbol	Factor	Level 1	Level 2	Level 3
A	Tool type	Turning I	Turning II	Turning III
B	Depth of cut (mm)	0.2	0.6	0.4
C	Feed (mm/rev)	0.1	0.2	0.3
D	Cutting speed (m/min)	300	450	600

Table 3. Milling parameters and their levels

Symbol	Factor	Level 1	Level 2	Level 3
A	Tool type	Milling I	Milling II	Milling III
B	Depth of cut (mm)	0.2	0.4	0.6
C	Feed (mm/rev)	0.2	0.4	0.6
D	Spindle speed (rpm)	2000	4000	6000

The turning processes took place in an OKUMA LB10ii CNC revolver turning machine with a maximum spindle speed of 10,000 rpm and a 10 HP drive motor. The milling processes were conducted on an OKUMA MX-45VAE five axis CNC milling machine with maximum spindle speed of 7,000 rpm and a drive motor of 15 HP. The workpiece during the turning was a bar with 300 mm length and 50 mm diameter, while for the milling a plate 200x50 mm with 5 mm width was used. The material for both workpieces was C45 (HB 180), a common steel in industry.

The cutting tool's characteristics are shown in Table 4 and Table 5 for turning and milling respectively. Common tools were used, made of General Carbide from SECO. The cut for turning was orthogonal, as long as back rake angle was kept at 0° in order to fit better the common cutting conditions.

Table 4. Turning tools characteristics

	Turning I	Turning II- Turning III
Tool Type	CDCB 04T002	DNMG 110404 M3
Coating	TiN – TiNAl	TiN – Al_2O_3 - TiC
Tool Width	3mm	5mm
Edge Radius	0.04mm	0.04mm
Side Rake Angle	-5°	-5°
Back Rake Angle	0°	0°
Lead Angle	-5°	-5°
Relief Angle	5°	5°

Table 5. Milling tools characteristics

	Milling I	Milling II – Milling III
Tool type	CCMX 060204T-MD06	XOEX 090304FR-E05
Coating	w/o coating	TiN – TiNAl
Tool Width	2mm	3.65mm
Edge Radius	0.04mm	0.04mm
Side Rake Angle	0°	5°
Back Rake Angle	0°	0°
Lead Angle	0°	0°
Relief Angle	7°	7°
Number of Teeth (inserts)	1	2
Cutter Diameter	12mm	25mm

For the cutting forces and the chip formations, analog data measurement equipment was used and the tool (for turning)/the workpiece (for milling) were adjusted on a dynamometer, while the chips were gathered at the end of each experiment. Each experiment was carried out twice. The cutting forces were measured in all the three directions. The measurements were done with a Kistler dynamometer 9257A. This is a three-component piezoelectric dynamometer platform. The force data were recorded by a specifically designed, very compact multi-channel microprocessor controlled data acquisition system with a single A/D converter preceded by a multiplexer.

In turning the x-axis is the main cutting force, y-axis the secondary cutting force and z-axis was parallel to the Axis rotational speed. In milling the dynamometer was on the workpiece, so x and y force components are forming the main cutting force.

During the experiments held on the laboratory, the workpiece was cut in both turning and milling for an adequate length in order to obtain clear results on the dynamometer. The length of cut at turning was 50-100mm and at milling was 200mm (full workpiece length).

The models provided are 3D turning and milling models developed with Third Wave AdvantEdge software, which integrates special features appropriate for machining simulation. The program menus are properly designed so that model preparation time is minimized. Furthermore, it possesses a wide database of workpiece and tool materials commonly used in cutting operations, offering all the required data for effective material modeling. The simulation was chosen to be 3D in order to have the ability to make a more accurate comparison between the experimental and the numerical results, especially for the chip morphology. Workpiece material, cutting tools and the processes' setup were modelled from the software menus and data library, with minimum intervention from the user for better results.

AdvantEdge is a Lagrangian, explicit, dynamic code which can perform coupled thermo-mechanical transient analysis. The program applies adaptive meshing and continuous remeshing for chip and workpiece, allowing for accurate results. For an analytical discussion on the numerical techniques used in the program and a comprehensive presentation of its functions Refs (Marusich and Ortiz, 1995; Mamalis, Kundrák, Markopoulos and Manolakos, 2008) are proposed.

The cutting conditions used as input of the FEM models were the same with the experimental work as it can be seen in Tables 6 and 7, for comparison reasons. Three different models for turning and three for milling were produced, for every cutting tool and machining setup and each model run with three different cutting conditions, according to the levels of the design of experiment procedure.

The maximum and minimum element size during the modelling may affect the simulation results and AdvantEdge offers the ability to manually change those values. The maximum element size in all the simulations of this paper was set for the tool as 0.1 mm and the minimum as 0.03 mm, while for workpiece, the maximum was 10 mm and the minimum 0.1 mm.

Minimum values where set at 0.1 mm in order to obtain acceptable computational time; simulation time was much longer while results were not affected considerably, when testing was performed with smaller values.

Simulation time varied at about 3-4 days for turning and 2-3 days for milling, by using an Intel Core2 Duo processor (2.66GHz). The length of the simulations was held on more than a full rotation, namely 400° for turning and 200° for milling, in order to obtain steady state force conditions.

The results of the average force for each experiment and the numerical results are tabulated in Table 6 and Table 7 for turning and milling respectively.

Table 6. Experimental and numerical results for turning

Number of experiment	Factors				Cutting Forces (N)					
					Experimental	Numerical	Experimental	Numerical	Experimental	Numerical
	Tool type	Depth of cut (mm)	Feed (mm/rev)	Cutting Speed (m/min)	F_x	F_x	F_y	F_y	F_z	F_z
1	Turning 1	0.2	0.1	300	150	15	230	15	175	15
2	Turning 1	0.6	0.2	450	290	300	300	300	350	60
3	Turning 1	0.4	0.3	600	210	300	240	100	195	90
4	Turning 2	0.2	0.2	600	35	30	17,5	50	15	10
5	Turning 2	0.6	0.3	300	210	200	25	50	55	50
6	Turning 2	0.4	0.1	450	145	130	35	40	45	200
7	Turning 3	0.2	0.3	450	300	250	50	10	50	50
8	Turning 3	0.6	0.1	600	90	40	40	25	35	30
9	Turning 3	0.4	0.2	300	310	250	80	30	60	90

Table 7. Experimental and numerical results for milling

Number of experiment	Factors				Cutting Forces (N)					
					Experimental	Numerical	Experimental	Numerical	Experimental	Numerical
	Tool type	Depth of cut (mm)	Feed (mm/rev)	Spindle Speed (rpm)	F_x	F_x	F_y	F_y	F_z	F_z
1	Milling 1	0.2	0.2	2000	600	500	125	250	175	300
2	Milling 1	0.4	0.4	4000	1050	500	275	300	275	350
3	Milling 1	0.6	0.6	6000	1950	1900	475	1100	375	400
4	Milling 2	0.2	0.4	6000	225	300	425	200	375	200
5	Milling 2	0.4	0.6	2000	1200	800	175	400	175	300
6	Milling 2	0.6	0.2	4000	400	500	125	200	125	300
7	Milling 3	0.2	0.6	4000	375	300	150	150	175	200
8	Milling 3	0.4	0.2	6000	1100	550	150	500	175	250
9	Milling 3	0.6	0.4	2000	1300	800	325	600	225	250

3. RESULTS AND DISCUSSION

A comparison between the experimental values and the numerical results in Tables 6 and 7 indicate that in general the results are in good agreement. More specifically, for turning the discrepancies are lower between experimental and numerical results; for milling the discrepancies tend to be higher. For certain experiments, e.g. number 1 for turning, the disagreement between the values is rather high. Although this high difference is present only in exceptional cases, inadequate modeling for this case may be concluded. In turn it may be attributed to inadequate workpiece material or contact conditions modeling. In most cases of FE metal cutting simulations, as well as in AdvantEdge, the simple Coulomb friction condition is adopted. Furthermore, the friction coefficient is assumed to be constant. Astakhov and Outeiro (2008) state that according to calculations there is a limiting value for the coefficient of friction above which no relative motion can occur at the tool-chip interface. Experimental data are in direct contradiction with this value but on the other hand FEM models always use a friction coefficient below the limiting value to suit the sliding condition at the interface. Still experiments and FEM models are in good agreement in the results they provide. It seems that there is an inherent difficulty in applying a suitable friction model, especially in the case of high speed machining where more complicated phenomena occur.

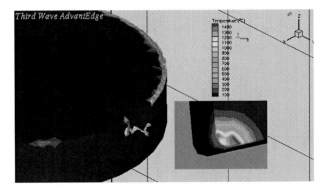

Figure 1. High speed turning experiment 2 with CDCB tool, depth of cut 0.6 mm, feed 0.2 mm/rev and cutting speed 450 m/min.

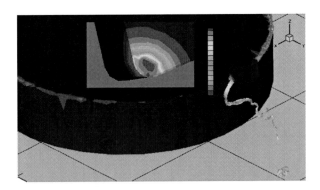

Figure 2. High speed turning experiment 3 with CDCB tool, depth of cut 0.4 mm, feed 0.3 mm/rev and cutting speed 600 m/min.

Figure 1 and Figure 2 depict the chip formation for experimental number 2 and 3 for high speed turning. Furthermore, temperatures on the tool, chip and workpiece can be seen. Figure 3 and Figure 4 shows the chip formation for experimental number 2 and 3 for high speed milling. Once again the temperatures on the tool, chip and workpiece can be observed. Temperatures are important for tool life and the estimation of areas where wear is more possible to show up. It can be seen that at the area close to the tip of the cutting tool temperatures up to 1400° C for turning and 1200° for milling may appear. In all the cases considered the maximum temperatures on the cutting tool have the same magnitude and appear in the same areas.

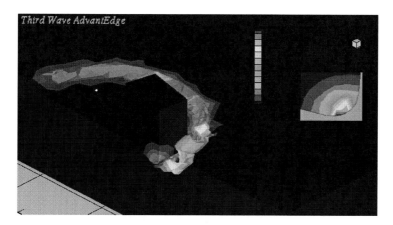

Figure 3. High speed milling experiment 2 with CCMX tool, depth of cut 0.4 mm, feed 0.4 mm/rev/tooth and spindle speed 4000 rpm.

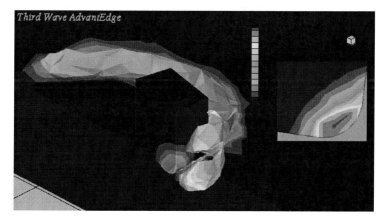

Figure 4. High speed milling experiment 3 with CCMX tool, depth of cut 0.6 mm, feed 0.6 mm/rev/tooth and spindle speed 6000 rpm.

In Figure 5, a comparison of the chips between the experimental and the numerical results for turning is made. More specifically, the chips for the first three cutting conditions are depicted. The pictures of the actual chips collected from the experiments indicate that in all three cases the chips are strips with side curling, typical for the process examined. On the other hand, the simulated chips seem to match well for experiments 2 and 3 while for experiment 1 the correlation is poor. This is attributed to the small depth of cut used in this case. Chip in experiment 3 is longer than that of experiment 2. This can be expalined by the fact that the chip of experiment 2 is thiner because of the smaller depth of cut used from experiment 3 and with more thermal load and because of that also more distorted; the chip breaks easier and forms shorter strips than the chip from experiement 3.

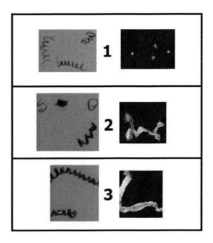

Figure 5. Chips as collected from the experiments and as prediced by the models for high speed turning experiments 1, 2 and 3.

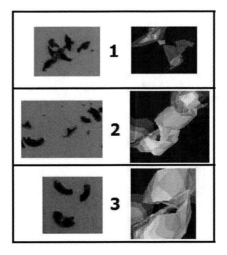

Figure 6. Chips as collected from the experiments and as prediced by the models for high speed milling.

Figure 6 presents the chips for the first three milling experiments. In all cases, in both experimental and numerical results, short side curled chips are seen. The correlation between experiments and simulation is quite satisfying.

Generally speaking, in most cases the chips were predicted accurately; in the cases where inconsistency between measured and predicted forces exists poor correlation between chips also appears.

Conclusions

In this paper 3D FEM simulations of the precision high speed turning and high speed milling are presented. The results are validated by experimental results performed with CNC machine tools. More specifically, the cutting forces and the chip morphology are examined and discussed. From the analysis performed it may be concluded that:

- 3D models need more time and computational power but offer more realistic results than 2D models.
- The results indicate that the proposed models can satisfactorily predict the cutting forces, in most of the examined cases.
- Temperatures may reach 1400° C for turning and 1200° C for milling at the area close to the tip of the cutting tool.
- The shape of the produced chip is accurately predicted by the simulations for most of the examined cutting conditions; in the cases where inconsistency between measured and predicted forces is observed, poor correlation between chips also appears.

In future works a more adequate friction model needs to be incorporated to the analysis to provide more reliable results. Nevertheless, the proposed models offer a useful and reliable tool for the analysis of high speed machining.

References

[1] Astakhov, V. P. and Outeiro, J. C. (2008), "Metal cutting mechanics, finite element modelling", in Davim, J. P. (Ed.), *Machining: Fundamentals and Recent Advances*, Springer-Verlag Limited.

[2] Bäker, M. (2006), "Finite element simulation of high-speed cutting forces", *Journal of Materials Processing Technology*, 176, 117-126.

[3] Balaji, A. K., Ghosh, R., Fang, X. D., Stevenson, R. and Jawahir, I. S. (2006), "Performance-based predictive models and optimization methods for turning operations and applications: Part 2 – Assessment of chip forms/chip breakability", *Journal of Manufacturing Processes*, 8 (2), 144-158.

[4] Davim, J. P. (2008), "*Machining: Fundamentals and Recent Advances*", Springer-Verlag Limited.

[5] Davim, J. P., Maranhão, C., Jackson, M. J., Cabral, G. and Grácio J. (2008), "FEM analysis in high speed machining of aluminium alloy (Al7075-0) using polycrystalline diamond (PCD) and cemented carbide (K10) cutting tools", *International Journal of Advanced Manufacturing Technology*, 39, 1093-1100.

[6] Dixit, P. M. and Dixit U. S. (2008), "*Modeling of Metal Forming and Machining Processes: by Finite Element and Soft Computing Methods*", London, Springer-Verlag London Limited.

[7] Erdel, B. P. (2003), "*High-speed machining*", Society of Manufacturing Engineers, Dearborn, USA.

[8] Hinkelmann, K. and Kempthorne, O. (2008), *"Design and analysis of experiments Vol. 1: Introduction to Experimental Design"*, Hoboken, New Jersey, John Wiley and Sons, Inc.

[9] Hortig, C. and Svendsen, B. (2007), "Simulation of chip formation during high-peed cutting", *Journal of Materials Processing Technology*, 186, 66-76.

[10] Kirby, E. D., Zhang, Z., Chen, J. C. and Chen, J. (2006), "Optimizing surface finish in a turning operation using the Taguchi parameter design method", *International Journal of Advanced Manufacturing Technology*, 30, 1021-1029.

[11] Koshy, P., Dewes R. C. and Asinwall D. K. (2002), "High speed end milling of hardened AISI D2 tool steel (~58 HRC)", *Journal of Materials Processing Technology*, 127, 266-273.

[12] Lin, H. M., Liao, Y. S. and Wei, C. C. (2008), "Wear behavior in turning high hardness alloy steel by CBN tool", *Wear*, 264, 679–684.

[13] Mackerle, J. (1999), "Finite-element analysis and simulation of machining: a bibliography (1976-1996)", *Journal of Materials Processing Technology*, 86, 17-44.

[14] Mackerle, J. (2003), "Finite element analysis and simulation of machining: an addendum A bibliography (1996-2002)", *International Journal of Machine Tools and Manufacture*, 43, 103-114.

[15] Mamalis, A. G., Kundrák, J., Markopoulos, A. and Manolakos, D. E. (2008), "On the finite element modeling of high speed hard turning", *International Journal of Advanced Manufacturing Technology*, 38 (5-6), 441-446.

[16] Marusich, T. D. and Ortiz, M. (1995), "Modelling and simulation of high-speed machining", *International Journal for Numerical Methods in Engineering*, 38, 3675-3694.

[17] Nalbant, M., Gökkaya, H. and Sur, G. (2007), "Application of Taguchi method in the optimization of cutting parameters for surface roughness in turning", *Materials and Design*, 28, 1379-1385.

[18] Nouari, M. and Ginting A. (2004), "Wear characteristics and performance of multi-layer CVD-coated alloyed carbide tool in dry end milling of titanium alloy", *Surface and Coatings Technology*, 200 (18-19), 5663-5676.

[19] Özel, T. and Altan, T. (2000), "Procss simulation using finite element method – prediction of cutting forces, tool stresses and temperatures in high-speed flat end milling", *International Journal of Machine Tools and Manufacture*, 40, 713-738.

[20] Peace, G. S. (1993), *"Taguchi methods: a hands-on approach"*, Addison-Wesley Publishing Company, Inc., Massachusetts, USA.

[21] Robinson, G. M. and Jackson M. J. (2005), "A review of micro and nanomachining from a materials perspective", *Journal of Materials Processing Technology*, 167, 316-337.

[22] Schulz, H. and Moriwaki, T. (1992), "High-speed machining", *Annals of the CIRP*, 41/2, 637-643.

[23] Tang, D. W., Wang, C. Y., Hu, Y. N. and Song, Y. X. (2009), "Finite-element simulation of conventional and high-speed peripheral milling of hardened mold steel", *Metallurgical and Materials Transactions A*, 40A, 3245-3257.

[24] Thamizhmanii, S. and Hasan, S. (2009), "Effect of tool wear and forces by turning process in hard AISI 440C and SCM 440 materials", *International Journal of Materials Forming*, 2 Suppl 1, 531–534.

[25] Tlusty, J. (1993), "High-speed machining", *Annals of the CIRP*, 42/2, 733-738.

[26] Trent, E. M. and Wright, P. K. (2000), *"Metal cutting"*, Woburn, Butterworth-Heinemann.

[27] Yang, W. H. and Tarng, Y. S. (1998), "Design optimization of cutting parameters for turning operations based on the Taguchi method", *Journal of Materials Processing Technology*, 84, 122-129.

[28] Yang, Y. and Li, J. F. (2010), "Study on mechanism of chip formation during high-speed milling of alloy cast iron", *International Journal of Advanced Manufacturing Technology*, 46, 43–50.

In: Machining and Forming Technologies, Volume 3
Editor: J. Paulo Davim

ISBN: 978-1-61324-787-7
© 2012 Nova Science Publishers, Inc.

Chapter 12

FINITE ELEMENTS MODELLING AND SIMULATION OF PRECISION GRINDING

A. P. Markopoulos[*]

Laboratory of Manufacturing Technology, National Technical University of Athens,
Athens, Greece

ABSTRACT

Grinding is a manufacturing process with a lot of applications in contemporary industry. Due to its importance, grinding is subjected to careful and systematic analysis. The Finite Elements Method (FEM), which is widely used for the modeling and simulation of manufacturing process, is also employed for grinding. In the present paper both thermal and coupled thermo-mechanical FEM models of very fine shallow surface grinding are presented. The models are based on previous work on this area; however, they incorporate new features that allow for the detailed and accurate simulation of the process. The data for the simulations are provided by experimental work conducted on three different steels. The models exhibit results on the maximum temperatures as well as the temperatures reached within the workpiece, revealing thus information on the heat affected zones and possible damage within the ground workpieces. Furthermore, the coupled analysis is able to supply extra information on the stresses appearing during grinding due to the grinding wheel and workpiece interaction.

Keywords: Grinding, Modelling and Simulation, Finite Elements Method, Coupled analysis

1. INTRODUCTION

Grinding is a precision manufacturing process, traditionally used as a finishing operation because of its ability to produce high workpiece surface quality. Improvements in its performance have allowed for the use of grinding in bulk removal of metal, maintaining at the same time its characteristic to be able to perform precision processing, thus opening new

---

[*]E-mail: amark@mail.ntua.gr

areas of application in today's industrial practice (Boothroyd and Knight, 2006). The ability of the process to be applied on metals and other difficult to machine materials such as ceramics and composites is certainly an advantage of this manufacturing method (Sun, Brandt and Dargusch, 2010). In a keynote paper by CIRP (Oliveira *et al.*, 2009), a thorough discussion on the applications of grinding, especially in the automotive industry, as well as the limitations and research conducted towards the improvement of the process was presented, justifying the great importance of grinding and the amount of researchers that show interest in analyzing its characteristics.

Especially, thermal aspects of grinding are investigated because of the damage induced to the workpiece due to excessive heat loading (Malkin and Guo, 2007). This heat input is responsible for a number of defects in the workpiece like metallurgical alterations, microcracks and residual stresses. The areas of the workpiece that are affected are described as *heat affected zones*. Thermal load is connected to the maximum workpiece temperature reached during the process and therefore the maximum temperature of the ground workpiece surface is of great importance (Mamalis *et al.*, 2003a). Nevertheless, certain difficulties arise when measuring surface temperatures during grinding, mainly due to the set-up of the process; a lot of research pertaining to grinding is performed through modeling and simulation instead of experimental investigation.

More specifically, the Finite Element Method has been employed for the thermal and mechanical modeling of surface grinding. An overview of the so far proposed models can be found in Snoeys, Maris and Peters (1978), Tönshoff *et al.* (1992), Brinksmeier *et al.* (2006), Markopoulos (2006) and Doman, Warkentin and Bauer (2009). From these references it can be concluded that most of the FEM models refer to thermal modelling that aims to predict temperature distribution within the workpiece. Coupled thermo-mechanical models are rarer because they are more complex and need not only more data but also more sophisticated modeling (Nélias and Boucly, 2008). Only a small portion of the cited works provides experimental validation (Anderson, Warkentin and Bauer, 2008). The above conclusions indicate that thermal phenomena are of greater interest than others due to their impact on the outcome of the process and that there is difficulty in providing adequate grinding temperature measuring methods.

The paper presents a Finite Element grinding model which is able to predict the temperature and stress distribution within the workpiece and finally determine the heat affected zones that may appear. The data used as input for the FEM models are gained from a series of experiments conducted for this reason. Additionally, model validation is performed with comparison with other models. Both experimentation and model validation will be described in detail in another section of the paper.

2. FINITE ELEMENT MODELLING

2.1. Thermal Modelling of Grinding

Almost all thermal models of grinding are based on the moving heat sources model suggested by Jaeger (1942). A two dimensional model is used, provided that the grinding width is large with respect to its length. In Jaeger's model the grinding wheel is represented by a heat source moving along the surface of the workpiece with a speed equal to the workspeed.

The heat source is characterized by a physical quantity, the heat flux, q, that represents the heat entering the workpiece per unit time and area and it is considered to be of the same density along its length, taken equal to the geometrical contact length, l_c, which is calculated from the relation:

$$l_c = \sqrt{a \cdot d} \tag{1}$$

where a is the depth of cut and d_s the diameter of the grinding wheel. The above can be seen in Figure 1.

It should be noted that other heat source profiles, besides the rectangular one, have been suggested and investigated, e.g. triangular, parabolic, oblique profile, by the author and other researchers (Mamalis *et al.*, 2004; Snoeys, Maris and Peters, 1978; Jin and Cai, 2001; Kuriyagawa, Syoji and Ohshita, 2003; Kruszynski and Pazgier, 2003; Tian *et al.*, 2009). However, it is concluded by most authors that in shallow surface grinding of steels, a typical heat source profile uniformly distributed over the contact length may be assumed without considerably affecting the maximum temperatures and the overall accuracy of the model.

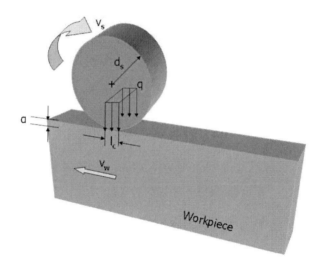

Figure 1. Jaeger's model in grinding.

The heat flux can be calculated from the following equation:

$$q = \varepsilon \frac{F_t' \cdot v_s}{l_c} \tag{2}$$

where ε is the percentage of heat flux entering the workpiece, F'_t the tangential force per unit width of the workpiece, v_s the peripheral wheel speed and l_c the contact length. The proportion of the heat flux entering the workpiece can be calculated by a formula suggested by Malkin (1978) for grinding with aluminum oxide wheels, by making assumptions on the partitioning of total specific grinding energy, u, required for grinding. The total specific grinding energy consists of three different components: the specific energy required for the formation and the removal of the chip, u_{ch}, the specific energy required for plowing, i.e. the plastic deformation in the regions where the grains penetrate the workpiece surface but no material is removed, u_{pl} and the specific energy required for making the flat wear grains slide on the workpiece surface, u_{sl}, thus:

$$u = u_{ch} + u_{pl} + u_{sl} \qquad (3)$$

It has been analytically and experimentally shown that approximately 55% of the chip formation energy and all the plowing and sliding energy are conducted as heat into the workpiece, i.e.

$$\varepsilon = \frac{0.55 \cdot u_{ch} + u_{pl} + u_{sl}}{u} = \frac{u - 0.45 \cdot u_{ch}}{u} \Rightarrow \varepsilon = 1 - 0.45 \frac{u_{ch}}{u} \qquad (4)$$

The component u_{ch} has a constant value of about 13.8 J/mm^3 for grinding for all ferrous materials (Malkin, 1989). The total specific grinding energy is calculated from the following equation:

$$u = \frac{F'_t \cdot v_s}{a \cdot v_w} \qquad (5)$$

where v_w is the workspeed and, consequently, as in Jaeger's model, the speed of the moving heat source. Note that, in both equations (2) and (5) the value of F'_t is needed in order to calculate the heat flux and the total specific grinding energy, respectively; it can be calculated from the power per unit width of the workpiece, P'_t, as follows:

$$F'_t = \frac{P'_t}{v_s} \qquad (6)$$

The last equation suggests that if the power per unit width of the workpiece is known, then the heat flux can be accurately calculated and used in the finite elements models described in the next paragraph. Indeed, the power per unit width of the workpiece can be experimentally measured (Mamalis *et al.*, 2003b) and accordingly used.

2.2. Finite Elements Thermal Modelling of Grinding

Following the previous paragraph's guidelines Finite Elements models can be constructed with the configuration presented in Figure 2.

On the top surface heat is entering the workpiece in the form of heat flux that moves along this surface. Cooling from the applied cutting fluid is simulated by means of convective boundary conditions. All the other sides of the workpiece are considered to be adiabatic, and so no heat exchange takes place in these sides. The cooling effect simulated refers to the flood method, where coolant at low pressure and room temperature fills the upper part of the workpiece, applying a uniform cooling in all the surface area.

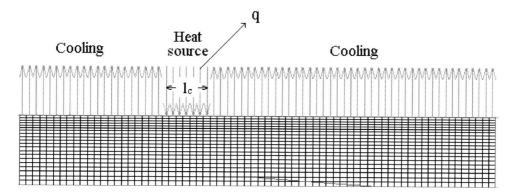

Figure 2. Thermal finite element model.

The model needs to have a sufficient enough length in order for the temperature fields to be deployed and observed in full length. A mesh, consisting of 4-noded rectangular full integration elements with one degree of freedom, namely the temperature, for the thermal models, is applied on the workpiece geometry for plane stress analysis. The mesh is denser towards the grinding surface, which is the thermally loaded surface, and, thus the most affected zone of the workpiece, allowing for greater accuracy to be obtained.

The kind of modelling suggested in this paper is suitable for grinding processes with very small depth of cut, as in precision grinding, since there is no modelling of the chip. The models of this paragraph use the method described in the previous section in order to simulate precision surface grinding. Nevertheless, the powerful Finite Elements method allows for certain improvements. First of all the presented models pertain to transient problems, where $\dot{T} = \partial T / \partial t \neq 0$, instead of steady state models, where $\dot{T} = 0$. Furthermore, the two coefficients of the workpiece material that are related to temperature, i.e. the thermal conductivity and the specific heat capacity, are considered to be temperature depended. Transient conditions and temperature depended material properties produce non-linear finite element problems, which are more difficult to be solved. Finally, as presented above, the cutting fluid effect is simulated; this is a feature that is not taken into account in the original Jaeger's model but adds considerably to the accuracy and the building of a sound model of grinding (Markopoulos, 2006).

2.3. Finite Elements Coupled Analysis

Coupled analysis incorporates thermal and mechanical features in the models, i.e. a coupled thermo-mechanical analysis is allowed. In principle, coupled analysis models are more complete than thermal since they provide all the data thermal models can but also stress parameters can be predicted. At the same time the complexity involved in making these models has not allowed for their wide spread. Most researchers are trying to correlate critical temperatures with the onset of unwanted phenomena, even if they are of mechanical nature, e.g. residual stresses.

From a modeling point of view thermal models are converted into thermo-mechanical models by adding more boundary and initial conditions, material properties and altering accordingly the elements' type, i.e. 4-noded rectangular full integration elements with four degrees of freedom. Boundary conditions include the constraining of the nodes on the vertical sides as well as the bottom side from movement and initially nodal displacement in the workpiece mesh is considered to be zero. Additionally, Young's modulus and thermal expansion coefficient are temperature dependent, providing thus a plane stress dynamic transient coupled analysis.

The results obtained from this model, e.g. stresses, are attributed only to thermal loading, because no mechanical interaction between the cutting wheel and the workpiece is simulated. This can be achieved if more boundary conditions with cutting forces are included in the analysis. The combined results are provided by superimposing the results of the models with and without cutting forces boundary conditions or, if it is allowed by the FEM software as is the case with MSC.Marc models presented hereafter, the coupled analysis can be performed simultaneously.

2.4. Model Validation

The aforementioned analysis is realised by the commercial Finite Elements software MSC.Marc Mentat. The software is able to perform the analysis using the Jaeger's model and furthermore include all the special features that improve the model, providing a novel more efficient and reliable simulation of precision grinding.

In order to investigate the validity of the model the sets of data from three different publications of the relevant literature are inserted to the model and the discrepancies of the original results from the results of the presented model are observed. More specifically, input data from models proposed by Snoeys, Maris and Peters (1978), Guo, Wu, Varghese and Malkin (1999) and Moulik, Yang and Chandrasekar (2001), are used and new models with the method described in paragraph 2 are constructed; the new models are referred hereafter as Model 1, Model 2 and Model 3 respectively. The grinding conditions and other data required for the construction of Models 1, 2 and 3 are presented in Table 1; these are the same data used in the corresponding initial publications from the literature. It can be seen that the provided data are quite different from each other, especially the ones referred to Model 2, mainly because of the grinding wheel composition but also due to different cutting conditions. Because of this, the results cannot be compared to each other but the only comparison is between the original model and the corresponding model constructed with the same input data.

In Figure 3 the maximum temperature for Models 1, 2 and 3 in comparison to the predicted maximum temperature from the corresponding original model are presented. It can be seen that the results for Model 1 and model 3 exhibit very small discrepancies, while the values for Model 2 are different. This can be attributed to the fact that the energy partition analysis presented in the previous paragraphs needs to be further investigated when CBN grinding wheels are considered, because of the different properties of the wheel material.

Table 1. Data from relevant references

	Model 1	Model 2	Model 3
Workpiece material	100Cr6	100Cr6	100Cr6
Depth of cut, $a(\mu m)$	18	300	25
Workspeed, v_w (mm/s)	80	125	100
Wheel speed, v_s (m/s)	24	60	30
Contact length, l_c (mm)	2,84	11	2,236
Additional data	$u=49$ J/mm^3	$P=11$ kW	$q=54.78$ W/mm^2
	$\varepsilon=85\%$	$\varepsilon=6.2\%$	$\varepsilon=88.7\%$
Grinding wheel	Aluminum Oxide	CBN	Aluminum Oxide

Nevertheless, the predicted 128° C are very close to the 110° C of the original paper. Furthermore, in Figure 4, the temperature versus distance from the workpiece edge at a certain step of the analysis is presented for Model 3 and the corresponding original model. The similarity of the two diagrams is obvious, showing once more that the presented model is in agreement with the models proposed from other investigators.

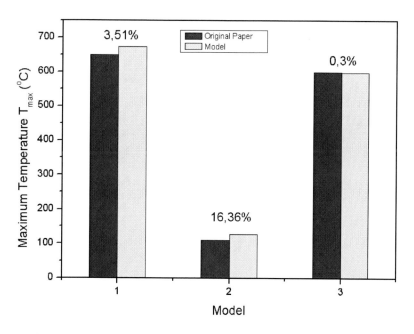

Figure 3. Comparison between the results of the model and results from the literature.

Note that each of the three models simulated with the corresponding model was constructed using the original configuration. For example, if the original model did not take into consideration the workpiece material dependency upon temperature, so was modeled with the new model. The good agreement of the results presented above allows for the use of the models for further investigation, if the appropriate input data can be attained. For this reason experimental work was conducted and is presented in the next section.

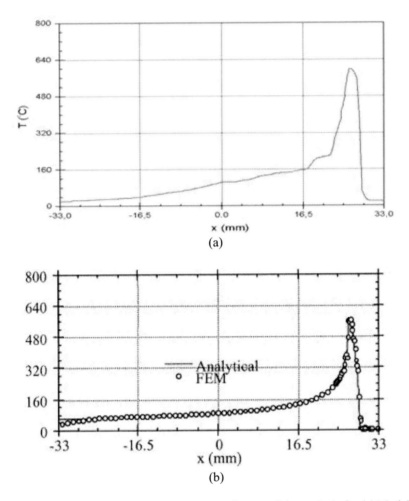

Figure 4. Temperature on the workpiece surface at a certain step of the analysis for (a) Model 3 and (b) model by Moulik, Yang and Chandrasekar (2001).

3. EXPERIMENTAL DATA

According to the analysis of section 2, in order to provide the appropriate input to the model, i.e. the heat flux, the power per unit width of the workpiece needs to be measured. For this reason, six aluminum oxide grinding wheels of the same diameter, d_s=250 mm and width b_s=20 mm with different bonding were used on a BRH 20 surface grinder. Four depths of cut were used, namely 10, 20, 30 and 50 μm while the workpiece speed was v_w=8 m/min and the wheel speed v_s=28 m/s kept constant for all sets of experiments, for all wheels. The workpiece materials were the 100Cr6, C45 and X210Cr12 steels; the characteristics of the materials are given in Table 2.

Finite Elements Modelling and Simulation of Precision Grinding

Table 2. Workpiece material data

Workpiece material	Hardness		Composition (% weight)							
			C	Cr	Mn	Si	Ni	Mo	P	S
100Cr6	62 HR$_C$	min	0.90	1.35	0.25	0.15	0.00	-	0.000	0.000
		max	1.05	1.65	0.45	0.35	0.30		0.030	0.025
C45	220 HB	min	0.42	0.00	0.60	0.00	0.00	0.00	0.000	0.000
		max	0.50	0.40	0.90	0.40	0.40	0.10	0.035	0.035
X210Cr12	64 HR$_C$	min	1.90	11.0	0.15	0.10	-	-	0.000	0.000
		max	2.20	12.0	0.45	0.40			0.030	0.030

Throughout the process the synthetic coolant Syntilo-4 was applied at 15 ℓ/min. For each grinding wheel, 10 passes of the same depth of cut were performed over the workpiece. The power per unit width of the workpiece was measured for each pass and its average value was calculated. For measuring the power, a precision three-phase wattmeter was used. First, the power of the idle grinding machine was measured and set as the zero point of the instrument. Then, the workpieces were properly ground and the power was registered on the measuring device. After 10 passes were performed the grinding wheel was dressed with a single point diamond dressing tool, with depth a_d=0.02 mm and feed of f_d=0.1-0.2 mm/wheel rev. In total, 72 measurements took place and are presented in Table 3. In Table 3, the calculations for heat flux, are also presented.

Table 3. Experimental data

Experiment Number	Grinding wheel	Workpiece material	α(μm)	P'_t (W/mm)	q (W/mm^2)
1	1	100Cr6	10	93.5	53.90
2		100Cr6	20	143.5	56.77
3		100Cr6	30	160.0	49.35
4		100Cr6	50	278.5	67.06
5	2	100Cr6	10	94.0	54.21
6		100Cr6	20	164.0	65.94
7		100Cr6	30	237.5	77.65
8		100Cr6	50	288.0	69.75
9	3	100Cr6	10	96.0	55.48
10		100Cr6	20	176.0	71.30
11		100Cr6	30	208.0	66.88
12		100Cr6	50	298.5	72.72
13	4	100Cr6	10	99.0	57.38
14		100Cr6	20	132.5	51.85
15		100Cr6	30	160.0	49.35
16		100Cr6	50	256.0	60.70
17	5	100Cr6	10	117.5	69.08

Table 3. (Continued)

Experiment Number	Grinding wheel	Workpiece material	α(μm)	P'$_t$ (W/mm)	q (W/mm^2)
18		100Cr6	20	170.5	68.84
19		100Cr6	30	176.0	55.20
20		100Cr6	50	277.5	66.78
21	6	100Cr6	10	102.5	59.59
22		100Cr6	20	141.0	55.65
23		100Cr6	30	202.5	64.87
24		100Cr6	50	378.5	95.35
25	1	C45	10	51.0	27.02
26		C45	20	96.5	35.75
27		C45	30	109.5	30.91
28		C45	50	224.0	51.65
29	2	C45	10	48.0	25.12
30		C45	20	81.0	28.82
31		C45	30	133.5	39.68
32		C45	50	194.5	43.30
33	3	C45	10	42.5	21.64
34		C45	20	90.5	33.07
35		C45	30	106.5	29.82
36		C45	50	208.0	47.12
37	4	C45	10	56.0	30.18
38		C45	20	117.5	45.14
39		C45	30	140.0	42.05
40		C45	50	234.5	54.62
41	5	C45	10	45.5	23.54
42		C45	20	102.5	38.43
43		C45	30	117.5	33.83
44		C45	50	216.0	49.38
45	6	C45	10	64.0	35.24
46		C45	20	103.0	38.66
47		C45	30	120.0	34.75
48		C45	50	266.5	63.67
49	1	X210Cr12	10	122.5	72.24
50		X210Cr12	20	206.0	84.72
51		X210Cr12	30	340.0	115.08
52		X210Cr12	50	368.0	92.38
53	2	X210Cr12	10	141.0	83.94
54		X210Cr12	20	271.5	114.01
55		X210Cr12	30	368.0	125.30
56		X210Cr12	50	426.5	108.92
57	3	X210Cr12	10	170.5	102.60
58		X210Cr12	20	330.5	140.40

Experiment Number	Grinding wheel	Workpiece material	α(μm)	P'$_t$ (W/mm)	q (W/mm^2)
59		X210Cr12	30	346.5	117.45
60		X210Cr12	50	450.5	115.71
61	4	X210Cr12	10	128.0	75.72
62		X210Cr12	20	219.5	90.76
63		X210Cr12	30	352.0	119.46
64		X210Cr12	50	378.5	95.35
65	5	X210Cr12	10	138.5	82.36
66		X210Cr12	20	261.5	109.54
67		X210Cr12	30	330.5	111.61
68		X210Cr12	50	421.5	107.51
69	6	X210Cr12	10	152.5	91.21
70		X210Cr12	20	241.5	100.60
71		X210Cr12	30	309.5	103.94
72		X210Cr12	50	501.5	130.14

According to the measured data the specific energy increases with decreasing depth of cut. This can be explained by the so-called size effect. The cross section of the chip which is smaller for smaller depths of cut, possess different mechanical properties at microscale, as compared with macroscale, due to the existence of fewer numbers of dislocations, grit faults and inclusions. Therefore, the micro hardness increases, resulting in an increase of the specific energy in grinding.

4. RESULTS AND DISCUSSION

As mentioned in section 2, MSC.Marc Mentat is used for the modelling and simulation of grinding. Four different types of models are introduced, one for each depth of cut and consequently for each contact length. This is done because for different contact lengths, different meshing is required. Model types I, II, III and IV are related to depths of cut 10, 20 30 and 50 μm correspondingly. It is reminded that the models take into consideration the material properties variation with temperature, the cooling effect of the cutting fluid and are transient state dynamic models.

The experimental data presented in Table 3 are used as input data for the models and the analysis is performed. Figure 5 presents four different increments, namely increment 3, 5, 7 and 9 from model type II, grinding wheel 6 for 100Cr6. The models although are transient state model, show an increase in temperature for only a few increments, indicating the passing from transient to steady-state condition. The heat input causes the rapid increase of temperature, as can be seen in Figure 5; note that temperatures are in degrees Celsius.

For all model types the isothermal bands and the temperature distribution on the workpiece surface show the same trends. The maximum temperature varies between different cutting conditions, material and cutting wheel. In Figure 6 the temperature variation on the workpiece surface for model type II and workpiece material 100Cr6 for all six wheels is presented and in Figure 7 the temperature variation for the same material but for model type IV is presented.

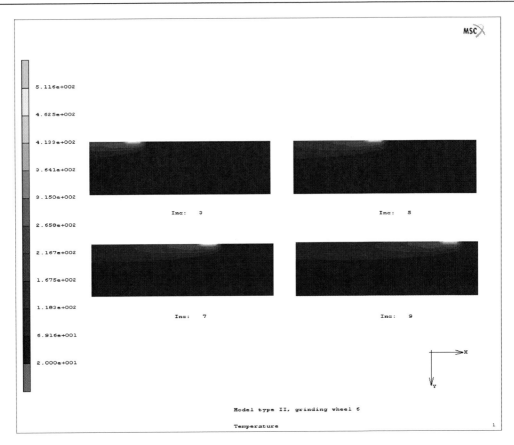

Figure 5. Isothermal bands for model type II, for grinding wheel 6 and workpiece material 100Cr6, in the third, fifth, seventh and ninth increment of the analysis.

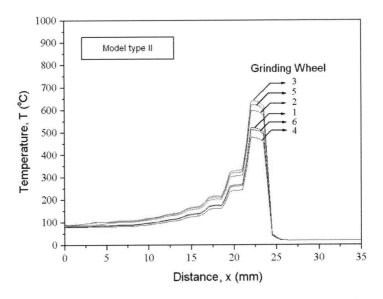

Figure 6. Temperature variation on the surface of the workpiece for model type II and workpiece material 100Cr6.

Finite Elements Modelling and Simulation of Precision Grinding 173

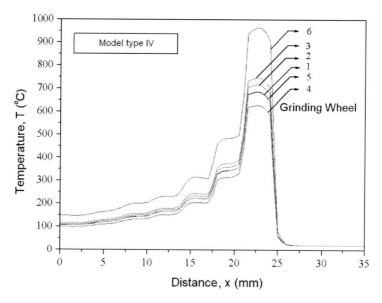

Figure 7. Temperature variation on the surface of the workpiece for model type IV and workpiece material 100Cr6.

From Figures 6 and 7 it can be concluded that the temperature fields appear to be the same for the same depth of cut and the only difference is the maximum temperature reached for each grinding wheel. Furthermore, it is revealed that the temperatures are higher in the regions on the back of the wheel, therefore, it seems that, it is more critical to direct the coolant to this side, in order to prevent the damage of the surface integrity due to the temperature rise. Furthermore, wheel 4 has the lower maximum temperature for both depths of cut, whilst when grinding with wheel 6 a significant deviation in temperature between the two cases is observed. This irregularity may be attributed either to the wheel specification being not suitable for the applied cutting conditions and so the grains are overloaded, or to the increase of the specific energy due to friction. The latter can be explained by the bigger radius in the workpiece-wheel contact zone, when the depth of cut is bigger, leading to the increase of the adhesive effect between the workpiece material and the wheel. The particles that are adhered to the wheel increase the friction surfaces, leading to the consumption of more energy. Such a phenomenon is not rare in grinding technology; it can be avoided by using the right wheel or coolant.

In Figures 8 and 9 the temperatures on the surface (y=0) and underneath it (y>0), for grinding wheel 4 and model types II and IV are presented. In these diagrams y (mm) is the distance of the nodes of the mesh of the finite element model from the surface, C is the center of the heat source and the region 0-l_{ci} is the current position of the heat source, where l_{ci}, i=I, II, III, IV, is the geometrical contact length between the grinding wheel and the workpiece for each model type. From Figures 8 and 9 it can be concluded that the maximum temperature appears to be reported very close to the center of the heat source and that the temperatures below the surface and especially for y>0.320 mm are approximately half than those on the surface of the workpiece. Furthermore, the regions in front of the heat source that are more than 1 l_{ci} away from the center of the heat source are not yet affected, while the regions even at a distance of 5.5 l_{ci} from the same point are still thermally loaded.

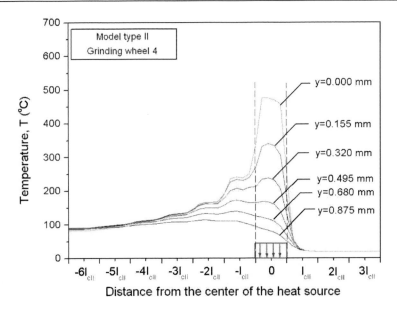

Figure 8. Temperature distribution within the workpiece for wheel 4, 100Cr6, model type II.

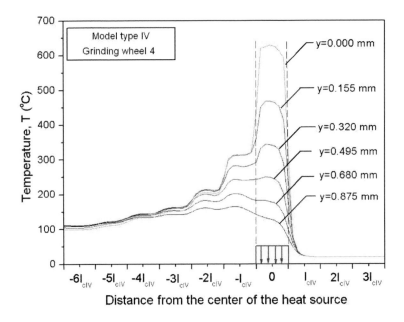

Figure 9. Temperature distribution within the workpiece for wheel 4, 100Cr6, model type IV.

Figure 10 and Figure 11 show the temperature variation on the surface of the workpiece, for grinding wheel 4, for all three workpiece materials, for model type II and IV respectively. It can be seen that the temperature distribution is similar for all materials and once again the only difference lies in the maximum temperature reached. The harder the material is the higher the maximum temperature is expected to be, since the process energy is higher for harder materials, as can be seen from the experimental data as well. As noted previously, the maximum temperature increases for increasing depth of cut.

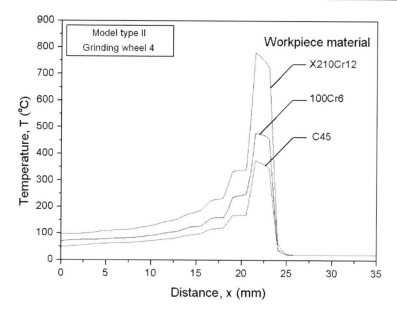

Figure 10. Temperature variation on the surface of the workpiece for model type II and grinding wheel 4 for the three workpiece materials.

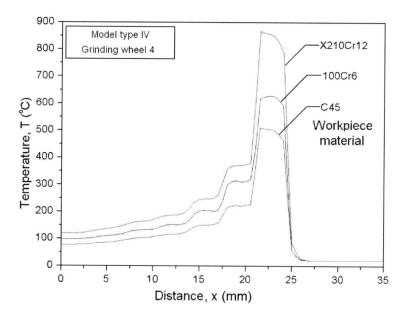

Figure 11. Temperature variation on the surface of the workpiece for model type IV and grinding wheel 4 for the three workpiece materials.

Figures 5-11 demonstrate maximum temperatures that vary from 450° to almost 1000° C, depending on the workpiece material, depth of cut and grinding wheel used. Such values for the maximum temperature, when grinding steels, are reported by other investigators, too (Snoeys, Maris and Peters, 1978; Hoffmeister and Weber, 1999; Kato and Fujii, 2000; Moulik, Yang and Chandrasekar, 2001).

The high temperatures that appear in grinding have a negative effect on the workpiece. The surface of the workpiece and also the layers that are near the surface and have been affected by the heat loading during the grinding process consist the heat affected zones of the workpiece. The excessive temperature in these zones contributes to residual stresses, microcracking and tempering and may cause microstructure changes, which result to hardness variations of the workpiece surface. Steels that cool down quickly from temperatures above the austenitic transformation temperature undergo metallurgical transformations; as a result, untempered martensite is produced in the workpiece. Excessive heat may also lead to metallurgical burn of the workpiece, which produces a bluish color on the surface of the processed material due to oxidation. If the critical temperatures at which these transformations take place are known, the size of the heat affected zones can be also predicted from the FEM model. The actual size of these zones and their composition depends on the duration of thermal loading, except the maximum temperature reached. The three critical temperatures for the 100Cr6 steel are $T_t=150$ °C for tempering, $T_m=250$ °C for martensitic and, $T_a=800$ °C for austenitic transformation and are related to hardness variation, residual stresses and the formation of untempered martensite layers within the workpiece (Rowe, Petit, Boyle and Moruzzi, 1988; Shaw and Vyas, 1994; Zhang and Mahdi, 1995; Chang and Szeri, 1998; Mamalis et al., 2003b).

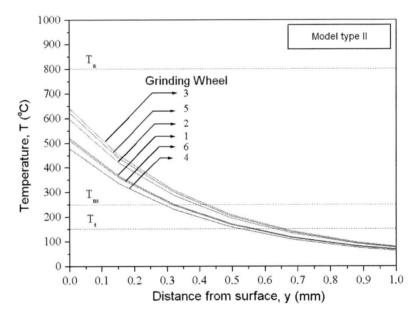

Figure 12. Variation of temperature versus distance from surface when grinding 100Cr6 steel with all grinding wheels for model type II.

In Figures 12 and 13 the variation of the temperatures within the workpiece with depth below the surface is shown, as calculated for all grinding wheels, for 100Cr6 for model types II and IV. These temperatures are taken underneath the grinding wheel where the maximum temperatures are reached. In the same diagrams the three critical temperatures for the 100Cr6 steel are also indicated. From these diagrams the theoretical depth of the heat affected zones, for each wheel used and depth of cut can be determined. In Figure 12 it can be seen that there is no exceeding of the austenitic transformation temperature since the temperatures are not high enough. On the contrary, in Figure 13, when grinding with grinding wheel 6, austenitic transformation temperature is exceeded in the layers with depth up to 0.1 mm below the surface.

Diagrams such as these can be provided for all the other data of Table 3. Note, however, that the critical temperatures are not the same for all the materials.

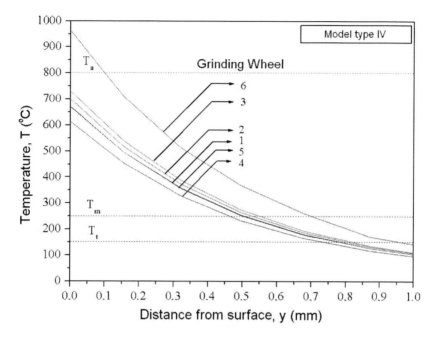

Figure 13. Variation of temperature versus distance from surface when grinding 100Cr6 steel with all grinding wheels for model type IV.

Figures 12 and 13 indicate that there is the possibility when machining hardened steels to exceed critical temperatures that may damage the workpiece and create heat affected zones through the metallurgical transformations that take place (Shaw and Vyas, 1994). This may be attributed to the use of unsuitable grinding wheel, non-effective cooling or inappropriate grinding conditions. On the other hand, it should be taken into account that the top surface of the workpiece, that is the part of the workpiece mostly affected by the thermal damage, is to some extend, depending on the depth of cut, carried away as a chip.

Finally, Figure 14 shows the σ_{xx}/Y and σ_{yy}/Y, a ratio consisting of the components of stress and the yield stress of the workpiece material, on the surface of the 100Cr6 workpiece versus the distance from the center of the heat source for grinding wheel 6 and model type IV. The diagram depicts the seventh increment of the analysis. The discontinuous lines show the results when no cutting forces boundary conditions are considered while the continuous lines include the mechanical boundary conditions. By comparison between previous diagrams and Figure 14 it can be concluded that the maximum stresses appear under the grinding wheel, where the maximum temperatures also appear. The stresses are quite high, reaching almost the yield stress of the material, and are compressive. Furthermore, it is worth noticing that the difference between the values that include the extra boundary conditions and those that do not include them are rather small, since stress is mainly attributed to thermal loading.

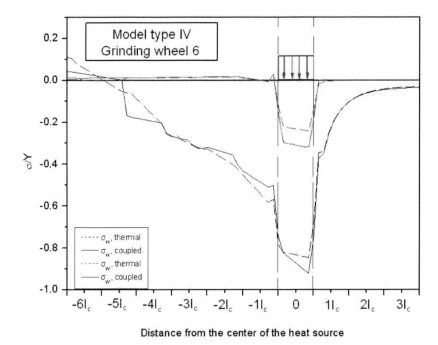

Figure 14. Stresses versus distance from the heat source for model type IV, grinding wheel 6 and workpiece material 100Cr6.

CONCLUSION

The FEM commercial sotware MSC.Marc Mentat was used for the construction of thermal and coupled thermo-mechanical models for the simulation of precision shallow grinding of steels. An analysis based on Jaeger's moving heat source theory was carried out, with extra features included that add to the reliability and the accuracy of the presented models. The aim of the models was to predict the maximum temperatures, the heat affected zones and the stresses within the ground workpieces. Experimental work was carried out in order to obtain the necessary input data for the models. However, the results of the analysis cannot be validated quantitatively by other published work, since there are no relevant papers with the same characteristics; only qualitative remarks can be made. Because of this, a validation procedure was adopted where three different models were constructed with the exact data from other researchers. In all cases, the constructed models where found to be in good agreement with the original results.

The results obtained from the models predict the maximum temperatures on the surface of the ground workpieces as well as the temperatures within the workpieces. In some cases, depending on the workpiece material, depth of cut and grinding wheel used, the temperature rise may be considerable. This in turn may lead to metallurgical transformations and the creation of heat affected zones. Responsible for this is the use of unsuitable grinding wheel, non-effective cooling or inappropriate grinding conditions. It is pointed out that grinding of steel and especially hardened steel should be applied with care in order to avoid these phenomena. Finally, the stresses during grinding can be evaluated and be taken into consideration, from the results obtained through the coupled thermo-mechanical models.

The models presented in this paper, although constructed with improvements in Jaeger's model, are simple to use and require little data. All the calculations of temperatures and stresses are based on kinematic and geometrical parameters, allowing for the monitoring of the process without using any temperature measurement devices. Furthermore, the models are fast, since it takes only a few seconds for running on a modern PC; the total time depends on the number of steps and the parameters applied. With the results and the simulations, the quality of the workpiece produced with precision grinding can be assessed and unwanted phenomena can be predicted and avoided.

REFERENCES

[1] Anderson, D., Warkentin, A. and Bauer, R. (2008), "Experimental validation of numerical thermal models for dry grinding", *Journal of Materials Processing Technology*, 204, 269–278.

[2] Boothroyd, G. and Knight, W. A. (2006), *"Fundamentals of machining and machine tools"*, 3rd edition, CRC Press, Boca Raton, USA.

[3] Brinksmeier, E., Aurich, J. C., Govekar, E., Heinzel, C., Hoffmeister, H. -W., Klocke, F., Peters, J., Rentsch, R., Stephenson, D. J., Uhlmann, E., Weinert, K. and Wittmann, M. (2006), "Advances in modeling and simulation of grinding processes", *Annals of the CIRP*, 55/2, 667-696.

[4] Chang, C. C. and Szeri, A. Z. (1998), "A thermal analysis of grinding", *Wear*, 216, 77-86.

[5] Doman, D. A., Warkentin, A. and Bauer, R. (2009), "Finite element modeling approaches in grinding", *International Journal of Machine tools and Manufacture*, 49, 109-116.

[6] Guo, C., Wu, Y., Varghese, V. and Malkin, S. (1999), "Temperatures and energy partition for grinding with vitrified CBN wheels", *Annals of the CIRP*, 48/1, 247-250.

[7] Hoffmeister, H.-W. and Weber, T. (1999), "Simulation of grinding by means of the finite element analysis", *3rd International Machining and Grinding SME Conference*, Ohio, MR99-234.

[8] Jaeger, J. C. (1942), "Moving sources of heat and the temperature at sliding contacts", *Journal and Proceedings of the Royal Society of New South Wales*, 76 (3), 203-224.

[9] Jin, T. and Cai, G. Q. (2001), "Analytical thermal models of oblique moving heat source for deep grinding and cutting", *Journal of Manufacturing Science and Engineering*, 123, 185-190.

[10] Kato, T. and Fujii, H. (2000), "Temperature measurement of workpieces in conventional surface grinding", *Transactions of the ASME - Journal of manufacturing Science and Engineering*, 122, 297-303.

[11] Kruszynski, B. and Pazgier, J. (2003), "Temperatures in grinding of magnetic composites - Theoretical and experimental approach", *Annals of the CIRP*, 52/1, 263-266.

[12] Kuriyagawa, T., Syoji, K. and Ohshita, H. (2003), "Grinding temperature within contact arc between wheel and workpiece in high-efficiency grinding of ultrahard cutting tool materials", *Journal of Materials Processing Technology*, 136, 39-47.

[13] Malkin, S. (1978), "Burning limit for surface and cylindrical grinding of steels", *Annals of the CIRP*, 27/1, 233-236.

[14] Malkin, S. (1989), "Grinding Technology: theory and applications of machining with abrasives", Society of Manufacturing Engineers, Michigan, USA.

[15] Malkin, S. and Guo, C. (2007), "Thermal Analysis of Grinding", *Annals of the CIRP*, 56/2, 760-782.

[16] Mamalis, A. G., Kundrák, J., Manolakos, D. E., Gyáni, K. and Markopoulos A. (2003a), "Thermal modelling of surface grinding using implicit finite element techniques", *International Journal of Advanced Manufacturing Technology*, 21, 929-934.

[17] Mamalis, A. G., Kundrák, J., Manolakos, D. E., Gyáni, K., Markopoulos A., and Horvath, M. (2003b), "Effect of the workpiece material on the heat affected zones during grinding: a numerical simulation", *International Journal of Advanced Manufacturing Technology*, 22, 761-767.

[18] Mamalis, A. G., Kundrák, J., Manolakos, D. E., Markopoulos A., and Tsimpidakis, K. (2004), "The effect of the heat source profile in the thermal modeling of surface grinding", *Proc. of the microCAD 2003 International Computer Science Conference*, Miskolc, Hungary, 133-139.

[19] Markopoulos, A. (2006), "Ultrprecision Material Removal Processes", *Ph. D. Dissertation*, Manufacturing Technology Division, School of Mechanical Engineering, National Technical University of Athens.

[20] Moulik, P. N., Yang, H. T. Y. and Chandrasekar, S. (2001), "Simulation of thermal stresses due to grinding", *International Journal of Mechanical Sciences*, 43, 831-851.

[21] Nélias, D. and Boucly, V. (2008), "Prediction of grinding residual stresses" *International Journal of Material Forming*, 1, 1115-1118.

[22] Oliveira, J. F. G., Silva, E. J., Guo, C. and Hashimoto, F. (2009), "Industrial challenges in grinding", *CIRP Annals - Manufacturing Technology*, 58, 663–680.

[23] Rowe, W. B., Petit, J. A., Boyle, A. and Moruzzi J. L. (1988), "Avoidance of thermal damage in grinding and prediction of the damage threshold", *Annals of the CIRP*, 37/1, 327-330.

[24] Shaw, M. C. and Vyas, A. (1994), "Heat affected zones in grinding steel", *Annals of the CIRP*, 43/1, 279-282.

[25] Snoeys, R., Maris, M. and Peters, J. (1978), "Thermally induced damage in grinding", *Annals of the CIRP*, 27/2, 571-581.

[26] Sun, S., Brandt, M. and Dargusch, M. S., (2010), "Thermally enhanced machining of hard-to-machine materials—A review", *International Journal of Machine Tools and Manufacture*, 50, 663–680.

[27] Tian, Y., Shirinzadeh, B., Zhang, D., Liu, X. and Chetwynd, D. (2009), "Effects of the heat source profiles on the thermal distribution for ultraprecision grinding", *Precision Engineering*, 33, 447–458.

[28] Tönshoff, H. K., Peters, J., Inasaki, I. and Paul, T. (1992) "Modelling and simulation of grinding processes", *Annals of the CIRP*, 41/2, 677-688.

[29] Zhang, L. and Mahdi, M. (1995), "Applied mechanics in grinding – IV. the mechanism of grinding induced phase transformation", *International Journal of Machine Tools and Manufacture*, 35, 1397-1409.

In: Machining and Forming Technologies, Volume 3
Editor: J. Paulo Davim

ISBN: 978-1-61324-787-7
© 2012 Nova Science Publishers, Inc.

Chapter 13

PREDICTION OF THIN WALL SURFACE SHAPE THROUGH SIMULATION OF THE MACHINING PROCESS FOR LIGHT ALLOY WORKPIECES

Serge St-Martin[*], Jean-François Chatelain[1], René Mayer[2], Serafettin Engin[3] and Stéphane Chalut[4]

[1]Department of Mechanical Engineering École de technologie supérieure, University of Quebec, Montreal, Canada
[2]Department of Mechanical Engineering École Polytechnique, Montreal, Canada
[3]Manufacturing Technology Development Pratt and Whitney Canada Corp., Longueuil, Canada
[4]CAM and Manufacturing Solutions, Business Transformation and Systems (BTS), Bombardier Aerospace, Dorval, Canada

ABSTRACT

This paper focuses on the prediction of the behaviour of the thin wall during the finishing process of pockets in order to predict surface shape of the workpiece. The clamped-free-clamped-clamped (CFCC) plate configuration, similar to the geometry of thin walls within milled mold, automobile parts and aircraft parts is covered. Since the thin wall usually exhibit a much lower rigidity than the machining system, the former deflect appreciably during the finishing process. The outcome is material in excess at the most flexible regions of the wall, close to the center the free edge. The proposed solution is based on an advanced algorithm integrating an adapted flexible force model for low radial immersion machining process of thin walls. It uses a superposition of analytical solutions derived from thin plate vibration theory.

Keywords: thin-walled components; finish machining; surface errors; surface shape; time domain simulation; geometry prediction

---

[*] Email: serge.st-martin@etsmtl.ca

1. INTRODUCTION

High quality product manufacturers largely benefit from the increase of high performance machining technology. This is particularly true for the manufacturing of airframe and complex aircraft parts that feature large amount of milled pockets. The machining process of these parts becomes delicate at the finishing step, as the occurrence of thin wall between pockets demands additional caution to avoid inaccuracies and form errors to the components.

For one part, finish machining is normally used to mill components to their designed dimensions within tolerances. It includes the cutting of the material in excess that have been left on the surfaces during the high removal rate machining due to tool static deflection. For low rigidity components such as thin walls, the local deflection of the workpiece as well has to be considered in order to reach dimensional and geometrical objectives. Secondly, the low tool immersion process will have a significant impact on the thin wall dynamic behaviour. At best, this will have an effect on the surface finish quality; at worst chatter could arise and generate cracking or breakage. In order to achieve dimensional accuracy of the component right on the first attempt, it is imperative to predict the surface profile generated by presently used cutting strategies. Thus, one is able to optimize the key parameters that will permit the compensation of the additional sources of inaccuracies inherent to thin wall machining.

Common works on the machining of thin wall components can be regrouped in two grand categories. On the first hand, there are these solutions that focus on the stability of the milling process using frequency domain analysis. Davies and Balachandran (2000) focused their studies on the dynamics of the flexible workpiece by the calculation of an ordinary differential equation derived from beam functions. The thin wall vibrations are considered largely influenced by impact dynamics, which integrates time delays terms. The cutting forces evaluation is based on a nonlinear model. Davies et al. (2000) (2002) studied the stability of thin wall milling process, but not for low radial immersion machining. Bravo et al. (2005) predetermine the occurrence of chatter by generating 3D stability lobes adapted to a machining structure and several workpieces having similar dynamic behaviours. Ismail and Ziaei (2002) developed an on and off lines cutting parameters control to reduce chatter occurrence during the machining of flexible components of turbine blades type. On the second hand, much knowledge has been acquired using temporal domain analysis in order to integrate the transitory phenomena to the analysis. This way, dimensional form error due to the particular machining conditions at the tool entry and exit at the ends of the workpiece can be evaluated. For an improved simulation of the thin wall behaviour, Sagherian and Elbestawi (1990) presented a removal of material technique that takes into account the deflection of the workpiece during the machining. Elbestawi and Sagherian (1991) proposed later the introduction of finite element analysis to add the deflection of both the cutter and the workpiece. Recent works propose solutions for the generation of the milled surfaces (Montgomery and Altintas, 1991), considering the occurrence of chatter (Sims, 2005) or focusing on low immersion milling (Campomanes and Altintas, 2001).

Temporal domain analysis can be clustered in two branches by consideration of the force model used. First, rigid force models have been seldom utilized since they have a direct use only to specific machining cases including tools of very basic geometry (no corner radius, no chip breaking groove). These models are base on the mechanics of orthogonal and oblique cutting. Secondly, the flexible force models, which are more accurate for actual machining tools, include the tool and workpiece deflections to determine useful cutting coefficients. Altintas (2000) and Budak (2006) presented the basic theory to evaluate the specific cutting force coefficients, which eliminates the need to conduct experimental cutting tests for every different pair of tool geometry and workpiece material. Engin and Altintas (2001) get more specific with the evaluation cutting pressure coefficients of widely use helical end mills, using the orthogonal to oblique cutting transformation method. However, these force models are not well adapted to thin wall machining since the shallow tool immersion process produces cutting forces that can no longer be considered linear to the uncut chip thickness (Lapujoulade et al., 2002, Paris et al., 2004).

Kline et al. (1982) proposed an adaptation of the machining cutting forces by considering the deflection of a thin clamped-free-clamped-clamped (CFCC) plate. Although the plate was represented by a finite element model, its behaviour had no impact on the uncut chip area. Sutherland and DeVor (1986) improved the model by including the effect of the tool immersion. The cutter was modeled as a cantilever beam and a single concentrated force summed up the force distribution at the area of contact between the part and the tool. They used Kline's (1982) dividing of the end mill into axial segments, in order to simplistically take into account the helix angle of the tool. They evaluated the instantaneous tooth immersion of the tool in the workpiece at each time step and considered each tooth runout. However, the solution didn't consider the milling of flexible workpiece. The use of slender end mills for the machining of cantilever plates was studied by Budak and Altintas (1995). The cutter was modeled as an elastic beam and they considered the influence of its dynamic deflection on the tool immersion, therefore modifying the cutting forces generated. In their case, Tsai and Liao (1999) modeled the cutting tool as a pre-twisted Timoshenko beam element. The workpiece as well as the tool deflections were included in order to obtain a more accurate surface shape of the thin cantilever plate.

As computing performances increase, better results are predicted through the use of numerical methods. Ever improving digital computer suggests the nearby and much broader utilisation of its capabilities to the time domain study of machining processes. Ratchev et al. (2003) (2004a) (2004b) (2004c) presented several results for the prediction of machining form errors, considering tool and workpiece deflections. An analytical flexible model based on an extended perfect plastic layer model is proposed. The finite element model determines the workpiece deflection. Although finite-element strategies have the advantage of accurately modeling the geometry and the physical characteristics of a workpiece, the calculation routines associated to a machining process are too time consuming for industries producing small number of identical parts or thin wall geometries. This paper focuses on the prediction the thin wall surface shape through time domain simulation of the finish machining process. It integrates the static and dynamic behaviour of the workpiece for a correct evaluation of the uncut chip geometry for each time increment. The plate vibration analysis is based on a superposition method of Lévy type solutions.

2. METHODOLOGY

In order to get a realistic prediction of the milled surface shape, an interactive algorithm was coded to take into account the behaviour of the thin wall during the machining. As the tool cuts material from the workpiece, the generated cutting forces can be evaluated in function of the immersion angle of each flute in the part. These forces modify continually the relative position of the wall in respect to the nominal position of the tool, thus influencing the immersion angle of the cutter. Therefore, a routine must be used to find the accurate uncut chip thickness at each time increment of the machining process.

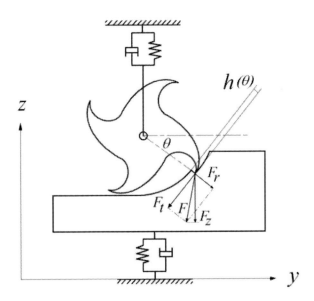

Figure 1. Milling dynamic model including instantaneous chip thickness.

2.1. Cutting Force Model

Finish machining is a low radial immersion machining process. The tool is without contact with the workpiece for most of the duration of the milling. In this case, rigid cutting force models, which average the cutting forces over one full rotation of the cutter, cannot be use. Instead, flexible cutting force models have been developed to consider the nonlinear characters involved in theses conditions. According to the immersion angle as shown on Figure 1, the radial and tangential component of the force vector is evaluated in function of the chip section (Altintas, 2000) as presented below:

$$F_t = K_{tc}bh + K_{te}b \qquad (1)$$

$$F_r = K_{rc}bh + K_{re}b \qquad (2)$$

where F_t and F_r are respectively the tangential and radial force components, K_{ic} and K_{ie} are the respective cutting and edge cutting constants, h is the thickness of the uncut chip and b is the depth of cut. The helix angle of the cutter is considered by Kline's (Kline et al., 1982) decomposition of the tool into several axial disks. Each disk has a zero helix angle and the angular position between them reproduces the tool geometry. Cutting pressure coefficients can be determined by well established orthogonal cutting force techniques and the orthogonal to oblique cutting transformation. As the milling simulation takes place, the contribution of each flute has to be calculated at each time period.

2.2. Part Deflection Model

As shown at Figure 2, most thin walls milled within aircraft parts have a CFCC configuration as all except one edge of the wall is integrated to the global geometry of the part. An analytical routine based on the free vibration of thin plates is used to determine the behaviour of the thin wall. Depending on the boundary conditions applied for each edge, one can select specific series solutions to express accurately the local deflection of the thin plate in response to a driving harmonic moment or vertical edge reaction distributed along a particular edge as Figure 3 shows an example. Superimposing wisely each solution according to the boundary conditions, the global displacement of the plate can be deduced. The technique is referred as the Superposition method (Gorman, 1982).

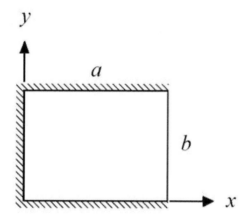

Figure 2. Schematic representation CFCC thin plate.

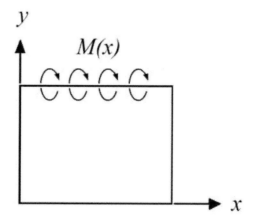

Figure 3. Distributed bending moment acting along a specific edge of a rectangular plate.

The final solution obtained is in perfect accordance with the plate governing differential equation presented below:

$$D\left[\frac{\partial^4 W}{\partial x^4} + 2\frac{\partial^4 W}{\partial x^2 \partial y^2} + \frac{\partial^4 W}{\partial y^4}\right] + \rho\frac{\partial^2 W}{\partial t^2} = F_{(x,y,t)} \tag{3}$$

$$D = \frac{Eh^3}{12(1-v^2)} \tag{4}$$

where D is the plate flexural rigidity, W is the plate lateral displacement, ρ is the mass of the plate per unit of area, F, null for the analysis of free vibrations, is the z-component of the force function applied, E is the Young's modulus of elasticity, h is the plate thickness and v is the Poisson's ratio of the plate.

In our case, the analytical solution is valid in the region defined by $0 < x < a$ and $0 < y < b$ for these specific boundary conditions:

$$W\Big|_{x=0} = 0 \qquad \frac{\partial W}{\partial x}\Big|_{x=0} = 0 \tag{5,6}$$

$$W\Big|_{y=0} = 0 \qquad \frac{\partial W}{\partial y}\Big|_{y=0} = 0 \tag{7,8}$$

$$W\Big|_{y=b} = 0 \qquad \frac{\partial W}{\partial y}\Big|_{y=b} = 0 \tag{9,10}$$

$$M_x\Big|_{x=a} = D\left(\frac{\partial^2 W}{\partial x^2} + v\frac{\partial^2 W}{\partial y^2}\right)\Bigg|_{x=a} = 0 \tag{11}$$

$$V_x\Big|_{x=a} = D\frac{\partial}{\partial x}\left(\frac{\partial^2 W}{\partial x^2} + v\frac{\partial^2 W}{\partial y^2}\right)\Bigg|_{x=a} = -F_z \tag{12}$$

where $M_x|_{x=a}$ is the bending moment and $V_x|_{x=a}$ the edge reaction of the plate at the proper boundary. Classical Lévy-type solutions used can be represented in the dimensionless form

$$W(\xi,\eta) = \lim_{k\to\infty}\sum_{m=1}^{k} Y_m(\eta)\sin m\pi\xi \tag{13}$$

which is a series solution comprising the Fourier trigonometric functions $\sin m\pi\xi$ and their associated coefficients Y_m function of the η, where ξ and η are the dimensionless form of variables x and y.

These solutions are commonly called building blocks. Once the solution of each building block is obtained, we can superimpose all theses solutions in the respect of the boundary conditions requirements by adjusting the relevant coefficients. For the free vibration of the CFCC thin plate, we obtain the building blocks represented at Figure 4.

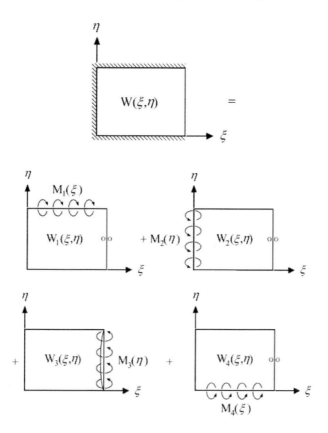

Figure 4. Building blocks of the resolution of the CFCC thin plate problem.

One can notice the pair of circles at the free edge $\xi = 1$ for all solution except W_3. These circles specify that the slope and the vertical reaction to that particular edge must be null. Those particular constraints are most valuable for the resolution of the problem and are broadly explained by Gorman (1982). We obtained the following solutions for the CFCC plate problem

$$W_1 = \sum_{m=1}^{k^*} E_m \theta_{11m} (\sinh \beta_m \eta + \theta_{1m} \sin \gamma_m \eta) * \sin((2m-1)\frac{\pi}{2}\xi)$$

$$+ \sum_{k^*}^{\infty} E_m \theta_{22m} (\sinh \beta_m \eta + \theta_{2m} \sinh \gamma_m \eta) * \sin((2m-1)\frac{\pi}{2}\xi)$$

(14)

$$W_2 = -\sum_{n=1}^{k^*} E_n \theta_{11n} (\cosh \beta_n (1-\xi) + \theta_{1n} \cos \gamma_n (1-\xi)) * \sin(n\pi\eta)$$

$$-\sum_{k^*}^{\infty} E_n \theta_{22n} (\cosh \beta_n (1-\xi) + \theta_{2n} \cosh \gamma_n (1-\xi)) * \sin(n\pi\eta) \qquad (15)$$

$$W_3 = \sum_{n=1}^{k^*} E_n \theta'_{11n} (\sinh \beta_n \xi + \theta'_{1n} \sin \gamma_n \xi) * \sin(n\pi\eta)$$

$$+\sum_{k^*}^{\infty} E_n \theta'_{22n} (\sinh \beta_n \xi + \theta'_{2n} \sinh \gamma_n \xi) * \sin(n\pi\eta) \qquad (16)$$

$$W_4 = -\sum_{m=1}^{k^*} E_m \theta_{11m} (\sinh \beta_m (1-\eta) + \theta_{1m} \sin \gamma_m (1-\eta)) * \sin((2m-1)\frac{\pi}{2}\xi)$$

$$-\sum_{k^*}^{\infty} E_m \theta_{22m} (\sinh \beta_m (1-\eta) + \theta_{2m} \sin \gamma_m (1-\eta)) * \sinh((2m-1)\frac{\pi}{2}\xi) \qquad (17)$$

where the quantities k^*, β, γ and θ are known functions dependant of the aspect ratio of the plate (b/a). Each solution satisfies completely the plate vibration governing differential equation and so does the combination of them. Depending on the number of terms used for each Fourier series, the coefficients E are adjusted so that the global effect of all the solutions respects the boundary conditions. Non-trivial solutions are resolved by matrix techniques documented by Gorman (1982) as well.

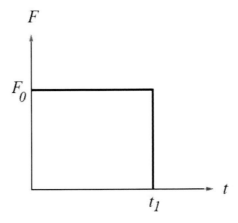

Figure 5. Square-pulse excitation.

It is possible to derive a solution in regard to a concentrated harmonic driving force located at an arbitrary location on the lateral surface of the plate. Since the equation of motion of the plate is assumed linear through space and time, the principle of convolution can be used to determine the global response of a combination of different forces based on their individual contribution. This convolution integral can be extended to support a force distribution case very similar to the proper component of the cutting force distribution. The

Fourier theorem permits the expression of an arbitrary force function by an infinite series of harmonic force terms. Concentrated force can consequently be estimated into a finite series of harmonic force, therefore representing the cutting forces of the milling process for each time step of the algorithm. The number of terms used for the harmonic force series determines the accuracy expected. In our case, a square-pulse excitation as shown at Figure 5 will be used to simulate the reaction of the plate to the cutting force. The excitation duration will be equivalent to the time increment of the algorithm.

2.3. Tool Deflection Model

The tool itself deflects under the cutting forces. The magnitude of this deflection can represent a serious source of inaccuracy. Therefore, a vast range of solutions has been developed to predict the impact of the tool flexibility on the surface profile accuracy. In most cases, the global rigidity of machining system far outweighs the thin wall counterpart. Consequently, the impact of the dynamic behaviour of the cutter on the form error of the plate is negligible. From an industrial point of view, the specific machine-holder-tool system on which a particular part will be milled is not necessarily known to the programmer of the machining process. There is even less chances that the static and dynamic properties of the machining system are known for the adaptation of the machining strategy employed. Therefore the solution presented here considers the tool infinitely rigid.

2.4. Material Removal Model

At each time step, the accurate position of the tool and the deflection of the thin wall are calculated. The tool immersion in the workpiece material is determined and the instantaneous surface topology generated.

Since high-speed machining usually takes place in the context under study, it is imperative to use this method of simulation since several teeth passages will occur on a specific local point of the surface of thin wall. An accurate evaluation of the tool immersion favours good simulation results.

As shown on Figure 6, the simulation contains several steps. The first step consist in the initialisation of all known variables such as cutting parameters, material properties, geometry and initial conditions of both workpiece and tool. Secondly, the position of each flute is calculated throughout the milling process in order to detect contact between the workpiece and the tool.

Iteration of the time step may begin. When cutting occurs, the instantaneous chip thickness is evaluated and used to determine the lateral cutting force applied. Thus, we obtain the dynamic behaviour of the thin plate under forced vibration. The position of the workpiece at the end of the time step is of first interest since it represents the reaction of the plate right after the square-pulse excitation. The uncut chip thickness can therefore be revaluated with the consideration of the motion of the plate. An iterative routine is used to obtain equilibrium between the position of the plate and the force generated. Once equilibrium is reached, the lateral position of the flute relative to the position of the plate is used in order to determine the amount of material removed. The geometry of the plate is updated consequently. When no cutting occurs, the plate vibrates freely until the next contact with the cutter or the end of the process.

As suggested by Peigne et al.(2003) and Paris et al. (2004), a second-order interpolation routine is use to simulate the accurate trochoidal path of the tooth in the workpiece material. This routine decreases the necessity of using a very fine space increment for a good evaluation of the uncut chip geometry. Its impact over of the simulated surface shape and finish is considerable.

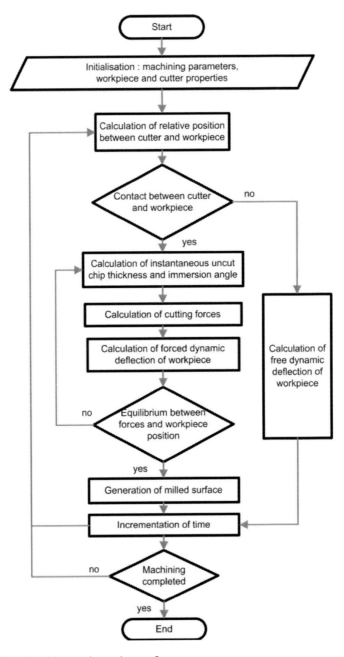

Figure 6. Simulation algorithm to determine surface errors.

3. EXPERIMENTAL ANALYSIS

3.1. Description of Tests

As shown in Figure 7, a test part has been designed to isolate and monitor the behaviour of one thin wall during a machining process. Rough machining is first realised to obtain general aspect of the part. Afterward only one side of the thin wall is finished. The wall is at its thinnest before machining the other side.

The machining was done with a 0.5" diameter carbide tool with three flutes. Its cutting length was of 1.0" and its stake out of 3". The helix angle was of 37°, the corner radius of 0.125" and the machining was done up milling. The spindle speed was 12,000 rpm and the feed rate of 216 ipm. The depth of cut has been programmed to 0.25" but set automatically by the CAM software at an average value of 0.196". The 0.040" and 0.020" widths of cut for the two finish passes are similar to general practice in the industry for finish machining of light aluminium alloys.

The wall is 0.090" thick before the machining. The ultimate thickness objective is therefore of 0.030". The radius of the light alloy component at the thin wall extremities are 0.28" laterally and the tool corner radius at the bottom. The part was mounted on a KISTLER dynamometer table in order to follow the progression of the cutting forces.

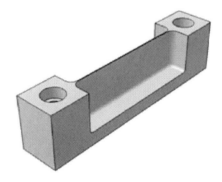

Figure 7. Test part presenting one thin wall.

To monitor the wall deflection, four proximity sensors have been strategically position on an adjacent support as shown on Figures 8 and 9. It was designed as rigid as possible in order for it to vibrate synchronously with the machining table, thus eliminating additional sources of error. Each sensor has a sensitivity of 10 V/mm with a resolution of 1 µm. An 8-channels data acquisition set-up was used to collect all the data.

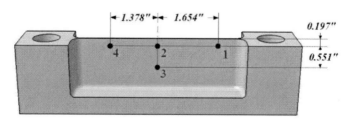

Figure 8. Proximity sensors locations.

Seven passes were necessary to cover the lateral area of the 4.45" wide by 1.375" high thin wall. Thus, 14 series of data has been collected in comparison with the simulation results. At this stage of the proposed solution, the removal of material has not been considered for the evaluation of the dynamic behaviour of the wall. Consequently, the simulation considers a constant 0.090" thick wall for the first seven passes and a 0.050" thick wall for the last seven. Figure 9 shows the direction and location of pass #1 and #7. Therefore, the tool passes in front of proximity sensor #4, then sensors #2 and #3 and finally proximity sensor #1 for all finishing passes (Figure 10).

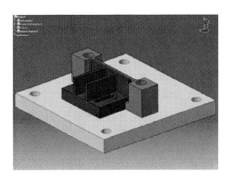 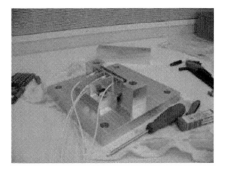

Figure 9. Adjacent support for proximity sensors.

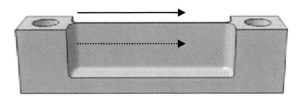

Figure 10. 1st (solid arrow) and 7th (dotted arrow) passes direction and location.

3.2. Results

The analysis of the part deflection is interesting. The raw data collected basically consist of the deflection of the wall at each sensor location against time, as shown on Figure 11. One set of four graphs is associated to one of the fourteen passes. In order to better understand the process, the graphs were placed with the same configuration as the sensors were located.

Figure 11 shows clearly that highest magnitude deflection at sensor #4 is of about .005" for pass #1. It occurs early in the pass since the tool passes in front of this sensor in the beginning of the process. For sensor #1, the tool passes in front of it at the end of the pass, reason why the greatest deflection occurs late in the process. The highest deflection magnitude of this sensor is lower than sensor's #4 since the sensor is located further away from the center of the thin wall.

If we compare the deflection magnitude at sensor #2 and #3, we notice the similarity of shape. The symmetry of both graphs come from the fact that the tool generates the highest deflection at the most flexible area of the wall, which is at the center, midway on the tool trajectory. Since sensor #2 is located closer to the free edge of the wall, at that location will be generated higher deflections than at sensor #3 for the present mode shape of the wall.

Looking at the four graphs, we get a basic idea of the behaviour of the wall during each finish pass.

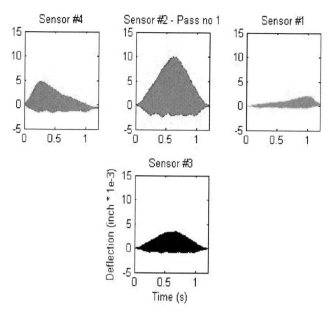

Figure 11. Deflections at Pass #1.

By a quick glance over the deflection graphs at sensor #2 for the first seven passes (Figure 12), we notice the similarities between some of them and observe the evolution of the global behaviour of the thin wall. The graphs have been separated in three categories:

- Category 1 : Passes # 1 and 2
- Category 2 : Passes # 3,4 and 5
- Category 3 : Passes # 6 and 7

For the Category 1, we see that the thin wall mainly deflects in the positive direction, which is away from the tool. It also deflects in the negative direction when the tool hasn't yet made contact with the workpiece. For the Category 2 and specifically for sensors #2 and #3, we can observe that the wall deflects away from the tool and is maintained at a fair distance from it. For these passes, the tool is positioned deeper on the lateral side of the wall and clearly generates different mode shapes of vibrations compared to Category 1 passes as seen on Figure 12. For the Category 3, we see again different mode shapes generated as the thin wall deflects in both positive and negative directions with closer magnitudes, compared to Category 1 passes. The graph for these 2 last passes suggest that the tool deflects enough to let the thin wall at sensor #2 position's to deflects in the negative direction. From these data, it is not possible to confirm the possibility that the thin wall hits the tool outside of the theoretical range of cut, which is the width of cut.

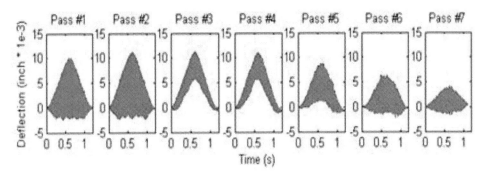

Figure 12. Deflections at sensor #2 for the first 7 passes.

For passes #1 through #4, we also notice at sensor #2 that the greatest deflection is greater than 0.010" midway in the passes. These deflections' magnitude represents more than 50% of the width of cut. Depending on the mode shape occurring at each pass, the deflections can have a major impact on the dimensional accuracy of the surface. These results therefore initiated the consideration of mode shape analysis in the study of thin wall finish machining.

At Figure 13 are represented the mean measured deflections at sensor #2 for the same seven passes. From this graph, it is not possible to appreciate the impact of the deflection's amplitude on the generated surface topology.

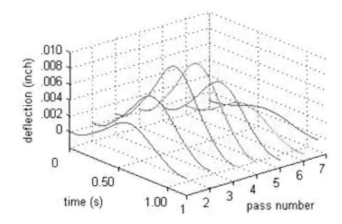

Figure 13. Mean measured deflections at sensor #2 (first 7 passes).

4. SIMULATION RESULTS

The workpiece has been measured in order to compare the profile to the simulation results as represented on Figure 14. The highest surface error measured on the workpiece from the ideal target surface is of 0.0143" compared to 0.0138" for the predicted surface. A large discrepancy at both end of the thin wall length can be observed. The negative values of the measured surface locations are mainly due to the corner shape tool paths and the rough machining cutting conditions required to produce the workpiece at its design geometry in an industrial context (Figure 7).

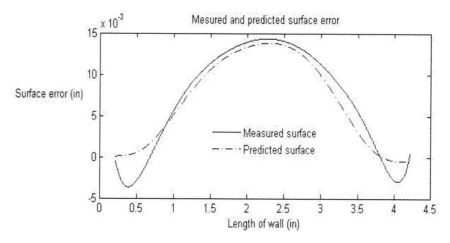

Figure 14. Variation of measured and predicted surface error along top of thin wall.

For most of the length, the predicted surface error is lower than the measured error. Therefore, the two curves cross themselves at values x = 0.881" and 3.806". Between these limits, the mean error between the measured and predicted values is 0.001" with a standard deviation of 0.0006". The average percentage difference is of 15% with a standard deviation of 32%. The simulation estimates the surface error of the thin wall in an appreciable accuracy.

Conclusion

The proposed model is based on the vibration analysis of thin plates by using the Superposition method in order to determine the thin wall behaviour during its machining. Mode shapes obtained with the algorithm are in accordance with the literature and show good convergence with the measured deflection of the workpiece. The model is integrated to a simulator predicting the shape of the surface once the finish machining is done.

The deflection of the wall is taken into account for the correct evaluation of the instantaneous teeth immersion in the workpiece material. From this, cutting forces can accurately be evaluated to determine the wall deflection at each time step of the simulation and therefore generate the corrected surface profile of the milled wall. Preliminary tests have been conducted to determine the specific cutting pressure involved in finish machining. The surface topography obtained show good accordance with experimental data collected.

The proposed model is adaptable to an industrial context since the algorithm is generic to a wide range of wall aspect ratio. The tool is considered rigid compared to the wall for simplicity. The model is part of an ongoing research on the development of accurate prediction tools of thin wall machining accuracy for the industry. The presented work does not take into account the change of dynamic properties of the wall caused by the removal of material.

REFERENCES

[1] Altintas, Y. (2000) '*Manufaturing Automation*', Cambridge, Cambridge University Press.

[2] Bravo, U., Altuzarra, O., Lopez De Lacalle, L. N., Sanchez, J. A. and Campa, F. J. (2005) 'Stability limits of milling considering the flexibility of the workpiece and the machine', *International Journal of Machine Tools and Manufacture*, Vol. 45, No. 15, pp. 1669-1680.

[3] Budak, E. (2006) 'Analytical models for high performance milling. Part I: Cutting forces, structural deformations and tolerance integrity', *International Journal of Machine Tools and Manufacture*, Vol. 46, No. 12-13, pp. 1478-1488.

[4] Budak, E. and Altintas, Y. (1995) 'Modeling and avoidance of static form errors in peripheral milling of plates', *International Journal of Machine Tools and Manufacture*, Vol. 35, No. 3, pp. 459-476.

[5] Campomanes, M. L. and Altintas, Y. (2001) 'An improved time domain simulation for dynamic milling at small radial immersions and large depths of cut'. 2001 *ASME International Mechanical Engineering Congress and Exposition*, Nov 11-16 2001. New York, NY, United States, American Society of Mechanical Engineers, New York, NY 10016-5990, United States.

[6] Davies, M. A. and Balachandran, B. (2000) 'Impact dynamics in milling of thin-walled structures', *Nonlinear Dynamics*, Vol. 22, No. 4, pp. 375-92.

[7] Davies, M. A., Pratt, J. R., Dutterer, B. and Burns, T. J. (2002) 'Stability prediction for low radial immersion milling', *Journal of Manufacturing Science and Engineering, Transactions of the ASME*, Vol. 124, No. 2, pp. 217-225.

[8] Davies, M. A., Pratt, J. R., Dutterer, B. S. and Burns, T. J. (2000) 'The stability of low radial immersion milling', *CIRP Annals - Manufacturing Technology*, Vol. 49, No. 1, pp. 37-40.

[9] Elbestawi, M. A. and Sagherian, R. (1991) 'Dynamic modeling for the prediction of surface errors in the milling of thin-walled sections', *Journal of Materials Processing Technology*, Vol. 25, No. 2, pp. 215-228.

[10] Engin, S. and Altintas, Y. (2001) 'Mechanics and dynamics of general milling cutters.: Part I: helical end mills', *International Journal of Machine Tools and Manufacture*, Vol. 41, No. 15, pp. 2195-2212.

[11] Gorman, D. J. (1982) '*Free vibration analysis of rectangular plates*', New York.

[12] Ismail, F. and Ziaei, R. (2002) 'Chatter suppression in five-axis machining of flexible parts', *International Journal of Machine Tools and Manufacture*, Vol. 42, No. 1, pp. 115-122.

[13] Kline, W. A., DeVor, R. E. and Shareef, I. A. (1982) '*Prediction of surface accuracy in end milling.*'. Design and Production Engineering Technical Conference. Washington, DC, USA, ASME, New York, NY, USA.

[14] Lapujoulade, F., Mabrouki, T. and Raissi, K. (2002) 'Prediction du comportement vibratoire du fraisage lateral de finition des pieces a parois minces: Vibratory behavior prediction of thin-walled parts during lateral finish milling', *Mecanique and Industries*, Vol. 3, No. 4, pp. 403-418.

[15] Montgomery, D. and Altintas, Y. (1991) 'Mechanism of cutting force and surface generation in dynamic milling', *Journal of Engineering for Industry, Transactions of the ASME*, Vol. 113, No. 2, pp. 160-168.

[16] Paris, H., Peigne, G. and Mayer, R. (2004) 'Surface shape prediction in high speed milling', *International Journal of Machine Tools and Manufacture*, Vol. 44, No. 15, pp. 1567-1576.

[17] Peigne, G., Paris, H. and Brissaud, D. (2003) 'A model of milled surface generation for time domain simulation of high-speed cutting', Proceedings of the Institution of Mechanical Engineers, *Part B: Journal of Engineering Manufacture,* Vol. 217, No. 7, pp. 919-930.

[18] Ratchev, S., Govender, E., Nikov, S., Phuah, K. and Tsiklos, G. (2003) 'Force and deflection modelling in milling of low-rigidity complex parts', *Journal of Materials Processing Technology,* Vol. 143-144, No. 1, pp. 796-801.

[19] Ratchev, S., Huang, W., Liu, S. and Becker, A. A. (2004a) 'Modelling and simulation environment for machining of low-rigidity components', *Journal of Materials Processing Technology,* Vol. 153-154, No., pp. 67-73.

[20] Ratchev, S., Liu, S., Huang, W. and Becker, A. A. (2004b) 'A flexible force model for end milling of low-rigidity parts', *Journal of Materials Processing Technology,* Vol. 153-154, No. 1-3, pp. 134-138.

[21] Ratchev, S., Liu, S., Huang, W. and Becker, A. A. (2004c) 'Milling error prediction and compensation in machining of low-rigidity parts', *International Journal of Machine Tools and Manufacture*, Vol. 44, No. 15, pp. 1629-1641.

[22] Sagherian, R. and Elbestawi, M. A. (1990) 'A simulation system for improving machining accuracy in milling', *Computers in Industry,* Vol. 14, No. 4, pp. 293-305.

[23] Sims, N. D. (2005) 'The self-excitation damping ratio: A chatter criterion for time-domain milling simulations', *Journal of Manufacturing Science and Engineering, Transactions of the ASME*, Vol. 127, No. 3, pp. 433-445.

[24] Sutherland, J. W. and DeVor, R. E. (1986) ' Improved method for cutting force and surface error prediction in flexible end milling systems.', Journal of Engineering for Industry, *Transactions ASME*, Vol. 108, No. 4, pp. 269-279.

[25] Tsai, J.-S. and Liao, C.-L. (1999) 'Finite-element modeling of static surface errors in the peripheral milling of thin-walled workpieces', *Journal of Materials Processing Technology,* Vol. 94, No. 2, pp. 235-246.

In: Machining and Forming Technologies, Volume 3
Editor: J. Paulo Davim

ISBN: 978-1-61324-787-7
© 2012 Nova Science Publishers, Inc.

Chapter 14

EXPERIMENTAL STUDY OF DRILLING CERAMIC MATRIX BRAKE PAD (CMBP)

*K.L. Kuo[1] and C.C. Tsao[2]**

[1]Department of Vehicle Engineering National Taipei University of Technology, Taipei, Taiwan, ROC

[2]Department of Automation Engineering Tahua Institute of Technology, Hsinchu, Taiwan, ROC

ABSTRACT

Ceramic matrix brake pads (CMBPs) are important parts of brakes in the automobile industry owing to their high temperature durability, wear resistance, corrosion resistance. The CMBPs are lighter than pads made with traditional asbestos and cast iron. The delamination damage for ceramic matrix brake pads caused by the tool thrust has been known as one of the major concerns during drilling. In order to investigate the degree of influence of drilling parameters (drill type, feed rate and spindle speed) in drilling ceramic matrix brake pads, a L_9 (3^4) orthogonal array was employed. Based on the experimental results the importance of feed rate and various drill types in assessing thrust force and delamination is highlighted. Moreover, the best combination to get lower thrust force and lower delamination in drilling CMBP is $A_3B_1C_1$ in this work, (i.e., drill type = twist drill (four-flute), feed rate = 0.01 mm/rev and spindle speed = 800 rpm) within the selected tested range.

Keywords: Twist drill, Candle stick drill, Ceramic matrix brake pads, Thrust force, Delamination factor

---

* Email:aetcc@msdb.thit.edu.tw

1. INTRODUCTION

Most developed countries have prohibited importing traditional brakes pads owing to the presence of asbestos in those products. The merits of the present brake pads, e.g. ceramic matrix brake pads, iron fiber brake pads and fiber magnesium brake pads, possess the high temperature durability, wear resistance and corrosion resistance. These brake pads are made of composite materials and have better mechanical properties than those made with traditional asbestos and cast iron. Moreover, the effective lifespan of the present brake pads can even reach 300,000 kilometers, which approaches the life expectancy of most vehicles (Shih, 2001). To use these brake pads, accurate, precise high quality holes need to be drilled to ensure proper and durable assemblies. Conventional drilling with twist drill still remains one of the most economical and, therefore, commonly adopted machining processes for drilling holes in structural parts. However, the geometry and the velocity of cutting edge for twist drill are not constant, but vary along the cutting edge. During drilling these brake pads, the problem of delamination often occurs at the entrance plane and the exit plane of brake pad, as shown in Figure 1. The drilling induced-delamination has been a number of studies with different methods of measurement (Tsao and Hocheng, 2005; Davim et al., 2007; Seif, 2007).

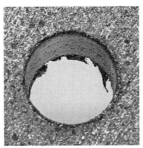 

Figure 1. During drilling ceramic matrix brake pad with a twist drill, delamination occurs at the entrance plane and the exit plane of CMBP.

Among the drilling parameters considered important for drilling operation, drill geometry and feed rate have significant effect on thrust force and delamination during drilling composite laminates. The chisel edge is a part of the intersection curve of two sequential flank surfaces of a twist drill. Galloway report that the chisel edge of twist drill mainly influences the thrust force and the hole quality is strongly dependent on drill geometry (Galloway, 1957). Jain and Yang developed a method for correlating the feed rate with the onset of delamination in composite drilling (Jain and Yang, 1993). Won and Dharan investigated the effect of the chisel edge on the thrust force (Won and Dharan, 2002). Tsao and Hocheng calculated an optimal range of diameter of pilot hole associated with chisel edge length (Tsao and Hocheng, 2003). Armarego and Zhao developed a model that can be used to predict thrust for point-thinned and circular centre edge twist drill designs for three different drill flank configurations (Armarego and Zhao, 1996). To reduce the chisel edge of twist drill from the flutes of both sides can decrease the induced-delamination in drilling composite materials. Koenig and Grab found a correlation between thrust force and delamination in drilling composite materials within a range of cutting conditions (Koenig and Grab, 1989). Doerr et al. designed a drill to cut materials toward the hole center and to shear at the hole edge (Doerr et al., 1982). Fujii et al. also investigated the effects of the chisel edge on conical drill performance for optimum output (Fujii et al., 1971). Koenig et al. studied the effects of processing variables on drilling damage (Koenig et al., 1984; Koenig et al., 1985). Salama and ElSawy investigated the tool and working geometry of a double plane sharpened twist drill point (Salama and ElSawy, 1996). They showed that the effect of feed must be taken into consideration when studying chisel edge action in a drilling operation. Chen proposed the idea of delamination factor to characterize the delamination in drilling carbon fiber reinforced plastic (CFRP) laminates (Chen, 1997). Mathew et al. reported the trepanning tool starts from the periphery of the cutting edge that puts the fibers in tension during the entire cutting operation (Mathew et al., 1999). Piquet et al. showed the capabilities of the specific cutting tool because several defects and damages often encountered in twist drilled holes were minimized or avoided (Piquet et al., 2000). This leads to several problems for the manufacturers, and there have been a number of studies on this problem (Lee and Chan, 1997; Tsao and Hocheng, 2004).

Generally, candle stick drills are used for generating holes in fiber reinforced plastic (FRP) and wooden products. Hocheng and Tsao (2003) developed the delamination model of the candle stick drill to explain the advantage of distributing the thrust force toward the drill periphery. However, the chisel edge of candle stick drill is smaller than twist drill. The crack propagation around the drilled holes is found to be gentler when the candle stick drill passes through the bottom laminates. Agapiou has shown that multi-flute drilling is an efficient means of generating highly accurate holes (Agapiou, 1993a; Agapiou, 1993b). Also, Ema et al. found that the frequent occurrence of whirling vibrations in conventional two-fluted drills disappeared upon using a three-fluted drill (Ema et al., 1988; Ema et al., 1991). Moreover, Davim studied the drilling of metal matrix composites based on Taguchi technique to find the influences of cutting parameters on tool wear, torque and surface finish (Davim, 2003). Tsao and Hocheng presents a prediction and evaluation of delamination factor in use of twist drill, candle stick drill and saw drill (Tsao and Hocheng, 2004). The approach is based on Taguchi method and the analysis of variance (ANOVA). Gaitonde et al. presented the methodology of Taguchi optimization method for simultaneous minimization of delamination factor at entry and exit of the holes in drilling medium density fibreboard panel (Gaitonde et al., 2008).

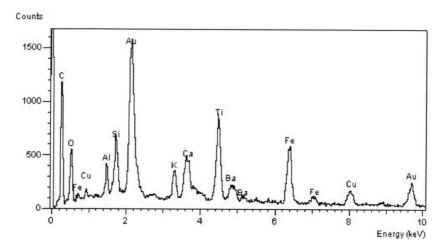

Figure 2. EDS analysis for ceramic matrix brake pad.

Based on the above studies, the objective of this study is to perform drilling on ceramic matrix brake pads with various drill types and drilling conditions to better understand the thrust force, delamination and processing parameters.

2. EXPERIMENTAL PROCEDURE

The results of ceramic matrix brake pads using energy dispersive spectrometer (EDS) and scanning electron microscope (SEM) are shown in Figures 2 and 3, respectively. The EDS analysis of ceramic matrix brake pad shows approximately 13.5 % of copper element. Since the heat conductivity for copper is excellent, the effect of heat emission on brakes is increased. At a temperature of 400-500 °F, the addition of oxidized copper will induce better braking. The thickness of ceramic matrix brake pad is 12 mm in all tests.

Figure 3. Structure of ceramic matrix brake pad under SEM.

(a) Candle stick drill   (b) Twist drill (three-flute)   (c) Twist drill (four-flute)

Figure 4. Photograph for various drill types.

The drill diameter, helix angle and drill material of the various drill types are 10 mm, 25° and submicron grade tungsten carbide, respectively. The photograph of various drill types is shown in Figure 4. The point angle for 3-flute and 4-flute of twist drill is 122°. To avoid the effect of chisel edge for twist drill on thrust force and delamination in drilling, the chisel edge was adjusted to zero in all used drill bits. Drilling tests are carried out on a 5.5 kW LEADWELL MCV-610AP machining center as shown in Figure 5. The mean thrust forces at the exit of the drill bits during drilling are measured with a Kistler 9273 piezoelectric dynamometer type and a Kistler 5019 charge amplifier. The electrical charge delivered from the measuring platform is converted by the amplifier into proportional voltages, which are further processed with the aid of a TEAC data acquisition system and digital recorder. All drilling tests were conducted coolant free.

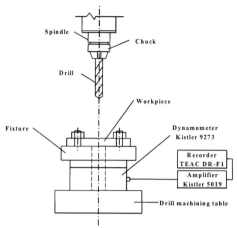

Figure 5. Schematic of experimental setup to measure the thrust force in drilling CMBP.

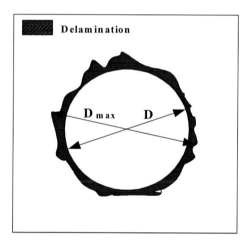

Figure 6. Schematic of drilling-delamination.

In this study, Matrox Imaging System (MIS) software was used to acquire the dimension of the drilling induced-delamination. The drilling delamination factor is determined by the ratio of the maximum diameter (D_{max}) of the delamination zone to the hole diameter (D). The delamination factor scheme is shown in Figure 6. The value of delamination factor (F_d) can be expressed as follows

$$F_d = \frac{D_{max}}{D} \qquad (1)$$

The processing parameters (drill types, feed rate and spindle speed) used in this experiment are shown in Table 1. The L_9 (3^4) orthogonal array table is used to determine the influence of each factor towards the error of thrust force and delamination factor when various drill bits work with the ceramic matrix brake pads. Each experiment is repeated three times to reduce the influence of uncontrolled factors (noise factors). These quality scores are further transformed to the signal-to-noise ratio (S/N ratio) via the following equation:

$$S/N = -10 \log \frac{1}{n} \sum_{i=1}^{n} y_i^2$$

(2)

where n is the number of repetitions of the experiment and y_i is the quality score with smaller-the-better value of experimental data i.

Table 1. Factors and levels in drilling ceramic matrix brake pads

Symbol	Control factor	Level 1	Level 2	Level 3
A	Drill type	Candle stick drill	Twist drill (three-flute)	Twist drill (four-flute)
B	Feed rate (mm/rev)	0.01	0.02	0.03
C	Spindle speed (rpm)	800	1000	1200

Table 2. L_9 (3^4) orthogonal array experimental results for the thrust force and delamination factor of various drill types in drilling ceramic matrix brake pads

Trail No.	Factor			Thrust force (N)	S/N (dB)	Delamination factor (mm/mm)	S/N (dB)
	A	B	C				
1	1	1	1	26.8	-28.6	1.167	-1.3
2	1	2	2	30.5	-29.7	1.197	-1.6
3	1	3	3	33.2	-30.4	1.218	-1.7
4	2	2	3	36.8	-31.3	1.182	-1.5
5	2	3	1	33.7	-30.6	1.197	-1.6
6	2	1	2	24.7	-27.9	1.170	-1.4
7	3	3	2	30.2	-29.6	1.167	-1.3
8	3	1	3	25.9	-28.3	1.152	-1.2
9	3	2	1	24.4	-27.8	1.167	-1.3

3. RESULTS AND DISCUSSIONS

The analysis of variance (ANOVA) is a statistical method, which used to understand the influence of control factors on the average variation of a product or process performance characteristic. By properly adjusting the average and reducing variation, the performance characteristic of the product or process can be improved or increased. In addition, the effect of control factor can also be estimated from ANOVA.

3.1. ANOVA of Thrust Force

The experimental results and ANOVA on the thrust force are shown in Tables 2 and 3. From the results in Table 3, it can be seen that the feed rate (*P*= 45.8 %) and various drill types (*P*= 30.0 %) are most significant variables affecting the thrust force, while the influence from spindle speed is trivial. The higher the levels value is, the better the overall performance. In other words, the factor levels with the highest value should always be selected. Accordingly, the average for each experimental level for thrust force was calculated using the highest value at the level for each parameter to produce the response graph, as shown in Figure 7. The best level for each control factor is the one with the highest S/N ratio. Figure 7 indicates that the best drill type level is A_3 (four-flute twist drill), the best feed rate level is B_1 (0.01 mm/rev) and the best spindle speed level is C_1 (800 rpm). Therefore, the best combination to get lower thrust force in drilling CMBP is expected to be $A_3B_1C_1$, (i.e., drill type = twist drill (four-flute), feed rate = 0.01 mm/rev and spindle speed = 800 rpm) within the selected tested range. The three-flute twist drill, however, produce better holes than the two-flute twist drill, but not as good as the four-flute twist drill shown by Agapiou, especially during interrupted cutting (Agapiou, 1993a; Agapiou, 1993b). The feed rate, on the other hand, is the significant factor in drilling CMBP. The thrust force increases with increasing feed rate. Tsao and Hocheng prove the effect of the chisel edge on thrust force and noticed that the thrust force increases with both chisel edge length and feed rate (Tsao and Hocheng, 2003). Basavarajappa et al. showed the dependent variables are greatly influenced by the feed rate rather than the speed for hybrid metal matrix composites (Basavarajappa et al., 2008).

Table 3. ANOVA for the thrust force of various drill types in drilling ceramic matrix brake pads

Factor	Level (S/N) 1	2	3	DF	SS	V	P (%)
A	-29.6	-29.9	-28.5	2	3.04	1.52	30.0
B	-28.2	-29.6	-30.2	2	6.07	3.03	45.8
C	-29.0	-29.1	-30.0	2	2.02	1.01	15.2
Error				2	2.12	1.06	16.0
Total				8	13.25		100

DF-degree of freedom, SS-sum of square, *V*-variance, *P*-percentage contribution

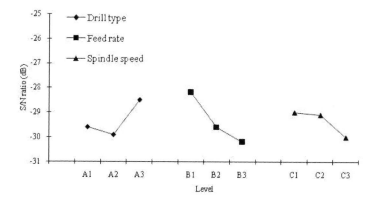

Figure 7. Response graph of thrust force for all control factors in drilling CMBP.

3. 2. ANOVA of Delamination Factor

The appearance after drilling with twist drill did not indicate any delamination at the entrance plane of CMBP, as shown in Figure 1. However, delamination does take place at the exit plane of CMBP because the uncut thickness at the bottom cannot withstand the drilling thrust as the drill approaches the exit plane. As the thrust exceeds the brake pad bond strength and delamination occurs. The experimental results and ANOVA on the delamination factor are shown in Tables 2 and 4. From the results of analysis in Table 4, various drill types ($P=$ 46.6 %) has a substantial effect on delamination factor, and the feed rate ($P=$ 43.0 %) also plays an important role in drilling induced-delamination. However, the influence of spindle speed is insignificant. The response graph of delamination factor for all control factors in drilling CMBP is shown in Figure 8. From Figure 8, it can be seen that the best drill type level is A_3 (four-flute twist drill), the best feed rate level is B_1 (0.01 mm/rev) and the best spindle speed level is C_1 (800 rpm). The results can be explained without conflict to the correlation between thrust force and delamination in drilling composite materials, as proposed by Koenig et al. (1985). However, the use a special drill can be operated at larger feed rate or in shorter cycle time without delamination damage at the drill exit side compared to the twist drill (Hocheng and Tsao, 2006). Besides, a combination of feed rate strategy and modified tool geometry can be used to avoid delamination in drilling (Jain and Yang, 1993).

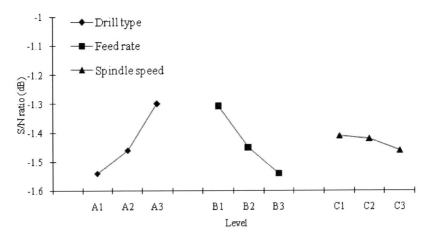

Figure 8. Response graph of delamination factor for all control factors in drilling CMBP.

Table 4. ANOVA for the delamination factor of various drill types in drilling ceramic matrix brake pads

Factor	Level (S/N) 1	2	3	DF	SS	V	P (%)
A	-1.54	-1.46	-1.30	2	0.087	0.044	46.6
B	-1.31	-1.45	-1.54	2	0.081	0.041	43.0
C	-1.41	-1.42	-1.46	2	0.004	0.002	2.4
Error				2	0.015	0.008	8.0
Total				8	0.187		100

DF-degree of freedom, *SS*-sum of square, *V*-variance, *P*-percentage contribution

CONCLUSION

An analysis of thrust force and delamination caused by various drill types in drilling ceramic matrix brake pads is presented in this study. From the results discussed above, the feed rate and drill type are found to be the most significant parameters among the three control factors (drill type, feed rate and spindle speed) that influence the thrust force and delamination in drilling CMBP. The influence of spindle speed was relatively insignificant. The experimental results indicated the best combination to get lower thrust and lower delamination factor is expected to be $A_3B_1C_1$. In other words, properly selected drill geometry and small feed rate should acquire low thrust force in drilling CMBP, which can decrease induced-delamination.

ACKNOWLEDGMENT

This work is partially supported by National Science Council, Taiwan, ROC, under contract NSC 95-2221-E-233-003.

REFERENCES

[1]   Agapiou, J. S. (1993a) 'Design characteristics of new types of drill and evaluation of their performance drilling cast iron - I: drill with four major cutting edges', *Int. J. Mach. Tools Manuf.*, Vol. 33, No. 3, pp. 321-341.

[2]   Agapiou, J. S. (1993b) 'Design characteristics of new types of drill and evaluation of their performance drilling cast iron - II: drills with three major cutting edges', *Int. J. Mach. Tools Manuf.*, Vol. 33, No. 3, pp. 343-365.

[3]   Armarego, E. J. A. and Zhao, H. (1996) 'Predictive force models for point-thinned and circular centre edge twist drill designs', *Ann. CIRP*, Vol. 45, No. 1, pp. 65-70.

[4]   Basavarajappa, S., Chandramohan, G. and Davim, J. P. (2008) 'Some studies on drilling of hybrid metal matrix composites based on Taguchi techniques', *J. Mater. Process. Technol.*, Vol. 196, pp. 332-338.

[5]   Chen, W. C. (1997) 'Some experimental investigations in the drilling of carbon fiber-reinforced plastic (CFRP) composite laminates', *Int. J. Mach. Tools Manuf.*, Vol. 37, No. 8, pp. 1097-1108.

[6]   Davim, J. P. (2003) 'Study of drilling metal matrix composites based on the Taguchi techniques', *J. Mater. Process. Technol.*, Vol. 132, pp. 250–254.

[7]   Davim, J. P., Compos Rubio, J. and Abrao, A. M. (2007) 'A novel approach based on digital image analysis to evaluate the delamination factor after drilling composite laminates', *Compos. Sci. Technol.*, Vol. 67, No. 9, pp. 1939-1945.

[8]   Doerr, R., Greene, E., Lyon, B. and Taha, S. (1982) 'Development of effective machining and tooling techniques for Kevlar composites', *Technical Report*, No. AD-A117853.

[9]   Ema, S., Fujii, H., Marui, E. and Kato, S. (1988) 'New type drill with three major cutting edges', *Int. J. Mach. Tools Des. Res.*, Vol. 28, No. 4, pp. 461-473.

[10] Ema, S., Fujii, H. and Marui, E. (1991) 'Cutting performance of drills with three cutting edges: effects of chisel edge shapes on the cutting performance', *Int. J. Mach. Tools Manuf.*, Vol 31, No. 3, pp. 361-369.

[11] Fujii, S., DeVries, M. F. and Wu, S. M. (1971) 'An analysis of the chisel edge and the effect of the d-theta relationship on drill point geometry', *ASME J, Eng, Ind.*, Vol. 93, pp. 1093-1105.

[12] Gaitonde, V. N., Karnik, S. R. and Davim, J. P. (2008) 'Taguchi multiple-performance characteristics optimization in drilling of medium density fibreboard (MDF) to minimize delamination using utility concept', *J. Mater. Process. Technol.*, Vol. 196, pp. 73-78.

[13] Galloway, D. F. (1957) 'Some experiments on the influence of various factors on drill performance', *Trans. ASME*, Vol. 79, pp. 191-237.

[14] Hocheng, H. and Tsao, C. C. (2003) 'Comprehensive analysis of delamination in drilling of composite materials with various drill bits', *J. Mater. Process. Technol.*, Vol. 140, pp. 335-339.

[15] Hocheng, H. and Tsao, C. C. (2006) 'Effects of special drill bits on drilling-induced delamination of composite materials', *Int. J. Mach. Tools Manufact.*, Vol. 46, pp. 1403-1416.

[16] Jain, S. and Yang, D. C. H. (1993) 'Effects of feedrate and chisel edge on delamination in composite drilling', *ASME J. Eng. Ind.*, Vol. 115, pp. 398-405.

[17] Koenig, W., Grass, P., Heintze, A., Okcu, F. and Schmitz-Justin, C. (1984) 'Developments in drilling, contouring composites containing Kevlar', *Prod. Eng.*, Vol. 9, pp. 56-61.

[18] Koenig, W., Wulf, C., Grass, P. and Willerscheid, H. (1985) 'Machining of fiber reinforced plastics', *Ann. CIRP*, Vol. 34, No. 2, pp. 538-548.

[19] Koenig, W. and Grab, P. (1989) 'Quality definition and assessment in drilling of fiber reinforced thermosets', *Ann. CIRP*, Vol. 38, No. 1, pp. 119-124.

[20] Lee, T. C. and Chan, C. W. (1997) 'Mechanism of the ultrasonic machining of ceramic composites', *J. Mater. Process. Technol.*, Vol. 71, No. 2, pp. 195-201.

[21] Mathew, J., Ramakrishnan, N. and Naik, N. K. (1999) 'Trepanning on unidirectional composites: delamination studies', *Compos. Part A: Appl. Sci. Manuf.*, Vol. 30, pp. 951-959.

[22] Piquet, R., Ferret, B., Lachaud, F. and Swider, P. (2000) 'Experimental analysis of drilling damage in thin carbon/epoxy plate using special drills', *Compos. Part A: Appl. Sci. Manuf.*, Vol. 31, pp. 1107-1115.

[23] Salama, A. S. and ElSawy, A. H. (1996) 'The dynamic geometry of a twist drill point', *J. Mater. Process. Technol.*, Vol. 56, pp. 45-53.

[24] Seif, M. A., Khashaba, U. A. and Rojas-Oviedo, R. (2007) 'Measuring delamination in carbon/epoxy composites using a shadow moiré laser based imaging technique', *Compos. Struc.*, Vol. 79, pp. 113-118.

[25] Shih, Y. H. (2001) 'Industrial evaluation on ceramic brake', Analytical Report on Industry from Economic Center, Industrial Technology Research Institute (ITRI).

[26] Tsao, C. C. and Hocheng, H. (2003) 'The effect of chisel length and associated pilot hole on delamination when drilling composite materials', *Inter. J. Mach. Tools Manuf.*, Vol. 43, No. 11, pp. 1087-1092.

[27] Tsao, C. C. and Hocheng, H. (2004) 'Taguchi analysis of delamination associated with various drill bits in drilling of composite material', *Int. J. Mach. Tools Manuf.*, Vol. 44, No. 10, pp. 1085-1090.

[28] Tsao, C. C. and Hocheng, H. (2005) 'Computerized tomography and C-Scan for measuring delamination in the drilling of composite materials using various drills', *Inter. J. Mach. Tools Manuf.*, Vol. 45, pp. 1282-1287.

[29] Won, M. S. and Dharan, C. K. H. (2002) 'Drilling of Aramid and carbon fiber polymer composites', *ASME J. Manuf. Sci. Eng.*, Vol. 124, pp. 778-783.

In: Machining and Forming Technologies, Volume 3
Editor: J. Paulo Davim

ISBN: 978-1-61324-787-7
© 2012 Nova Science Publishers, Inc.

Chapter 15

TURNING HARDENED STEEL USING CARBIDE INSERT UNDER HIGH-PRESSURE COOLANT CONDITION

Prianka B. Zaman, S. K. Dey and N. R. Dhar[*]
Department of Industrial & Production Engineering, Bangladesh University
of Engineering & Technology (BUET) Dhaka, Bangladesh

ABSTRACT

Machining hardened steel and other difficult to cut materials is associated with generating large amount of heat as well as high cutting temperature. Additionally, high production machining inherently produce high cutting zone temperature. So, it requires instant heat transfer from the cutting edge of the tool to improve its life. The efficiency of metal cutting operations depends on the thermal frictional conditions of the chip-tool interface. Application of the high-pressure coolant shows better result than other conventional lubrication process. The present work deals with experimental investigation in the role of high-pressure coolant jet on chip formation (chip breakability), cutting temperature, cutting force, tool wear and surface roughness in turning of hardened medium carbon steel at industrial speed-feed combination by carbide insert. The results have been compared with dry machining. The results indicate that the use of high-pressure coolant jet leads to reduced surface roughness, delayed tool flank wear and lower cutting temperature and cutting forces and favorable chip tool interaction.

Keywords: Hard turning, chip, temperature, cutting force, wear and surface roughness

1. INTRODUCTION

Hard turning of harder material differs from conventional turning because of its larger specific cutting forces requirements. Typically, in the machining of hardened steel materials, no cutting fluid is applied in the interest of low cutting forces and low environmental impacts.

---

[*] Email: nrdhar@ipe.buet.ac.bd

Higher temperatures are generated in the cutting zone, and because cutting is typically done without coolant, hard turned surfaces can exhibit thermal damage in the form of microstructural changes and tensile residual stresses. The potential economic benefits of hard turning can be offset by rapid tool wear or premature tool failure if the brittle cutting tools required for hard turning are not used properly. Even, progressive tool wear can result in significant changes in cutting forces, residual stresses, and microstructural changes in the form of a rehardened surface layer [Dawson and Kurfess, 2000].

Machining hardened steels using advanced tool materials, such as CBN or PCBN, or with multilayer coated carbide tools at high cutting speeds has certain advantages compared to the traditional machining sequence of processes, i.e., soft machining, heat treatment and grinding. Lower cutting force, residual stress, reduced cycle time and mainly low energy consumption, are some of those advantages [Matsumoto 1991 and Panov 1989]. Research in this area has often focused on the choice of appropriate cutting tool materials, with results typically indicating that CBN tools perform better than carbides or alumina based tools [Matsumoto et al. 1986, Luo et al. 1999 and Thiele and Melkote 1999]. Under proper conditions, CBN tooling can easily pay for its expensive initial cost with substantial tool life. However, short tool life is not the only result of rapid tool wear. Flank wear has been found to be the most significant factor affecting the depth of white layer [Kevin and Hui 2004]. Similar detrimental effects on residual stresses and white layers have been found by others [Chou and Evans 1999; Luo et al. 1999].

In general, high-speed machining for a given material can be defined as that speed above which shear-localization develops completely in the primary shear zone. Due to shear localization, a huge amount of heat generates at the chip tool interface, which leads a very high cutting temperature [Tonshoff and Brinkomeier 1986]. Such high cutting temperature not only reduces dimensional accuracy and tool life but also impairs the surface integrity of the product [Sales et al. 2002]. The application of conventional cutting fluid during machining is believed to reduce this cutting temperature either by removing heat as coolant or by reducing the heat generation as a lubricant and increase tool life. But it has been experienced that lubrication is effective at low cutting velocities when it is accomplished by diffusion through the workpiece and by forming solid boundary layers from the extreme pressure additives, but at high cutting velocities no sufficient lubrication effect is evident [Merchant 1958 and Sales et al. 2002].

Under these considerations, the concept of high-pressure coolant presents itself as a possible solution for high speed machining in achieving slow tool wear while maintaining cutting forces/power at reasonable levels, if the high pressure cooling parameters can be strategically tuned. With the use of high-pressure coolant during machining under normal cutting conditions, the tool life and surface finish are found to improve significantly due to the decrease in heat generated and cutting forces [Kovacevic et al. 1994; Lindeke 1991].

Mazurkiewicz [1989] reported that a coolant applied at the cutting zone through a high-pressure jet nozzle could reduce the contact length and coefficient of friction at chip-tool interface and thus could reduce cutting forces and increase tool life to some extent. High-pressure coolant injection technique not only provided reduction in cutting forces and temperature but also reduced the consumption of cutting fluid by 50% [Ezugwu and Bonney 2004; Aronson 2004 and Senthil et al. 2002].

Dhar et al. [2007] reported that the machining of alloy steel with uncoated carbide insert under high-pressure coolant supplies improve tool life up to 4 folds, especially at high speed conditions. Tool life tends to improve with increasing coolant pressure. There is also evidence that once a critical pressure has been reached any further increase in coolant pressure may only result to a marginal increase in tool life. Lower cutting forces are recorded with increasing coolant supply pressure when machining alloy steel with uncoated carbide inserts. The reduction in cutting forces observed is also partly due to the chip segmentation when machining with high-pressure coolant supplies [Sultana and Dhar 2010]. It has been reported that the drilling of steel with high pressure coolant jet improves the roundness of the hole and surface integrity in respect of lowering cutting temperature and torque [Dhar et al. 2006].

The review of the literature suggests that high pressure coolant jet provides several benefits in machining. The objective of the present work is to experimentally investigate the influence of high pressure coolant jet on cutting temperature, chip formation mode, cutting force, tool wear and surface roughness in turning hardened steel at industrial speed–feed conditions by carbide insert and compare the effectiveness of high pressure coolant with that of dry machining.

2. EXPERIMENTAL INVESTIGATION

The experiment was carried out on Okuma LB15 CNC lathe, which has a 7.5 kW spindle and maximum spindle speed of 4200 rpm. The photographic view of the experimental set-up is shown in Figure 1.The work material was medium carbon steel, hardened to nominal values of 56 HRC having external diameter of 87 mm, internal diameter of 61 mm and length of 230 mm. The cutting tool used was carbide insert (Widia SNMG 120408). The tool holder provided negative 6° side and back rake angles and 6° side cutting-edge and end cutting-edge angles. The ranges of the cutting speed (V_c) and feed rate (f) were selected based on the tool manufacturer's recommendation and industrial practices. Depth of cut (a_p), being less significant parameter, was kept fixed. The experimental conditions are given in Table-1.

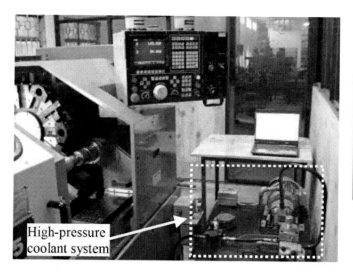

Figure 1. Photographic view of the experimental set-up.

High-pressure coolant jet impinged at the chip-tool interface zone for removing temperature through the nozzle at an angle from a suitable distance. The cutting fluid needs to be drawn at high pressure from the coolant tank and impinged at high speed through the nozzle. Considering the conditions required for the present research work and uninterrupted supply of coolant at pressure around 80 bar over a reasonably long cut, a coolant tank has been designed, fabricated and used. The photographic view of the experimental setup along with high pressure coolant system which contain motor-pump assemble, flow control valve, relief valve and directional control valve is shown in Figure 1. The positioning of the nozzle tip with respect to the turning tool has been settled after a number of trials [Dhar, 2006]. The high-pressure coolant jet is directed in such a way that it reaches at the rake and flank surface and to protect auxiliary flank to enable better dimensional accuracy. The application of high-pressure coolant jet is expected to affect the various machinability characteristics mainly by reducing the cutting temperature.

Table-1. Experimental conditions

Machine tool	:	CNC lathe (Okuma LB15), 7.5 kW
Work materials	:	Hardened steel
Hardness (HRC)	:	56
Size	:	External dia.=87, internal dia.=61 mm and length=230 mm
Cutting insert	:	Carbide, SNMG 120408 (P30 grade ISO specification), WIDIA 
Tool holder	:	PSBNR 2525 M12 (ISO specification), WIDIA
Working tool geometry	:	Inclination angle : -6°
		Orthogonal rake angle : -6°
		Orthogonal clearance angle : 6°
		Auxiliary cutting edge angle : 15°
		Principal cutting edge angle : 75°
		Nose radius : 0.8 mm
Process parameters		
Cutting speed, V_c	:	70, 100, 130, 156 m/min
Feed, f	:	0.12, 0.16, 0.20, 0.24 mm/rev
Depth of cut, a_p	:	1.5 mm
High-pressure Coolant	:	Pressure: 80 bar, Coolant: 6 l/min through external nozzle having 0.5 mm tip diameter.
Environment	:	Dry and High-pressure coolant

The average chip-tool interface cutting temperature was measured under both high pressure coolant and dry conditions by simple but reliable tool-work thermocouple technique with proper calibration. Figure 2 shows the calibration technique employed for the tool-work thermocouple used in the present investigation. The thermocouple junction was constructed using a long continuous chip of the concerned work material and a tungsten carbide insert to be used in actual cutting. To avoid generation of parasitic emf, a long carbide rod was used to extend the insert. A graphite block embedded with an electrically heated porcelain tube served as the heat sink. A chromel-alumel thermocouple was used as a reference in the vicinity of the tool-work thermocouple for measuring the temperature of the graphite block. The junction temperature measured by the reference thermocouple was recorded using a digital temperature readout meter (Eurotherm, UK) while, the emf generated by the tool-work thermocouple was recorded by a digital multimeter (Rish Multi, India).

Figure 2. Tool-work thermocouple calibration set up.

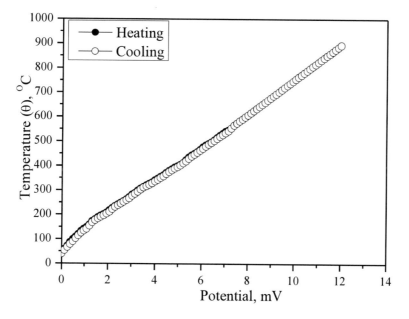

Figure 3. Tool-work thermocouple calibration curve.

Figure 3 shows the calibration curves obtained for the tool-work pair with tungsten carbide (P30 Grade, Widia) as the tool material and the hardened steel undertaken as the work material. In the present case, almost linear relationships between the temperature and emf have been obtained with correlation coefficients of 0.994. In the present work the average cutting temperature has been measured by tool-work thermocouple technique as indicated in Figure 4 taking care of the aforesaid factors.

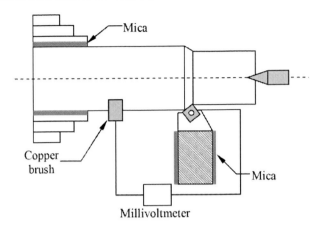

Figure 4. Schematic view of the tool-work thermocouple loop.

The machining chips were collected during all the treatments for studying their shape, colour and nature of interaction with the cutting insert at its rake surface. An importantmachinability index is chip reduction coefficient, ξ (ratio of chip thickness after and before cut). For given tool geometry and cutting conditions, the value of chip reduction coefficient depends upon the nature of chip-tool interaction, chip contact length, curl radius and form of the chips all of which expected to be influenced by high-pressure coolant in addition to the level of cutting speeds and feed rates. The thickness of the chips was repeatedly measured by a digital slide caliper to determine the value of chip reduction coefficient. The schematic view of the formation of chip is shown in Figure 5.

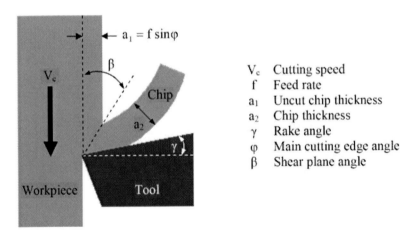

Figure 5. Schematic view of the formation of chip.

The tangential or main component, P_z of the cutting force was monitored by a dynamometer (Kistler, model: 9257B) and recorded in a PC using a data acquisition system during hard turning under different V_c and f. The charge signal generated at the dynamometer was amplified using charge amplifiers. The amplified signal is acquired and sampled by using data acquisition on a computer at a sampling frequency of 2000 Hz per channel. Time-series profiles of the acquired force data reveal that the forces are relatively constant over the length of cut and factors such as vibration and spindle run-out were negligible.

The cutting insert was withdrawn at regular intervals to study the pattern and extent of wear on main and auxiliary flanks for all the trials. The average width of the principal flank wear, V_B and auxiliary flank wear, V_S were measured using in inverted metallurgical microscope (Olympus, Model: MG) fitted with micrometer of least count 1.0 μm. The surface roughness was monitored by a Talysurf (Surtronic 3P, Rank Taylor Hobson) using a sampling length of 0.8mm. At the end of full cut, the cutting inserts were inspected under scanning electron microscope (Model: JSM 5800, JEOL, Japan).

3. EXPERIMENTAL RESULTS AND DISCUSSION

During machining heat is generated at the primary deformation zone, secondary deformation zone and the flank (clearance) surfaces but temperature becomes maximum at the chip-tool interface. The cutting temperature is measured in the present work refers mainly to that chip-tool interface temperature. The machining temperature at the cutting zone is an important index of machinability and needs to be controlled as far as possible. Cutting temperature increases with the increase in specific energy consumption and material removal rate (MMR). Such high cutting temperature adversely affects, directly and indirectly, chip formation, cutting forces, tool life and dimensional accuracy and surface integrity of the products. Besides, this high temperature invites environmental problems when tried to be controlled by conventional cutting fluid application.

The average cutting temperature measured by tool-work thermocouple technique during hard turning at different cutting velocities and feeds under both dry and high-pressure coolant (HPC) condition is shown in Figure 6. It shows how and to what extent average cutting temperature has decreased due to high-pressure coolant application under the different experimental conditions. With the increase in V_c and f, the average temperature increased as usual, even under HPC condition, due to increase in energy input. It can clearly be observed from Figure 6 that HPC is able to reduce average cutting temperature up to 11% compared to dry machining. At $V_c \leq 100$ m/min and all feed range, reduction in cutting temperature varies within 6~11%. Again, at $V_c > 100$ m/min and all feed range reduction in average temperature varies within 3~6%. It is evident from Figure 6 that as the cutting velocity and feed rate increases, the percentage reduction in average cutting temperature decreases. It may be for the reason that, the bulk contact of the chips with the tool with the increase in V_c and f did not allow significant entry of high-pressure coolant jet. Only possible reduction in the chip-tool contact length by the high-pressure coolant jet, particularly along the auxiliary cutting edge, could reduce the temperature to some extent. When the chip velocity was high due to higher speed-feed combinations, reduction in average cutting temperature is quite significant in pertaining tool life and surface finish.

Chip types in metal cutting are primarily dominated by the combined effects of cutting conditions, work material properties and tool geometry. Therefore, the understanding of chip formation, control of chip flow and chip breaking usually play an important role in machining process optimization, dimensional accuracy, surface integrity and the product performance. The chip samples collected during hard turning under both dry and high-pressure coolant condition have been visually examined and categorized as per ISO standard 3685 with respect to their shape [Shaw 2005]. The result of such categorization of the produced chips during hard turning under both the conditions is in Table 2. It is quite interesting and important to note in Table 2 that the chips were segmented more distinctly and sharply when the same material was turned at same V_c-f combination but with high-pressure coolant jet. From Table 2, it can be stated that the chips that are generated under high-pressure coolant condition are in favorable condition whereas the chips generated under dry condition fall under unfavorable condition though chips generated at lower velocity and higher feed rate under dry condition shows favorable condition. It can be clearly understood from Table 2 that by applying high-pressure coolant, chip breakability increases to a great extent. All the chips produced during hard turning are discontinuous and this discontinuity increases with the increase of speed-feed. This is desirable in generating better surface finish.

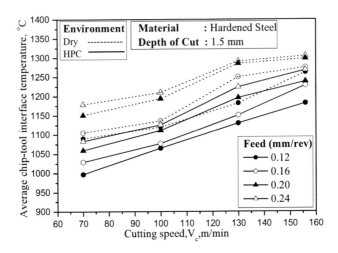

Figure 6. Variation in average chip-tool interface temperature with V_c and f under dry and high-pressure coolant conditions.

Almost all the parameters involved in machining have direct and indirect effect on the thickness of the chips during deformation. The degree of chip thickness which is assessed by the chip reduction coefficient (ξ) plays an important role on the cutting force and hence on cutting energy requirements and cutting temperature. The variation in value of ξ with change in V_c and f as well as machining environment evaluated for hard turning have been plotted and shown in Figure 7. Figure 7 shows that high-pressure coolant jet has reduced the value of ξ particularly at lower values of V_c and f. By high-pressure coolant applications, ξ is reasonably expected to decrease for reduction in friction at the chip-tool interface and reduction in deterioration of effective rake angle by built-up edge formation and wear at the cutting edges mainly due to reduction in cutting temperature.

The magnitude and pattern of the cutting forces is one of the most important machinability indices because that plays vital roles on power and specific energy consumption and product quality. Therefore it is reasonably required to study and assess how the cutting forces and tool life are affected by high-pressure coolant jet. The nature of variation in the main cutting force (P_z) during hard turning at different V_c and f under both dry and HPC condition is shown in Figure 8. Figure 8 clearly shows that P_z has uniformly decreased with the increase in V_c more or less under all the feeds, for both the tool and environments undertaken as usual due to favourable change in the chip-tool interaction resulting in lesser friction and intensity or chances of built-up edge formation at the chip-tool interface. It is evident from Figure 8 that P_z decreased significantly due to high-pressure coolant jet more or less at all the V_c-f combinations. This improvement can be reasonably attributed to reduction in the cutting temperature particularly near the main cutting edge where seizure of chips and formation or tendency of formation of built-up edge is more predominant. In this respect, the high-pressure coolant jet impinged along the auxiliary cutting edge seems to be more effective in cooling the neighbourhood of the auxiliary cutting edge.

Table 2. Influence of environment on chip during hard turning

Feed rate, f (mm/rev)	Cutting speed, V_c (m/min)	Environment			
		Dry		HPC	
		Shape	Effect	Shape	Effect
0.12	70	snarled ribbon	□	short conical helical	■
	100	snarled ribbon	□	short tubular	■
	130	snarled ribbon	□	short conical helical	■
	156	snarled ribbon	□	short tubular	■
0.16	70	snarled ribbon	□	loose arc	■
	100	snarled ribbon	□	short conical helical	■
	130	snarled ribbon	□	short conical helical	■
	156	connected arc	□	flat spiral	■
0.20	70	connected arc	■	loose arc	■
	100	connected arc	■	loose arc	■
	130	snarled ribbon	□	loose arc	■
	156	connected arc	□	loose arc	■
0.24	70	connected arc	■	loose arc	■
	100	snarled ribbon	■	loose arc	■
	130	snarled ribbon	□	connected arc	■
	156	connected arc	□	loose arc	■
Effect	■ Favorable			□ Unfavorable	
Chip shape					
Group	short tubular	connected arc	snarled ribbon	loose arc	flat spiral	shortconical helical

The cutting tools in conventional machining, particularly in continuous chip formation processes like turning, generally fails by gradual wear by abrasion, adhesion, diffusion, chemical erosion, galvanic action etc. depending upon the tool-work materials and machining condition. Tool wear initially starts with a relatively faster rate due to what is called break-in wear caused by attrition and micro-chipping at the sharp cutting edges. Cutting tools may also often fail prematurely, randomly and catastrophically by mechanical breakage and plastic deformation under adverse machining conditions caused by intensive pressure and temperature and/or dynamic loading at the tool tips particularly if the tool material lacks strength, hot-hardness and fracture toughness. However, in the present investigations with the tool and work material and the machining conditions undertaken, the tool failure mode has been mostly gradual wear.

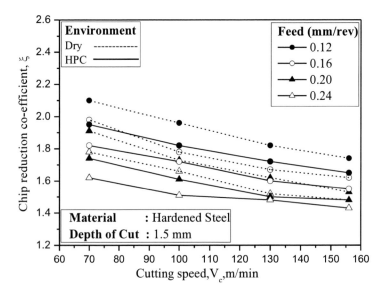

Figure 7. Variation in chip reduction coefficient with V_c and f under dry and high-pressure coolant conditions.

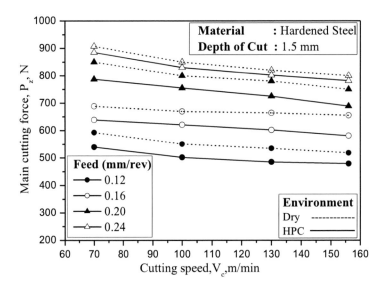

Figure 8. Variation in main cutting force (P_z) with V_c and f under dry and high pressure coolant conditions.

Among the aforesaid wears, the principal flank wear is the most important because it raises the cutting forces and the related problems. The life of carbide tools, which mostly fail due to wear, is assessed by the actual machining time after which the average value of its principal flank wear (V_B) reaches a limiting value, like 300 μm. Therefore, attempts should be made to reduce the rate of growth of flank wear (V_B) in all possible ways without sacrificing

MRR. The cutting insert was withdrawn at regular intervals to study the pattern and extent of wear on main and auxiliary flanks under both dry and high-pressure coolant conditions.

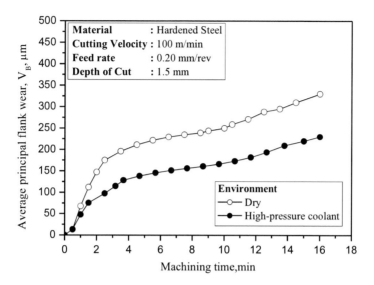

Figure 9. Growth of principal flank wear, V_B in the insert under dry and high-pressure coolant conditions.

The growth of principal flank wear (V_B) with progress of machining was recorded during hard turning at moderately high cutting speed, feed and depth of cut under both dry and high-pressure coolant condition have been shown in Figure 9. Figure 9 clearly shows that flank wear (V_B) particularly its rate of growth, decreases substantially by high-pressure coolant jet. The cause behind reduction in V_B observed may reasonably be attributed to reduction in the flank temperature by high-pressure coolant jet impinged along the auxiliary cutting edge, which helps in reducing abrasion wear by retaining tool hardness and also adhesion and diffusion types of wear which are highly sensitive to temperature.

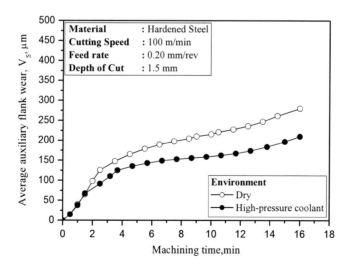

Figure 10. Growth of auxiliary flank wear, V_S in the insert under dry and high-pressure coolant conditions.

Because of such reduction in rate of growth of flank wear, the tool life would be much higher if high-pressure coolant is properly applied. The auxiliary flank wear affects dimensional accuracy and surface finish has also been recorded at regular intervals of machining under all the conditions undertaken. The growth of average auxiliary flank wear (V_S) with machining time under both dry and high-pressure coolant conditions have been shown in Figure 10. It appears from Figure 10 that auxiliary flank wear (V_S) has also decreased significantly due to application of high-pressure coolant jet.

The SEM views of the worn out insert after being used under both dry and high-pressure coolant conditions are shown in Figure 11 and Figure 12. Under all the environments, abrasive scratch marks appeared in the flanks. The examination of the craters revealed deep scratches left by the backside of the chip on the rake surface of the tool. There have also been some indications of adhesive wear especially under dry condition, which produced unfavorable chips as compared to high pressure coolant condition, which produced favorable chips.

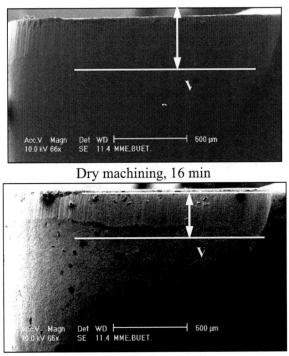

Figure 11. SEM images of principal flank of cutting insert under dry and high-pressure coolant conditions.

The flank wear occurred quite fast due to rapid attrition wear followed by adhesion and diffusion in addition to usual abrasion particularly at the tool tip where stresses and temperature are high. Rapid start of flank wear causes more intimate contact at the work-tool interface and initiates severe rubbing which again aggravates flank wear further. Flank wear grow so fast in hard turning by carbide insert that notching and grooving type wear do not appear separately.

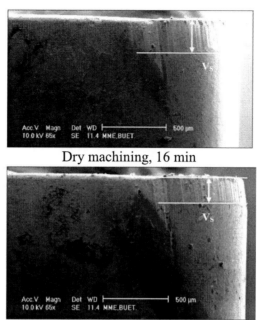

Dry machining, 16 min

HPC machining, 16 min

Figure 12. SEM images of auxiliary flank of cutting insert under dry and high-pressure coolant conditions.

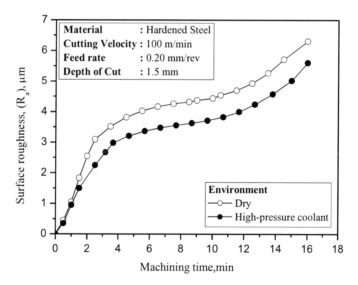

Figure 13. Surface roughness developed with progress of machining under dry and high-pressure coolant conditions.

But it is clearly evident from Figure 11 and Figure 12 that high-pressure coolant jet machining caused lesser wear than that produced by dry machining. Such reduction in wear is seemingly indebted to reduction of the cutting temperature sensitive wear phenomenon like diffusion and adhesion enabled by direct and indirect cooling by high-pressure coolant jet.

The resulting surface finishes under both dry and high-pressure coolant conditions are shown in Figure 13. As high-pressure coolant jet reduced average auxiliary flank wear (V_S)

on auxiliary cutting edge, surface roughness also grew very slowly under high-pressure coolant condition. The effect of high-pressure coolant is to improve the surface finish under the given set of cutting conditions.

CONCLUSIONS

i. The present high-pressure coolant system enabled reduction in average chip tool interface temperature upto 11% and even such apparently small reduction, unlike common belief, enabled significant improvement in the major machinability indices.

ii. Due to high-pressure coolant jet application, the form and colour of the chips became favourable for more effective cooling and improvement in nature of interaction at the chip-tool interface.

iii. High-pressure coolant jet reduced the cutting force by about 5% to 12%. Such reduction has been more effective for higher value of chip reduction coefficient (ξ) but adverse chip-tool interaction causing large friction and built-up edge formation at the chip-tool interface. Favourable change in the chip-tool interaction and retention of cutting edge sharpness due to reduction of cutting zone temperature seemed to be the main reason behind reduction of cutting forces by the high-pressure coolant jet.

iv. The most significant contribution of application of high-pressure coolant jet in machining hardened steel by the carbide insert undertaken has been the high reduction in flank wear, which would enable either remarkable improvement in tool life or enhancement of productivity (MRR) allowing higher cutting speed and feed. Such reduction in tool wear might have been possible for retardation of abrasion and notching, decrease or prevention of adhesion and diffusion type thermal sensitivity wear at the flanks and reduction of built-up edge formation which accelerates wear at the cutting edges by chipping and flaking. Deep notching and grooving, which are very detrimental and may cause premature and catastrophic failure of the cutting tool, are remarkably reduced by high-pressure coolant jet.

v. Surface finish also substantially improved mainly due to significant reduction of wear and damage at the tool tip by the application of high-pressure coolant jet.

ACKNOWLEDGMENT

This research work has been funded by Bangladesh University of Engineering and Technology (BUET), Dhaka, Bangladesh. The help extended by the Head, Department of Materials & Metallurgical Engineering, BUET, Dhaka for obtaining the scanning electron micrograph is sincerely acknowledged. Okuma Corporation, Japan, is also gratefully acknowledged for providing the CNC Lathe.

REFERENCES

[1] Aronson, R. B., (2004), "Using high-pressure fluids, cooling and chip removal are critical", *Manufacturing Engineering*, 132(6), 34-46.

[2] Chou, Y. K. and Evans, C. J., (1999), "White layers and thermal modeling of hard turned surfaces", *Int. J. Mach. Tools Manuf.*, 39, 1863-1881.

[3] Dawson, T. G. and Kurfess, T. R., (2000), "An investigation of tool wear and surface quality in hard turning", *Trans. North American Manuf. Research Institution of SME*, 28, 215-220.

[4] Dhar, N. R.,Kamruzzaman, M. and Rashid, M. H., (2007), "On Effect of High-Pressure Coolant on Tool Wear and Surface Finish in Turning AISI-8740 Steel", *Proceedings of the International Conference on Manufacturing Automation (ICMA-2007)*, Singapore, 695-703.

[5] Dhar, N. R., Rashid, M. H. and Siddiqui, A. T., (2006), "Effect of High-Pressure Coolant on Chip, Roundness Deviation and Tool Wear in Drilling AISI-4340 Steel" *ARPN Journal of Engineering and Applied Sciences*, 1(3), 53-59.

[6] Ezugwu, E. O. and Bonney, J., (2004), "Effect of high-pressure coolant supply when machining nickel-base, Inconel 718, alloy with coated carbide tools", *J. Mater. Process. Technology*, 153–154, 1045-1050.

[7] Kevin, Chou, Y. and Hui, Song., (2004), "Tool nose radius effects on finish hard turning", *Journal of Materials Processing Technology*, 148, 259-268.

[8] Kovacevic, R., Cherukuthota, C. and Mohan, R. (1994), "Improving the surface quality by high pressure waterjet cooling assistance", *Geomechanics*, 93, 305-310.

[9] Lindeke, R. R, Schoeing Jr F. C., Khan, A. K. and Haddad, J., (1991), "Cool your jets", *Cutting Tool Engineering*, 31-37.

[10] Luo, S. Y., Liao, Y. S. and Tsai, Y. Y., (1999), "Wear characteristics in turning high hardness alloy steel by ceramic and CBN tools", *Journal of Materials Processing Technology*, 88, 114-121.

[11] Matsumoto, Y., (1991), "Effect of machining process on the fatigue strength of hardened AISI3040 steel", Transactions of ASME, *Journal of Engineering for Industry*, 113,154-159.

[12] Matsumoto, Y., Barash, M. M. and Liu, C. R., (1986), "Effect of hardness of surface integrity of AISI4340 steel", *J. Eng. Ind.*, 108 (3), 169-175.

[13] Mazurkiewicz, M., Kubala, Z. and Chow, J., (1989), "Metal machining with high-pressure water jet cooling assistance- a new possibility", *J. Eng. Ind.*, 111, 7-12.

[14] Merchant, M. E., (1958), "The physical chemistry of cutting fluid action", *Am. Chem. Soc.*,Preprint, 3(4A), 179-189.

[15] Panov, A. A., (1989), "Intensifying components machining by means of tools provided with synthetic superhard materials and ceramics", *Soviet Engineering Research*, 9(11), 45-49.

[16] Sales, W. F., Guimaraes, G., Machado, A. R. and Ezugwu, E. O., (2002), "Cooling ability of cutting fluids and measurement of the chip-tool interface temperatures", *Ind. Lub. Tribol.*, 54(2), 57-68.

[17] Senthil Kumar, A., Rahman, M. and Ng., S. L., (2002), "Effect of high-pressure coolant on machining performance",*Int. J. Adv. Manuf. Technol*, 20, 83-91.

[18] Shaw, M. C., (2005), *"Metal Cutting Principal"*, 2nd Edition, Oxford University Press, 479-498.

[19] Sultana, I. and Dhar, N. R., (2010), "Ga Based Multi Objective Optimization of the Predicted Models of Cutting Temperature, Chip Thickness Ratio and Surface Roughness in Turning AISI 4320 Steel by Uncoated Carbide Insert Under Hpc Condition", *Proceedings of the International Conference on Mechanical, Industrial and Manufacturing Technologies (MIMT-2010)*, China, 161-168.

[20] Thiele, J. D. and Melkote, S. N., (1999), "Effect of cutting edge geometry and workpiece hardness on surface generation in the finish hard turning of AISI 52100 steel", *J. Mater. Process Technology,* 94, 216–226.

[21] Tonshoff, H. K. and Brinkomeier, E., (1986), "Determination of the mechanical and thermal influences on machined surface by microhardness and residual stress analysis", *Annals of CIRP,* 29(2), 519-532.

In: Machining and Forming Technologies, Volume 3
Editor: J. Paulo Davim

ISBN: 978-1-61324-787-7
© 2012 Nova Science Publishers, Inc.

Chapter 16

OPTIMIZATION OF CNC TURNING OF AISI 304 AUSTENITIC STAINLESS STEEL USING GREY BASED FUZZY LOGIC WITH MULTIPLE PERFORMANCE CHARACTERISTICS

C. Ahilan[1], S. Kumanan[1] and N. Sivakumaran[2]*
[1]Department of Production Engineering
[2]Department of Instrumentation and Control Engineering, National Institute of Technology, Tiruchirappalli, India

ABSTRACT

This paper presents the effect of computerised numerical control (CNC) turning parameters of AISI 304 austenitic stainless steel on surface roughness and power consumption using grey based fuzzy logic approach. This approach integrates both grey relational analysis and fuzzy logic for optimizing the complicated multiple performance characteristics. This approach converts the optimization of multiple performance characteristics in to a single grey-fuzzy reasoning grade. Optimum level of parameters has been identified based on grey-fuzzy reasoning grade. In this study, CNC turning parameters namely cutting speed, feed rate, depth of cut and nose radius are optimized with consideration of performance characteristics such as surface roughness and power consumption. The significant contributions of parameters are estimated using Analysis of Variance (ANOVA). Confirmation test is conducted for validation and reported.

Keywords: Power consumption, Surface roughness, Optimization, Grey based Fuzzy logic, ANOVA, Turning process

---

* Email: ahilan@nitt.edu

1. Introduction

In engineering industries, turning is one of the important and extensively used machining processes (Davim, 2008). The cutting conditions such as cutting speed, feed rate and depth of cut, features of tools, work piece materials affects the process efficiency and performance characteristics (Boothroyd and Knight, 1898 & Shaw, 1984). Performance evaluation of CNC turning is based on the performance characteristics like surface roughness, material removal rate (MRR), tool wear, tool life and power consumption. Surface quality is an important performance to evaluate the productivity of machine tools as well as machined components. Hence achieving desired surface quality is of great importance for the functional behavior of the mechanical parts (Benardos and Vosniakos, 2003). Surface roughness is used as the critical quality indicator for the machined surfaces. Very few research attempts have been done to estimate the significance of energy required for the machining process. Recent increase in energy demand and constraints in supply of energy becomes a priority for the manufacturing industry. Now in manufacturing industries, special attention is given to surface finish and power consumption. Austenitic stainless steels are a high work hardening rate, low thermal conductivity and resistance to corrosion (Groover, 1996). Stainless steels are known for their resistance to corrosion but their machinability is more difficult than the other alloy steels due to reasons such as having low heat conductivity, high BUE tendency and high deformation hardening (Kopac and Sali, 2001). Work to date has shown that little work has been carried on the determination of optimum machining parameters when machining austenitic stainless steels. The influences of cutting fluids on tool wear and surface roughness during turning of AISI 304 are investigated (Xavior and Adithan, 2009). The optimum cutting speed for turning of AISI 304 austenitic stainless steel based on tool wear and surface roughness are determined (Ihsan et al 2004) and it needs further investigation. The high cost of CNC machine tools compared to their conventional counterparts, there is an economic need to operate these machines as efficiently as possible in order to obtain the required payback (Davim, 2003). The desired cutting parameters are determined based on experience or by use of a handbook which does not guarantee optimal performance. It is mandatory to select the most appropriate machining settings in order to improve cutting efficiency, process at low cost and produce high-quality products.

Optimization of cutting parameters through experimental methods and mathematical models has grown substantially over time to achieve a common goal of improving higher machining process efficiency (Montgomery, 1997). The experimental Taguchi method can optimize the performance characteristics through the settings of process variables and reduce the sensitivity of the system performance to sources of variation (Nalbant et al, 2007). Hence, the Taguchi method has become a powerful design of experiments method (Lin et al., 2002). The traditional Taguchi method can solve only single objective problems. But in the real world most of the engineering applications consist of multiple objectives (Chua et al., 1993). In complex processes, optimization of multiple objectives is complicated by using single objective method and engineering decision is principally used to resolve such difficult problems. Engineering decisions often raise the degree of uncertainty during decision-making process.

The grey relational analysis theory initialized by Deng (1989) makes use of this to handle uncertain systematic problem with only partially known information. This theory is used to solve the intricate interrelationships among the multiple objectives. Through this approach optimization of multiple objectives can be converted into optimization of single grey relational grade. This approach is implemented in optimization of multiple objectives in electrical discharge machining (Lin and Lin, 2002), chemical-mechanical polishing process (Ho and Lin, 2003) and drilling operation (Tosun, 2006). Fuzzy logic theory was initiated by Zadeh (1965) has been proven to be useful for dealing with uncertain and vague information. The performance or objective definition such as lower-the-better, higher-the-better and nominal-the-best contains certain degree of uncertainty and vagueness. Hence this approach is applied to find the optimal setting of parameters for multiple performance characteristics. This approach was implemented in optimizing the multi-objective problems in electrical discharge machining process (Lin et al., 2000 and Lin et al., 2002).

The grey based fuzzy logic approach has merits of grey relational analysis and fuzzy logic method. Optimization of complicated multiple performance characteristics can be converted into single objective through this method i.e. grey-fuzzy reasoning grade. The optimum level of process parameters are the level with higher value of mean grey-fuzzy reasoning grade. This approach was successfully implemented in optimizing the multiple performance characteristics of complicated problems in manufacturing process (Lin and Lin, 2005), design parameters of pin-fin heat sink (Chiang et al., 2006), side milling of SUS304 (Chang and Lu 2007), die casting process (Chiang et al, 2008) and rough cutting process in side milling (Lu et al, 2008) . This work presents the application of grey based fuzzy logic in optimization of surface roughness and power consumption in CNC turning of AISI 304 austenitic stainless steel. ANOVA (Montgomery, 1997) is used to find the significant contribution of each cutting parameters on the multiple performance characteristics.

2. GREY BASED FUZZY LOGIC APPROACH

The grey based fuzzy logic approach combines grey relational analysis and fuzzy logic to determine the optimum process parameters for multiple performance characteristics.

2.1. Grey Relational Analysis

The grey relational analysis based on the grey system theory can be used to solve the complex interrelationships among the multiple performance characteristics effectively (Deng, 1989). In grey relational analysis, system has a level of information between black and white. In a white system, the relationships among factors in the system are certain; in a grey system, the relationships among factors in the system are uncertain. In other words, in a grey system, some information is known and some information is unknown.

Data pre-processing is a means of transferring the original sequence to a comparable sequence. It is normally required since the range and unit in one data sequence may differ from the others. It is also necessary when the sequence scatter range is too large, or when the directions of the target in the sequences are different. Depending on the characteristics of a data sequence, there are various methodologies of data pre-processing available for the grey relational analysis. Experimental data y_{ij} is normalized as Z_{ij} $(0 \leq Z_{ij} \leq 1)$ for the ith performance characteristics in the jth experiment can be expressed (Haq et al, 2008) as:

For Larger-the-better condition

$$Z_{ij} = \frac{y_{ij} - \min\left(y_{ij}, i = 1, 2, ...n\right)}{\max\left(y_{ij}, i = 1, 2, ...n\right) - \min\left(y_{ij}, i = 1, 2, ...n\right)}_! \tag{1}$$

For smaller-the-better

$$Zij = \frac{\max\left(y_{ij}, i = 1, 2, ...n\right) - y_{ij}}{\max\left(y_{ij}, i = 1, 2, ...n\right) - \min\left(y_{ij}, i = 1, 2, ...n\right)} \tag{2}$$

For nominal-the-best

$$Zij = \frac{(y_{ij} - \text{Target}) - \min(|y_{ij} - \text{Target}|, i = 1, 2, ...n)}{\max(|y_{ij} - \text{Target}|, i = 1, 2, ...n) - \min(|y_{ij} - \text{Target}|, i = 1, 2, ...n)} \tag{3}$$

Then, calculate grey relational co-efficient γ_{ij} for the normalized values.

$$\gamma_{ij} = \frac{\Delta \min + \xi \Delta \max}{\Delta_{oj}(k) + \xi \Delta \max} \tag{4}$$

Where

$j=1,2…n$; $k=1,2…m$, n is the number of experimental data items and m is the number of responses.

$y_o(k)$ is the reference sequence $(y_o(k)=1, k=1,2…m)$; $y_j(k)$ is the specific comparison sequence.

$\Delta_{oj} = \left\| y_o(k) - y_j(k) \right\|$ = The absolute value of the difference between $y_o(k)$ and $y_j(k)$.

$\Delta \min = \min\limits_{\forall j \in i} \min\limits_{\forall k} \left\| y_o(k) - y_j(k) \right\|$ is the smallest value of $y_j(k)$.

$\Delta \max = \max\limits_{\forall j \in i} \max\limits_{\forall k} \left\| y_o(k) - y_j(k) \right\|$ is the largest value of $y_j(k)$.

ζ is the distinguishing coefficient which is defined in the range $0 \leq \zeta \leq 1$ (the value may adjusted based on the practical needs of the system).

This grey relation co-efficient γ_{ij} is applied to show the relationship between the optimal (best=1) and actual normalized results. The higher value of γ_{ij} represents, the corresponding experimental result is closer to the optimal (best) normalized value for the single objective.

2.2. Fuzzy Logic

Fuzzy logic imitates human reasoning about information (Ross, 1997). The interesting fact about fuzzy logic is that fuzzy inferences make it possible to deduce a proposition similar to the consequence from some proposition that is similar to the antecedent (Zadeh, 1995). It is an effective mathematical model of resolving problems in a simple way which contain the uncertain and huge information. In fuzzy logic analysis, the fuzzifier uses membership functions (MFs) to fuzzify the grey relational coefficient. The fuzzy inference engine then performs a fuzzy inference on fuzzy rules in order to generate a fuzzy value. Finally, the defuzzifier converts the fuzzy value into a grey-fuzzy reasoning grade. **Fuzzy expert system has 3** simple steps shown in Figure 1 and defined below

Step 1: Fuzzification: In a fuzzy expert system application, each input variable's crisp value is first fuzzified into linguistic values before the inference engine proceeds in processing with the rule base.

Step 2: Inference engine: The collection of fuzzy IF-THEN rules is stored in the fuzzy rule base which is referred to by the inference engine when processing inputs. Once all crisp input values have been fuzzified into their respective linguistic values, the inference engine will access the fuzzy rule base of the fuzzy expert system to derive linguistic values for the intermediate as well as the output linguistic variables. The grey relational coefficients x_1, x_2, ... x_n and a multi-objective output y that is of the form

Rule 1: IF x_1 is A_{11} and x_2 is A_{21} ...and x_n is A_{n1} THEN y is C_1 else

Rule 2: IF x_1 is A_{12} and x_2 is A_{22} ...and x_n is A_{n2} THEN y is C_2 else

$\vdots$

Rule i: IF x_1 is A_{1i} and x_2 is A_{2i} ...and x_n is A_{ni} THEN y is C_i else

$\vdots$

Rule n: IF x_1 is A_{1n} and x_2 is A_{2n} ...and x_n is A_{nn} THEN y is C_n else

where A_{1i}, A_{2i},..., A_{ni} and C_i are fuzzy subsets defined by the corresponding MFs, i.e., μA_{1i}, μA_{2i},...μA_{ni} and $\mu C_{i.}$ The fuzzy multiple objectives output y is provided from those above rules by employing the max-min interfaced operation. Inference results in a fuzzy set with MF for the y can be expressed as following

$$\mu C_0(y) = (\mu A_{11}(x_1) \wedge \mu A_{21}(x_2)...\mu C_1(y))... \vee (\mu A_{1n}(x_1) \wedge \mu A_{2n}(x_2)...\mu C_n(y)) \tag{5}$$

Where $\wedge$ and $\vee$ are the minimum operation and maximum operation.

Step 3: Defuzzification: The last phase is the defuzzification of the linguistic values of the output linguistic variables into crisp values. The most common techniques for defuzzification are Center-of-Maximum (CoM) and Center-of-Area (CoA). The fuzzy inference output $\mu C_0(y)$ transferred to a non-fuzzy value y_0 by using the centroid defuzzification method, i.e.

$$y_0 = \frac{\sum y \mu C_0(y)}{\sum \mu C_0(y)} \tag{6}$$

This non-fuzzy value y_0 is called as grey-fuzzy reasoning grade. The grey-fuzzy reasoning grade is used to optimize the multiple performance characteristics and the relational degree between main factor and other factors for each performance characteristic. The higher grey-fuzzy reasoning grade indicates that the experimental result closer to the ideally normalized value. The mean value of grey-fuzzy reasoning grade for each level of parameters is used to construct response table and response graph.

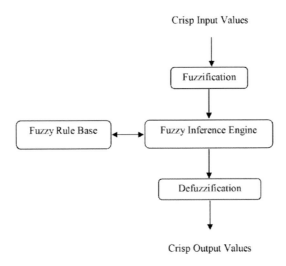

Figure 1. Fuzzy Expert System Model.

The optimum level of turning parameters is the level with higher value of mean grey-fuzzy reasoning grade for each factor concerning power consumption and surface roughness. To identify the contribution of process parameters on multiple performance characteristics ANOVA is carried out. Finally the confirmation test is conducted to validate the optimum parameter setting.

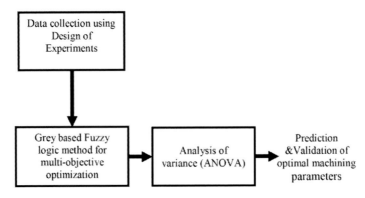

Figure 2. The proposed grey based fuzzy logic method.

The scheme of grey based fuzzy logic method is shown in Figure 2 and steps are as follows

Stage I: Data collection by using design of experiments
  Step 1: Design an appropriate orthogonal array to plan the experimental design and determining the level of parameters
  Step 2: Conduct the experiment based on the arrangement of the orthogonal array
Stage II: Grey based Fuzzy logic method- To optimize multiple performance characteristics
  Step 3: Data pre-processing of experimental results using Eqs. (1) – (3)
  Step 4: Compute the grey relational coefficient for each response by using Eq. (4)
  Step 5: Establishing the membership function and fuzzy rule to fuzzify the inputs (grey relational coefficient of each response) of fuzzy logic.
  Step 6: The fuzzy logic output (grey-fuzzy reasoning grade for multiple performance characteristics) is calculated by defuzzification of the output linguistic variables into crisp values.
  Step 7: Mean response for each level of process parameters are calculated and presented in the form of response table and response graph.
  Step 8: Select the optimal level of machining parameters by which level of each individual parameter having higher value of mean grey-fuzzy reasoning grade in response table and response graph.
Stage III: Analysis of process parameters contribution
  Step 9: Analyze the grey-fuzzy reasoning grade with ANOVA
Stage IV: Validate the optimum parameters through confirmation test
  Step 10: Verify the optimal process parameters through Confirmation test

3. DETERMINATION OF OPTIMAL MACHINING PARAMETERS

3.1. Experimental Details

A CNC lathe with maximum spindle speed of 4500 rpm and spindle power of 7.5KW is used to perform the turning operation. A schematic diagram and photographic view of the experimental set-up is shown in Figure 3 and Figure 4. FLUKE 43B Power Quality Analyzer is connected to the power supply of CNC turning center for measuring the power consumption (in watts) of cutting process is shown in Figure 5. The surface roughness Ra (in μm) is measured using Taylor-Hobson Talysurf which is a stylus and skid type instrument is working on carrier modulating principle is shown in Figure 6. The work material is AISI 304 stainless steel of 70 HRC in the form of round bars with 50mm diameter and 200 mm cutting length. Chemical composition of work material is as follows: 0.08 C, 18-20 Cr, 2 Mn, 8-10.5 Ni, 0.045 P, 0.03 S, 1 Si and remaining Fe. It is majorly used for chemical equipments, food processing equipment, dairy equipment, textile dyeing equipment, cryogenic vessels and hospital surgical equipment. Carbide tool inserts of standards CNMG120404, CNMG120408 and CNMG120412 is used for machining. These inserts are recommended for machining stainless steel by Kennametal and had CNMG120404, CNMG120408 and CNMG120412 Kennametal designations.

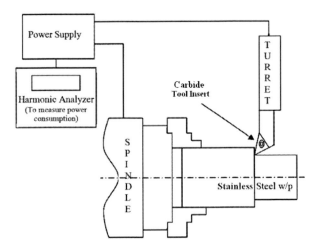

Figure 3. Experimental set-up in CNC turning operation.

Figure 4. Photographic view of experimental set-up in CNC turn.

Figure 5. Photographic view of FLUKE 43B Power Quality Analyzer.

Figure 6. Photographic view of Taylor-Hobson Talysurf to measure surface roughness.

Table 1. Cutting parameters and their levels

Factor	Cutting Parameter	Unit	Level 1	Level 2	Level 3
A	Cutting speed 'V'	m/min	100	125	150
B	Feed rate 'f'	mm/rev	0.05	0.1	0.15
C	Depth of cut 'd'	mm	0.20	0.35	0.50
D	Nose radius 'NR'	mm	0.4	0.8	1.2

To perform the experimental design, three levels of turning parameters cutting speed, feed rate, depth of cut and nose radius are selected and are shown in Table 1. An appropriate orthogonal array for the experiments must have the degrees of freedom greater than or atleast equal to those for the process parameters. In this study, an L_{27} (3^4) orthogonal array is used because it has 26 degrees of freedom more than the 8 degrees of freedom in the machining parameters. The experimental combinations of the turning parameters using the L_{27} orthogonal array are presented in Table 2.

Based on the designed orthogonal array combination turning operations are performed in CNC lathe. Power consumption and surface roughness are taken as performance characteristics. Power consumption is measured for each setting of turning operation and idle running operation by using FLUKE 43B Power Quality Analyzer. The difference between the turning operation and idle running operation power consumption is the actual power needed for turning operation. Surface roughness Ra is measured using Taylor-Hobson Talysurf instrument. The performance characteristic for power consumption and surface roughness has lower-the-better manner. The experimental results are summarized in Table 3.

Table 2. Experimental layout using L_{27} orthogonal array

Test	X1(Col1)	X2(Col2)	X3(Col5)	X4(Col9)
1	1	1	1	1
2	1	1	2	2
3	1	1	3	3
4	1	2	1	2
5	1	2	2	3
6	1	2	3	1

Table 2. (continued)

Test	X1(Col1)	X2(Col2)	X3(Col5)	X4(Col9)
7	1	3	1	3
8	1	3	2	1
9	1	3	3	2
10	2	1	1	2
11	2	1	2	3
12	2	1	3	1
13	2	2	1	3
14	2	2	2	1
15	2	2	3	2
16	2	3	1	1
17	2	3	2	2
18	2	3	3	3
19	3	1	1	3
20	3	1	2	1
21	3	1	3	2
22	3	2	1	1
23	3	2	2	2
24	3	2	3	3
25	3	3	1	2
26	3	3	2	3
27	3	3	3	1

Table 3. Experimental design using L_{27} orthogonal array their results of Power consumption and Surface Roughness

Exp. No.	A Cutting speed (m/min)	B Feed rate (mm/rev)	C Depth of Cut (mm)	D Nose radius (mm)	Power consumption (Watts)	Surface roughness (Ra in µm)
1	100	0.05	0.2	0.4	213	2.04
2	100	0.05	0.35	0.8	320	1.74
3	100	0.05	0.5	1.2	332	2.02
4	100	0.1	0.2	0.8	283	1.25
5	100	0.1	0.35	1.2	340	1.1
6	100	0.1	0.5	0.4	393	1.02
7	100	0.15	0.2	1.2	275	1.5
8	100	0.15	0.35	0.4	350	1.12
9	100	0.15	0.5	0.8	620	1.35
10	125	0.05	0.2	0.8	392	1.82
11	125	0.05	0.35	1.2	438	1.52
12	125	0.05	0.5	0.4	441	1.78
13	125	0.1	0.2	1.2	391	1.04
14	125	0.1	0.35	0.4	570	0.84

Exp. No.	A Cutting speed (m/min)	B Feed rate (mm/rev)	C Depth of Cut (mm)	D Nose radius (mm)	Power consumption (Watts)	Surface roughness (Ra in μm)
15	125	0.1	0.5	0.8	668	1.02
16	125	0.15	0.2	0.4	394	1.16
17	125	0.15	0.35	0.8	617	1.26
18	125	0.15	0.5	1.2	760	1.48
19	150	0.05	0.2	1.2	448	2.02
20	150	0.05	0.35	0.4	516	1.54
21	150	0.05	0.5	0.8	585	1.94
22	150	0.1	0.2	0.4	476	1.08
23	150	0.1	0.35	0.8	625	1.16
24	150	0.1	0.5	1.2	765	1.42
25	150	0.15	0.2	0.8	528	1.46
26	150	0.15	0.35	1.2	706	1.38
27	150	0.15	0.5	0.4	873	1.64

3.2. Optimization of Machining Parameters

The data pre-processing value of each performance characteristic (power consumption and surface roughness) is found by using one of the Eqs (1), (2) and (3) depends on their type of quality characteristic. Compute the grey relational co-efficient by using Eq (4). The value for ξ is taken as 0.5 in Eq (4) since all the process parameters are of equal weighting. Calculated data pre-processing value for each performance characteristic and their grey relational coefficient are shown in Table 4.

The grey relational grade can be determined using fuzzy logic analysis. The fuzzy model inputs are grey relational coefficients of power consumption and surface roughness and the output is grey-fuzzy reasoning grade. Developed fuzzy logic model using MATLAB 2007b is shown in Figure 7. The most popular defuzzification method centroid calculation is used, which returns the centre of area under the curve. Defuzzifier converts the fuzzy value into non-fuzzy value which is called as grey-fuzzy reasoning grade.

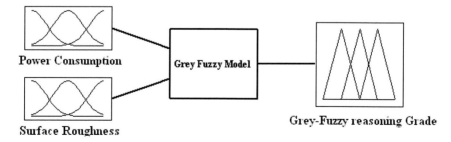

Figure 7. Developed Fuzzy logic model for CNC turning responses.

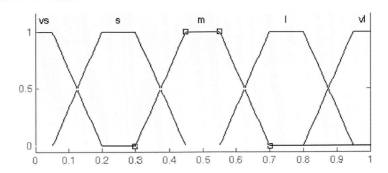

Figure 8. Membership function for power consumption and surface roughness.

Table 4. The data pre-processing of each performance characteristic (Power Consumption and Surface Roughness) and their Grey relational coefficient

Exp. No	Data pre-processing value of Power Consumption	Data pre-processing value of Surface Roughness	Grey relational coefficient of Power Consumption	Grey relational coefficient of Surface Roughness
1	1.0000	0.0000	1.0000	0.3333
2	0.8379	0.2500	0.7551	0.4000
3	0.8197	0.0167	0.7350	0.3371
4	0.8939	0.6583	0.8250	0.5941
5	0.8076	0.7833	0.7221	0.6977
6	0.7273	0.8500	0.6471	0.7692
7	0.9061	0.4500	0.8418	0.4762
8	0.7924	0.7667	0.7066	0.6818
9	0.3833	0.5750	0.4478	0.5405
10	0.7288	0.1833	0.6483	0.3797
11	0.6591	0.4333	0.5946	0.4688
12	0.6545	0.2167	0.5914	0.3896
13	0.7303	0.8333	0.6496	0.7500
14	0.4591	1.0000	0.4803	1.0000
15	0.3106	0.8500	0.4204	0.7692
16	0.7258	0.7333	0.6458	0.6522
17	0.3879	0.6500	0.4496	0.5882
18	0.1712	0.4667	0.3763	0.4839
19	0.6439	0.0167	0.5841	0.3371
20	0.5409	0.4167	0.5213	0.4615
21	0.4364	0.0833	0.4701	0.3529
22	0.6015	0.8000	0.5565	0.7143
23	0.3758	0.7333	0.4447	0.6522
24	0.1636	0.5167	0.3741	0.5085
25	0.5227	0.4833	0.5116	0.4918
26	0.2530	0.5500	0.4010	0.5263
27	0.0000	0.3333	0.3333	0.4286

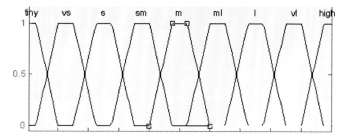

Figure 9. Membership function for grey-fuzzy reasoning grade.

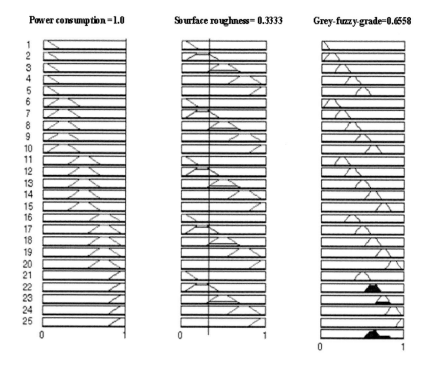

Figure 10. Fuzzy logic rules viewer for the model Exp 1.

Table 5. Fuzzy rule for grey-fuzzy reasoning grade in a matrix form

CNC Turning Process: grey-fuzzy reasoning grade rules		Membership functions for grey relational coefficient of surface roughness				
		VS	S	M	L	VL
Membership functions for grey relational coefficient of power consumption	VS	T	VS	S	SM	M
	S	VS	S	SM	M	ML
	M	S	SM	M	ML	L
	L	SM	M	ML	L	VL
	VL	M	ML	L	VL	H

The trapezoidal membership function is used in this study, which has a flat top and really is just a truncated triangle curve. The objectives for the turning process have been designed as membership function of the developed fuzzy model shown in Figure 8. There are five fuzzy sets for variables of grey relational coefficient of power consumption and surface roughness:

very small (VS), small (S), medium (M), large (L), and very large (VL). The same way output variable grey-fuzzy reasoning grade there are nine fuzzy sets shown in Figure 9: tiny (T), very small (VS), small (S), small medium (SM), medium (M), medium large (ML), large (L), very large (VL), and huge (H).The fuzzy rules in a matrix form used for the fuzzy logic are shown in Table 5.

The graphic representation of fuzzy logic reasoning procedure for Exp. No.1 result of L $_{27}$ orthogonal array is shown in Figure 10, in which rows represents the 25 rules, and columns are the two inputs and one output variable. The locations of trapezoidal indicates the determined fuzzy sets for each input and output value. The height of the darkened area in each trapezoidal corresponds to the fuzzy membership value for that fuzzy set. For Exp. No. 1, the input value of grey relational coefficient of power consumption and surface roughness are 1.0 and 0.3333 respectively. The defuzzified output for the Exp. No. 1 gives the grey-fuzzy reasoning grade value as 0.6558 from the combined darkened areas shown in the last column of Grey-fuzzy-grade in Figure 10. The entire results of the calculated grey-fuzzy reasoning grade for the experiments are shown in Table 6. Finally, the grades are considered for optimizing the multiple performance characteristics parameters selection.

Table 6. Grey-fuzzy reasoning grades and their ranks

Exp. No.	Grey-fuzzy reasoning Grade	Rank
1	0.6558	8
2	0.5811	12
3	0.5342	16
4	0.6958	6
5	0.7475	2
6	0.7035	5
7	0.6627	7
8	0.7319	3
9	0.4976	20
10	0.5160	17
11	0.5402	14
12	0.4878	21
13	0.7053	4
14	0.7500	1
15	0.5974	11
16	0.6529	9
17	0.5346	15
18	0.4383	24
19	0.4507	23
20	0.5000	18
21	0.4211	26
22	0.6315	10
23	0.5751	13
24	0.4367	25
25	0.5000	18
26	0.4564	22
27	0.3843	27

Based on the value of grey-fuzzy reasoning grade in Table 6, the main effects are tabulated in Table 7 and the factors effects are plotted in Figure 11. Considering the parameter level with higher value of mean grey-fuzzy reasoning grade for each parameter in Table 7 and Figure 11, the optimal parameter conditions $A_1B_2C_1D_1$ is obtained.

Table 7. Response table for the grey-fuzzy reasoning grade

Factors	Level-1	Level-2	Level-3
A	0.6456	0.5803	0.4840
B	0.5208	0.6492	0.5399
C	0.6079	0.6019	0.5001
D	0.6109	0.5465	0.5524

4. ANALYSIS OF EXPERIMENTAL RESULTS AND CONFIRMATION TESTS

The steep slope of response graph indicates the more influence of turning parameter to the performance characteristics in the Figure 11. Results show that the turning parameter (A) the cutting speed and (B) the feed rate have greater value of steep slope and will have great influence for affecting the multiple performance characteristics. From the grey-fuzzy reasoning grade value, ANOVA is formulated for identifying the significant factors. It is calculated by the sum of the squared deviations from the total mean of the grey-fuzzy reasoning grade, into contributions by each of the process parameter and the error. The total sum of the squared deviations is decomposed into two sources: the sum of the squared deviations due to each process parameter and the sum of the squared error. The percentage contribution by each of the process parameter in the total sum of the squared deviations can be used to evaluate the importance of the process parameter change on the process response. The results of ANOVA test are given in Table 8. Based on the ANOVA results, it is clear that cutting speed (37.57 %) influences more on CNC turning of AISI 304 austenitic stainless steel followed by feed rate (27.30%), depth of cut (20.85%) and nose radius (7.19%).

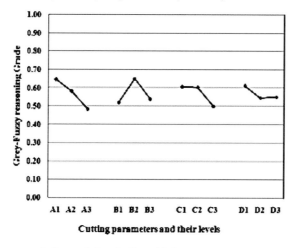

Figure 11. The response graph for each level of machining parameters.

Table 8. Results of the analysis of variance for the grey-fuzzy reasoning grade

Factor	DOF	SS	MS	F value	% Contribution
A	2	0.1189	0.0595	47.6429	37.57
B	2	0.0864	0.0432	34.6269	27.30
C	2	0.0660	0.0330	26.4409	20.84
D	2	0.0228	0.0114	9.1155	7.19
Error	18	0.0225	0.0012	-	7.10
Total	26	0.3166	-	-	100.00

4.1 Determination of Predicted Optimum Value

In order to predict the optimum condition, the expected mean at the optimal settings (μ) is calculated by using the following model.

$$\mu = \overline{A}_1 + \overline{B}_2 + \overline{C}_1 + \overline{D}_1 - 3 \times \overline{T}_{gg} \tag{8}$$

where, $\overline{A}_1$, $\overline{B}_2$, $\overline{C}_1$ and $\overline{D}_1$, are the mean values of the grey-fuzzy reasoning grade with the parameters at optimum levels and $\overline{T}_{gg}$ is the overall mean of average grey-fuzzy reasoning grade. The expected mean (μ) at optimal setting is found to be 0.8038.

Confidence interval (CI) is calculated as

$$CI = \sqrt{F_\alpha(1, f_e) V_e \left[\frac{1}{n_{eff}} + \frac{1}{R} \right]} \tag{9}$$

$$= \pm 0.0840$$

Where, $F_\alpha(1, f_e)$ is the F ratio at a significance level of α %, α is the risk, f_e is the error degrees of freedom, V_e is the error mean square, n_{eff} is the effective total number of tests and R is the number of confirmation tests

$$n_{eff} = \frac{\text{Total number of observations}}{1 + \text{Total degrees of freedom associated with items used in estimating } \mu} \tag{10}$$

Therefore 95% confidence interval of the predicted optimum condition is given by following model, where μ = the grey-fuzzy reasoning grade value after conducting the confirmation experiments with optimal setting point, i.e., $A_1 B_2 C_1 D_1$.

$$(0.8038 - 0.0840) < \mu < (0.8038 + 0.0840)$$

$$(0.7198) < \mu < (0.8878)$$

4.2. Confirmation Tests

The final step is to predict and verify the improvement of the total performance characteristics using the optimal level of process parameters setting $A_1B_2C_1D_1$. The result of the confirmation experiment is expressed by the estimated grey-fuzzy reasoning grade. Table 9 shows the multiple performance characteristics for initial and optimal machining parameters. The initial designated levels of machining parameters are A_1, B_1, C_2 and D_2 which is the experimental No. 2 in the Table 3. As noted from Table 9, the surface roughness Ra is decreased from 1.74 μm to 1.14 μm and the power consumption is decreased from 320 watts to 245 watts respectively. The estimated grey-fuzzy reasoning grade is increased from 0.5811 to 0.7992, which is the largest value obtained in all the experimental results in Table 6. It is clearly shown that the multiple performance characteristics in the turning process are collectively improved by using this method.

Table 9. The comparison results of initial and optimal turning performance

Initial turning parameters		Optimal turning parameters	
		Prediction	Experiment
Levels	$A_1B_1C_2D_2$	$A_1B_2C_1D_1$	$A_1B_2C_1D_1$
Power consumption (Watts),	320		245
Surface roughness (Ra in μm)	1.74		1.14
Grey-Fuzzy reasoning grade	0.581	0.8038	0.7992
Improvement of grey-fuzzy reasoning grade	1	0.2227	0.2181

CONCLUSION

The effect of machining parameters and the optimum machining parameters for CNC turning of AISI 304 austenitic stainless steel on multiple performance characteristics are systematically investigated by grey based fuzzy logic with orthogonal arrays is presented. The following conclusions are obtained

- Grey relational coefficient analyzes the relational degree of the multiple responses (surface roughness and power consumption). Fuzzy logic is used to perform a fuzzy reasoning of the multiple performance characteristics. Consequently, this technique can greatly simplify the optimization procedure of the complicated machining responses.
- From the ANOVA results of multiple performance characteristics together (grey-fuzzy reasoning grade), it is identified that cutting speed and feed rate are predominant factors which affect the turning of AISI 304 austenitic stainless steel. Cutting speed influences more, followed by feed rate, depth of cut and nose radius.

- By this method the optimal level of cutting parameters were acquired and verified by confirmation test. The results proved that the multiple performance characteristics were improved simultaneously.

ACKNOWLEDGMENT

The authors express their sincere thanks to the National Institute of Technology Tiruchirappalli, India for the sponsorship under its research grant scheme. The authors are grateful to the reviewers for their valuable comments and suggestions in reviewing and improving this paper.

REFERENCES

[1] Boothroyd, G. and Knight, W. A. (1989). *Fundamentals of Machining and Machine Tools*.2nd ed. Marcel-Dekker Inc, New York.

[2] Benardos, P. G. and Vosniakos, G. C. (2003) *Predicting surface roughness in machining: a review Int J Machine Tool Manuf* 43:833-844.

[3] Chang, C. K. and Lu, H. S. (2007) The optimal cutting-parameter selection of heavy cutting process in side milling for SUS304 stainless steel. *Int J Adv Manuf Technol* 34:440–447.

[4] Chua, M. S., Rahman, M., Wong, Y. S. and Loh, H. T. (1993) Determination of optimal cutting conditions using design of experiments and optimization techniques. *Int J Machine Tool Manuf* 33:297-305.

[5] Chiang, K. T., Chang, F. P. and Tsai, T. C. (2006) Optimum design parameters of Pin-Fin heat sink using the grey-fuzzy logic based on the orthogonal arrays. *Inter Comm Heat Mass Trans* 33:744-752.

[6] Chiang, K. T., Liu, N. M. and Chou, C. C. (2008) Machining parameters optimization on the die casting process of magnesium alloy using the grey-based fuzzy algorithm. *Inter J Adv Manuf Technol* 38:229-237

[7] Deng, J. L. (1989) Introduction to grey system theory. *J Grey Sys* 1:1-24.

[8] Davim, J. P. (2003) Design of optimization of cutting parameters for turning metal matrix composites based on the orthogonal arrays. *J Mater Process Technol* 132:340-344.

[9] Davim, J. P., Gaitonde, V. N. and Karnik, S. R. (2008) Investigation into the effect of cutting conditions on surface roughness in turning of free machining steel by ANN models. *J Mater Process Technol* 205:16–23.

[10] Groover, M. P. (1996) *Fundamentals of Modern Manufacturing—Materials Processing and Systems*. Prentice-Hall, Englewood Cliffs, NJ., pp. 85–96.

[11] Haq, A. N., Marimuthu, P. and Jeyapaul, R. (2008) Multiresponse optimization of machining parameters of drilling Al/SiC metal matrix composite using grey relational analysis in the Taguchi method. *Inter J Adv Manuf Technol* 37:250-255.

[12] Ho, C. Y. and Lin, Z. C. (2003) Analysis and application of grey relational and ANOVA in chemical-mechanical polishing process parameters. *Inter J Adv Manuf Technol* 21:10-14.

[13] Ihsan, K., Mustafa, K., Ibrahim, C. and Ulvi, S. (2004) Determination of optimum cutting parameters during machining of AISI 304 austenitic stainless steel. *Mater Des* 25:303– 305.

[14] Kopac, J. and Sali, S. (2001) Tool wear monitoring during the turning process. *J Mater Process Technol* 113:312–316.

[15] Lin, C. L. and Lin, J. L. (2005) The use of grey-fuzzy logic for the optimization of the manufacturing process. *J Mater Process Technol* 160:9-14.

[16] Lin, C. L. and Lin, J. L. and Chiang, K. T. (2002) Optimization of the EDM Process based on the orthogonal array with fuzzy logic and Grey relational analysis method. *Inter J Adv Manuf Technol* 19:271-277.

[17] Lin, C. L. and Lin, J. L. (2002) The use of orthogonal array with Grey relational analysis to optimize the EDM Process with multiple performance characteristics. *J Mater Process Technol* 42:237-244.

[18] Lin, J. L., Wang, K. S., Yan, B. H. and Tarng, Y. S. (2000) Optimization of the electrical discharge machining process based on the Taguchi method with fuzzy logics. *J Mater Process Technol* 102: 48-55.

[19] Lu, H. S., Chen, J. Y. and Chung, Ch T. (2008) The optimal cutting parameter design of rough cutting process in side milling. *J of Achievements Mater and Manuf Engg* 29: 183-186.

[20] Montgomery, D. C. (1997) *Design and analysis of experiments.* 4th ed. New York: Wiley.

[21] Mukherjee, I. and Ray, P. K. (2006) A review of optimization techniques in metal cutting processes. *Comp & Indust Engg* 50:15-34.

[22] Nalbant, M., Gokkaya, H. and Sur, G. (2007) Application of Taguchi method in the optimization of cutting parameters for surface roughness in turning. *Materials and Design* 28: 1379- 1385.

[23] Ross, T. J. (1997) *Fuzzy logic with engineering applications.* McGraw-Hill Inc., Singapore.

[24] Shaw, M. C. (1984) *Metal cutting principles.* Oxford University Press, New York.

[25] Tosun, N. (2006) Determination of optimum parameters for multi-performance characteristics in drilling by using grey relational analysis. *Inter J Adv Manuf Technol* 28:450-455.

[26] Xavior, M. A.and Adithan, M. (2009) Determining the influence of cutting fluids on tool wear and surface roughness during turning of AISI 304 austenitic stainless steel. *J Mater Process Technol* 209:900–909.

[27] Zadeh, L. A. (1965) *Fuzzy sets, Information Control* 8:338-353.

[28] Zadeh, L. A. (1995) *The MathWorks User guide*, Version 2, Fuzzy Logic Toolbox.

In: Machining and Forming Technologies, Volume 3
Editor: J. Paulo Davim

ISBN: 978-1-61324-787-7
© 2012 Nova Science Publishers, Inc.

Chapter 17

OPTIMIZATION OF ROLLER BURNISHING PROCESS ON TOOL STEEL MATERIAL USING RESPONSE SURFACE METHODOLOGY

M. R. Stalin John[*] and B. K. Vinayagam
SRM University, Kattankulathur, Kanchipuram District,
Tamil Nadu, India

ABSTRACT

Burnishing, a plastic deformation process, is becoming more popular as a finishing process. Selection of the burnishing parameters to reduce the surface roughness and to increase the surface hardness is especially crucial because of non linear characteristics between burnishing process parameters and surface characteristics. The paper analyses a roller burnishing process on CNC lathe to super finish the turning process. The tool and work piece materials are tungsten carbide (69 HRC) and HCHCr Tool steel (35 HRC) respectively. The input parameters are burnishing force, feed, speed and number of passes. The mathematical model is developed for the surface characteristics (such as surface roughness and surface hardness) using surface response methodology and validated using Pearson product moment correlation coefficient. The output parameters are optimized using response surface methodology (RSM). The optimum surface roughness and its surface hardness are 0.17 µm and 35.99 HRC respectively. The surface roughness has reduced by 88.8% and hardness has improved by 33.14%.

Keywords: Roller burnishing; response surface methodology; optimization technique; CNC lathe

---

[*] Email: m_r_stalin@yahoo.com

1. INTRODUCTION

Conventional machining process such as turning has inherent irregularities and defects like tool marks and scratches that cause energy dissipation and surface damage. Conventional finishing processes such as grinding, honing and lapping are used to overcome these defects. But these methods essentially depend on chip removal to attain the desired surface finish and also skill of the workers. To resolve these problems, burnishing process is applied for better surface finish on the post machined components due to its chip-less and relatively simple operations.

Burnishing is a cold working, surface treatment process in which plastic deformation of surface irregularities occurs by exerting pressure through a very hard and smooth roller on a surface to generate a uniform and work hardened surface. When the applied burnishing pressure exceeds the yield strength of the material, the asperities of the surface will deform plastically and spreads out permanently to fill the valleys so that the desired surface finish is achieved. (Shankar et al., 2008).

Klocke and Liermann (1998) investigated a hard roller burnishing operation using a hydrostatically borne ceramic ball rolls over the component surface (100 Cr6 steel (62 HRC)) under high pressures. The surface finish was improved to 30% to 50%. The white layer which exists after hard turning is still present as an even layer after hard roller burnishing.

Hassan et al. (1998) optimized the ball burnishing process using response surface methodology. The authors analyzed and optimized the ball burnishing process on brass materials for the effect of burnishing force and number of pass.

El-Axir (2000) studied the effect of roller burnishing process on the center lathe. Mathematical models were presented using response surface methodology for predicting the surface microhardness and roughness of Steel-37 caused by roller burnishing under lubricated conditions. The author discussed various problems of burnishing process such as flaking, structural in homogeneity and Compressive residual stress.

Ne´mat and Lyons (2000) investigated the burnishing process on A113, grade A of mild steel and AA6463 E of Aluminium alloy using the lathe. Depending on the burnishing conditions, burnishing process decreased surface roughness up to 70%.

El-Khabeery and El-Axir (2001) investigated the effect of roller-burnishing upon surface roughness, surface micro-hardness and residual stress of 6061-T6 aluminum alloy. A Group Method of Data Handling Technique was used to develop mathematical models correlating three process parameters: burnishing speed, burnishing depth of penetration and number of passes, were established.

Yung-Chang Yen (2004) developed finite element analysis (FEA) simulation models (2-D and 3-D) for analyzing residual stress of the ball burnishing process with model parameter as burnishing pressure, ball diameter, speed, and feed rate. The validation of the simulation was done with limited experiments.

Hamadache et al. (2006) investigated the effect of ball and roller burnishing process on Rb40 steel in lathe. The authors analyzed the characteristic of ball and roller burnishing process for the change in the burnishing force, feed, speed and number of pass.

El-Tayeb et al. (2007) designed burnishing tool, with interchangeable adapter for roller and fabricated to perform roller burnishing processes on Aluminum 6061 under different parameters and different burnishing orientations. The authors investigated the impact of burnishing speed, burnishing force and burnishing tool dimensions on the surface qualities and tribological properties.

Paiva et al. (2007) has proposed hybrid approach, combining response surface methodology and principal component analysis to optimize multiple correlated responses in a turning process.

El-Taweel and Ebeid (2008) investigated a novel finishing process, which integrated the merits of electrochemical smoothing and roller burnishing for minimizing the roundness error and increasing surface micro-hardness of cylindrical parts. Surface microhardness considerably increased about 31.5% compared to the initial surface micro-hardness, and about 2.32 μm roundness error was achieved using the optimum combination of process parameters.

Shankar et al. (2008) has studied the roller burnishing process on Al-(SiC)p metal matrix composites in a lathe machine. The material used for the roller was tungsten carbide. The author studied the effects of different lubricants such as mineral oil, soluble oil, kerosene and kerosene with graphite. The author has also achieved a surface finish improvement of 95% .

Yeldose and Ramamoorthy (2008) presented an investigation for the comparison of the effect of the uncoated and TiN coating by reactive magnetron sputtering on EN31 rollers in burnishing with varying process parameters such as burnishing force, feed, speed, number of passes upon surface roughness of EN24 steel work material.

Hudayim and Haldun (2009) optimized the burnishing parameter using fuzzy logic on Aluminum alloy (Al 7075 T6). The number of revolution, feed, number of passes, and pressure force as model parameter in fuzzy logic to predict surface characteristics.

Many researchers in the area of burnishing process stated the optimizing parameters from the characteristic of the process and they have not used any statistical methods to arrive the results (Hassan et al. (1998), El-Axir (2000), El-Khabeery and El-Axir (2001), El-Taweel and Ebeid (2008)). Very few analytical models are available. Hence, this paper deals with RSM optimisation technique for a roller burnishing process parameters on tool steel under different burnishing parameters such as burnishing feed, force, speed and number of pass against surface roughness and surface hardness.

2. EXPERIMENTAL WORKS

The work piece material is High Carbon High Chromium Tool Steel (HCHCr). The chemical compositions are as follows: 2.03 % C, 0.33 Mn, 12.01% Cr, 0.22 % Si, 0.023% S and 0.02 % P. The surface hardness is 35 HRC. The Jyoti Turning Centre DX200 CNC machine is used for machining. The positioning and accuracy of the machine used are 0.015 mm and ± 0.003 mm respectively. The work piece is initially turned to 32 mm diameter of 150mm length. The following machining parameters are used for rough turning: speed: 1200 rpm, feed 0.22 mm/rev, depth of cut: 1 mm, cutting speed: 120 mm/min and tool insert: TNMG 160408. The following machining parameters are used for finish turning: speed: 1200 rpm, feed: 0.1 mm/rev and depth of cut: 1mm. Figure 1 shows the schematic drawing of the work piece. The turning is done up to a length of 83.80mm. The remaining work piece area is used for holding on the chuck. Once turning is done with the same tool three grooves are done at 20 mm apart on the turned surface. This is done to identify the tool pass on all the four different sections of the work piece during burnishing. The surface roughness and surface hardness are measured from TR-200 hand held surface roughness indicator and Rockwell hardness testing machine respectively. The range of the surface roughness tester used is 0.01 - 40 μm for Ra value. ISO 4288 and IS 1516 standards are used to measure roughness and Rockwell hardness respectively. The surface roughness and surface hardness after turning operation are 1.52 μm and 24 HRC respectively.

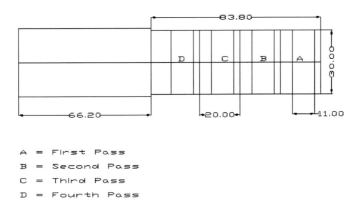

Figure 1. The schematic drawing of the work piece.

Figure 2 (a). Photographic view of the roller burnishing tool.

Figure 2 (b). Photographic view of the roller burnishing process on CNC lathe.

Figure 2 (c). Photographic view of TR 200 hand held roughness tester.

The dimensions of the tungsten carbide roller are a 42 mm diameter roller with a 4 mm land of 16 mm overall width of roller. A spring enclosed within the shank for providing the burnishing force or depth of cut. This spring is provided with a dowel pin indicator on the shank for indicating load movements. Two springs are designed based on load calculations, one spring to withstand load up to 500N and the other spring to withstand up to 1200N. The photographic view of the roller burnishing tool, roller burnishing process and TR 200 roughness tester is shown in Figures 2(a), 2(b) and 2(c) respectively. The soluble oil is used as lubricant. From the literature survey, the main affecting burnishing parameters are identified as speed, feed, force and number of pass on surface characteristics. Hence, the input parameters are speed (200rpm, 400rpm and 600rpm), feed (0.1mm/rev, 0.2mm/rev and 0.3mm/rev), burnishing force (300N, 600N and 900N) and number of pass (1, 2, 3 and 4). The output parameters are surface roughness and surface hardness.

3. EXPERIMENTAL DESIGN AND ANALYSIS

The main objective of this work is to optimize the stated parameters of the roller burnishing process on surface characteristics. The experimental design of central component factorial design is used in the response surface methodology (RSM). The values of the levels of each burnishing parameter used in this work are coded and it is shown in Table 1. The range of each parameter is coded in three levels (-1, 0, 1) using the transformation equations. The transformation equations are given in Eqs. 1-4.

$$\text{Burnishing Force (N)} = (\text{Burnishing Force } (X1) - 600)/300 \tag{1}$$

$$\text{Burnishing Feed (mm/rev)} = (\text{Burnishing Feed } (X2) - 0.2) / 0.1 \tag{2}$$

$$\text{Burnishing Speed (rpm)} = (\text{Burnishing Speed } (X3) - 400) / 200 \tag{3}$$

$$\text{Number of Pass} = (\text{No. of Pass } (X4) - 3) / 1 \tag{4}$$

Table 1. Coded levels of the burnishing process parameters

Coded Levels	-1	0	1
Burnishing Force, N (X_1)	300	600	900
Burnishing Feed, mm/rev (X_2)	0.1	0.2	0.3
Burnishing Speed, rpm (X_3)	200	400	600
Number of Pass (X_4)	2	3	4

4. MATHEMATICAL MODELS

The surface roughness and surface hardness are modeled for the input parameters of burnishing force (X_1), burnishing feed (X_2), burnishing speed (X_3) and number of pass (X_4)

using response surface methodology. The general regression equation for surface roughness and surface hardness is given in Eq. 5.

$$Y = \beta_0 + \beta_1 X_1 + \beta_2 X_2 + \beta_3 X_3 + \beta_4 X_4 + \beta_{11} X_1^2 + \beta_{12} X_2^2 + \beta_{13} X_3^2 + \beta_{14} X_4^2 + \beta_{21} X_1 X_2 + \beta_{22} X_1 X_3 + \beta_{23} X_1 X_4 + \beta_{24} X_2 X_3 + \beta_{25} X_2 X_4 + \beta_{26} X_3 X_4 \tag{5}$$

Where,

β_0	Constant of RSM
$\beta_1, \beta_2, \beta_3, \beta_4$	Coefficient of linear variables X_1, X_2, X_3 and X_4 respectively
$\beta_{11}, \beta_{12}, \beta_{13}, \beta_{14}$	Coefficient of squares of linear variables X_1, X_2, X_3 and X_4 respectively
$\beta_{21}, \beta_{22}, \beta_{23}, \beta_{24}, \beta_{25}, \beta_{26}$	Coefficient of interaction of linear variables X_1, X_2, X_3 and X_4 respectively

Table 2. Experimental design of the burnishing process

s.no	Force (N)	Feed (mm/rev)	Speed (rpm)	No. of passes	Roughness (micron)	Hardness (HRC)
1	1	1	-1	1	0.289	36.0
2	1	-1	1	-1	0.2453	49.0
3	1	-1	-1	1	0.2206	60.0
4	0	0	0	1	0.273	42.0
5	-1	1	1	1	0.7946	37.0
6	1	1	1	1	0.2666	45.0
7	-1	1	-1	-1	0.5726	32.0
8	-1	-1	-1	-1	0.8923	32.0
9	1	-1	1	1	0.1533	54.0
10	0	0	0	0	0.3376	40.0
11	1	1	-1	-1	0.565	33.0
12	-1	1	-1	1	0.4886	35.0
13	-1	-1	1	1	0.8396	38.0
14	1	1	1	-1	0.4033	39.0
15	0	0	-1	0	0.368	37.5
16	0	0	1	0	0.337	39.0
17	-1	1	1	-1	0.842	34.0
18	0	-1	0	0	0.4246	48.0
19	-1	0	0	0	0.762053	30.5
20	0	1	0	0	0.253	34.0
21	1	0	0	0	0.3633	49.0
22	-1	-1	1	-1	0.9113	32.0
23	0	0	0	-1	0.366	36.0
24	1	-1	-1	-1	0.3226	50.0
25	-1	-1	-1	1	0.783	35.0

Experimental design of the burnishing process is given in Table 2. From the experimental values, the coefficient for the regression equation and analysis of variance (ANOVA) table of surface roughness and surface hardness are calculated using MiniTAB software and it is

Optimization of Roller Burnishing Process on Tool Steel Material Using ...   253

given in Table 3 and 4 respectively. The significant coefficients are selected based on the value of P. The parameters are modeled using confidence level of 95%. Hence, less than 0.05 value of P are considered as significant. The regression equation of surface roughness and surface hardness are given in Eq. 6 and Eq. 7 respectively. In Table 4, it shows all the values of P for linear, square and interaction variables of regression equation for both surface roughness and surface hardness are less than 0.005. Hence the fit is obtained with less error. The Pearson product moment correlation coefficient is used to validate the predicted response from Eq. 6 and Eq. 7. The values are 0.944 and 0.947 for the surface roughness and surface hardness respectively. The output of response surface methodology and experimental values are plotted in Figures 3 and 4 for surface roughness and surface hardness respectively.

$$\text{Surface Roughness } (\mu m) = 0.34 - 0.225\, X_1 - 0.056\, X_4 + 0.217\, X_1^2 + 0.08\, X_1 X_2 - 0.06\, X_1 X_3 \tag{6}$$

$$\text{Surface Hardness } (HRC) = 39.44 + 6.1\, X_1 - 4.055\, X_2 + 2.5\, X_4 - 3.812\, X_1 X_2 \tag{7}$$

Table 3. Estimated Regression Coefficients for Surface Roughness, μm and Surface hardness, HRC – RSM

Term	Surface Roughness				Surface Hardness				Significant Coefficient of Surface Roughness	Significant Coefficient of Surface Hardness
	Coefficient	Std. Error of the Coefficient	T	P	Coefficient	Std. Error of the Coefficient	T	P		
β_0	0.344	0.02718	12.661	0.000	39.44	1.2342	31.960	0.000	0.344	39.44
β_1	-0.225	0.01483	-15.194	0.000	6.1	0.6737	9.027	0.000	-0.225	6.08
β_2	-0.017	0.01483	-1.191	0.261	-4.055	0.6737	-6.020	0.000		-4.05
β_3	0.0161	0.01483	1.091	0.301	0.916	0.6737	1.361	0.204		
β_4	-0.056	0.01483	-3.790	0.004	2.5	0.6737	3.711	0.004	-0.056	2.5
β_{11}	0.217	0.03944	5.516	0.000	0.412	1.7911	0.230	0.822	0.217	
β_{12}	-0.006	0.03944	-0.161	0.876	1.647	1.7911	0.920	0.379		
β_{13}	0.0073	0.03944	0.187	0.856	-1.102	1.7911	-0.615	0.552		
β_{14}	-0.025	0.03944	-0.650	0.530	-0.352	1.7911	-0.197	0.848		
β_{21}	0.081	0.01573	5.206	0.000	-3.812	0.7146	-5.335	0.000	0.081	-3.812
β_{22}	-0.061	0.01573	-3.892	0.003	0.062	0.7146	0.087	0.932	-0.061	
β_{23}	-0.018	0.01573	-1.169	0.270	0.562	0.7146	0.787	0.449		
β_{24}	0.028	0.01573	1.828	0.097	1.437	0.7146	2.012	0.072		
β_{25}	-0.010	0.01573	-0.672	0.517	-0.562	0.7146	-0.787	0.449		
β_{26}	0.013	0.01573	0.888	0.396	0.062	0.7146	0.087	0.932		
	Standard error of the regression = 0.063 Percentage of response variable variation (R- Sq)= 97.21% Predicted R- Sq = 79.65% Adjusted R- Sq = 93.31%				Standard error of the regression = 2.85 Percentage of response variable variation(R- Sq) = 94.39% Predicted R- Sq = 62.29% Adjusted R- Sq = 86.54%					
The coefficient are significant, when p≤0.05										

Table 4. Analysis of Variance for Surface Roughness, μm and Surface Hardness, HRC –RSM

	Source	Degrees of freedom	Sequential sum of squares	Adjusted sum of squares between factors	Adjusted mean squares	F	P
Surface Roughness	Regression	14	1.381	1.381	0.098	24.92	0.000
	Linear	4	0.981	0.981	0.245	61.96	0.000
	Square	4	0.209	0.209	0.052	13.21	0.001
	Interaction	6	0.190	0.190	0.031	8.03	0.002
	Residual Error	10	0.039	0.0396	0.003		
	Total	24	1.421				
Surface Hardness	Regression	14	1374.89	1374.89	98.2	12.02	0.000
	Linear	4	1089.45	1089.45	272.36	33.34	0.000
	Square	4	9.57	9.57	2.39	0.29	0.876
	Interaction	6	275.87	275.87	45.97	5.63	0.009
	Residual Error	10	81.70	81.70	8.17		
	Total	24	1456.59				

Table 5. Conformal Test of RSM for Surface Roughness, μm and Surface Hardness, HRC

S. No.	Force (N)	Feed (mm /rev)	Speed (rpm)	No. of passes	Roughness (micron)	Hardness (HRC)	Calculated Roughness (micron) using RSM	Calculated Hardness (HRC) using RSM	% of Error in Roughness	% of Error in Hardness
1	-1	-1	-1	0	0.846	33.0	0.802	33.58	5.23	-1.77
2	-1	-1	0	-1	0.92	32.5	0.918	31.08	0.22	4.36
3	-1	-1	0	1	0.861	37.0	0.806	36.08	6.45	2.48
4	-1	-1	1	0	0.882	34.5	0.922	33.58	-4.54	2.66
5	-1	0	-1	-1	0.722	32.0	0.778	30.84	-7.69	3.54
6	-1	0	0	-1	0.809	29.0	0.838	30.84	-3.46	-6.35
7	-1	0	1	-1	0.959	31.5	0.898	30.84	6.39	2.10
8	-1	0	1	0	0.915	35.0	0.842	33.34	7.98	4.74
9	-1	1	1	0	0.798	35.0	0.762	33.10	4.58	5.44
10	0	0	1	1	0.283	41.0	0.284	41.94	-0.35	-2.29
11	1	-1	-1	0	0.299	58.0	0.312	53.41	-4.35	7.92
12	1	-1	1	0	0.191	50.0	0.192	53.41	-0.52	-6.81
13	1	0	0	-1	0.403	44.0	0.388	43.04	3.79	2.18

The conformal test is carried out with new experiments and the results are given in Table 5. The test shows that the error in prediction of surface roughness is vary from 0.22% to 7.98% and surface hardness is 1.77% to 7.92% and product moment correlation coefficient is

0.989 and 0.972 respectively. The percentage of error and correlation coefficient shows that the surface response of the model is accurately predicted.

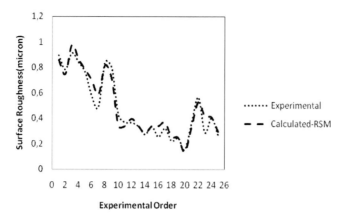

Figure 3. Comparison of RSM and Experimental values of surface roughness.

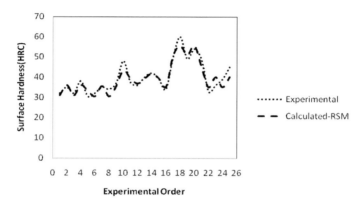

Figure 4. Comparison of RSM and Experimental values of surface hardness.

5. RESPONSE OPTIMIZATION

The parameters are optimized with an objective function where the surface roughness is to be minimized and surface hardness is to be maximized. The constraints for the objective functions are given Eqs. 8 – 10.

$$-1 <= \text{surface roughness (Ra)} => 1 \tag{8}$$

$$-1 <= \text{surface hardness (HRC)} => 1 \tag{9}$$

and

$$Ra, HRC => 0 \tag{10}$$

In MiniTAB software, response surface optimization module is used to optimize the coded parameters. The optimized coded parameters are found and the values are: burnishing force=0.2525, feed = 1, speed = -0.9633 and number of pass = 1. The corresponding optimized values of the burnishing process are: burnishing force = 675.75 N, burnishing feed = 0.3 mm/rev, burnishing speed = 207 rpm and fourth pass. Figure 5 shows the optimized plot of the burnishing parameters. The optimization of the surface roughness and surface hardness is found out with desirability factor of 0.983 and 0.999 respectively and the values are 0.1656 µm and 35.99 HRC respectively. By adjusting vertical chain line (red color line for online version (or) dash line in balck & white version), variation in the output is evaluated. The experiment is conducted for the optimized values of RSM and the values are: Surface roughness = 0.173 µm and Surface hardness = 38 HRC.

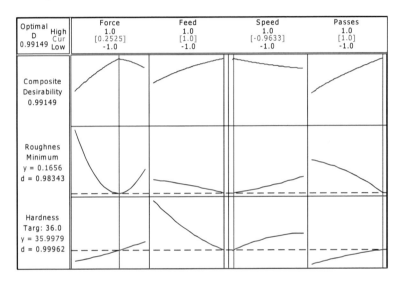

(a)

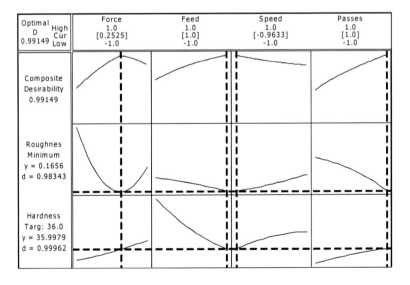

(b)

Figure 5a. Optimized plots of the burnishing parameters. (for online version). 5b. Optimized plots of the burnishing parameters. (for printing in Black and White).

Optimization of Roller Burnishing Process on Tool Steel Material Using ... 257

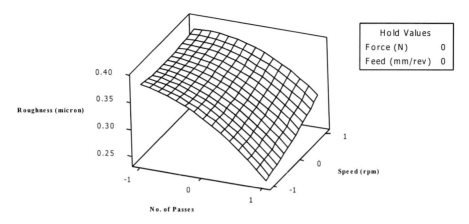

Figure 6. Surface plot of Surface Roughness Vs Coded Speed, No. of Pass for the mathematical model using RSM.

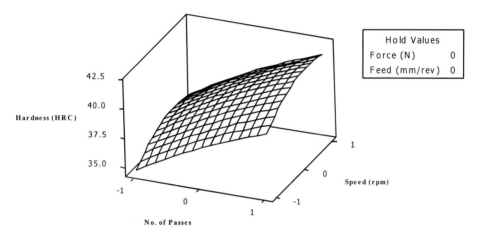

Figure 7. Surface plot of Surface Hardness Vs Coded Speed, No. of Pass for the mathematical model using RSM.

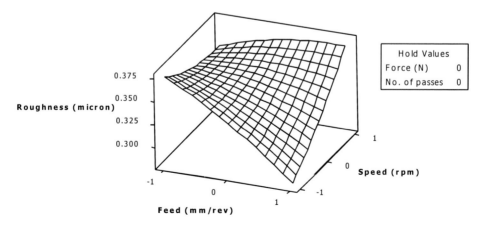

Figure 8. Surface plot of Surface Roughness Vs Coded Burnishing Feed, Speed for the mathematical model using RSM.

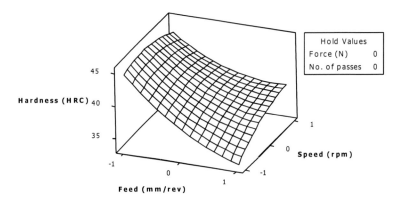

Figure 9. Surface plot of Surface Hardness Vs Coded Burnishing Feed, Speed for the mathematical model using RSM.

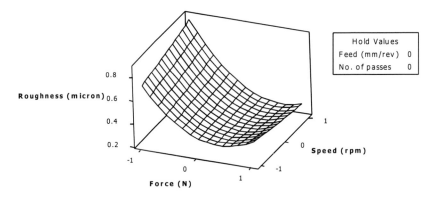

Figure 10. Surface plot of Surface Roughness Vs Coded Burnishing Force, Speed for the mathematical model using RSM.

6. RESULTS AND DISCUSSIONS

The deterioration of the surface finish in the burnishing process at high burnishing speeds (as shown in Figures 6, 8 and 10) and it is believed to be caused by the chatter that results in instability of the burnishing tool across the work piece surface. In addition, the low surface finish can be interpreted by the low deforming action of the rollers at high speeds and also because the lubricant loses its effect due to the insufficient time for it to penetrate between the roller and the work piece surfaces. It is better then to select low speeds (200 rpm) because the deforming action of the burnishing tool is greater and metal flow is regular at low speed. The corresponding surface hardness is more due to compression action at the surface asperities (as shown in Figures 7, 9 and 11).

From Figures 6, and 12, the surface roughness is decreases due to increase in number of passes. This is because of the repetition of the burnishing on the same work piece that leads to an increase in the structural homogeneity which results in a decrease in mean surface roughness and increase in surface hardness (as shown in Figures 7 and 13). However an increase in the number of passes when burnishing force is increasing, it deteriorates the surface quality as a result of excessive compression action on the work piece surface (as shown in Figure 14).

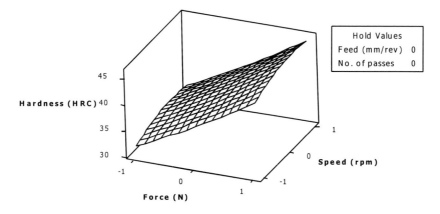

Figure 11. Surface plot of Surface Hardness Vs Coded Burnishing Force, Speed for the mathematical model using RSM.

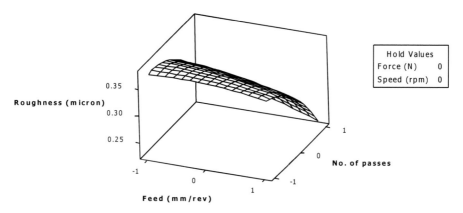

Figure 12. Surface plot of Surface Roughness Vs Coded Burnishing Feed, No. of Pass for the mathematical model using RSM.

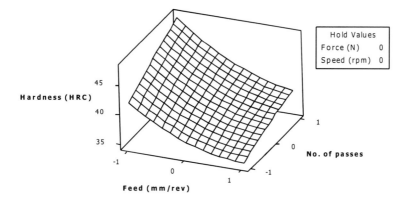

Figure 13. Surface plot of Surface Hardness Vs Coded Burnishing Feed, No. of Pass for the mathematical model using RSM.

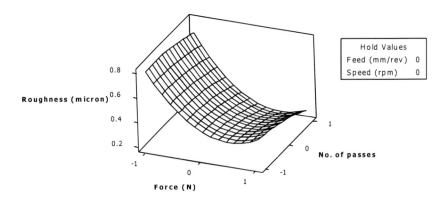

Figure 14. Surface plot of Surface Roughness Vs Coded Burnishing Force, No. of Pass for the mathematical model using RSM.

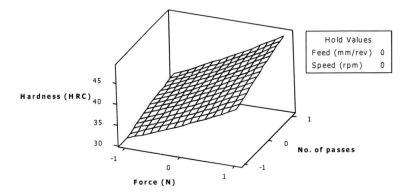

Figure 15. Surface plot of Surface Hardness Vs Coded Burnishing Force, No. of Pass for the mathematical model using RSM.

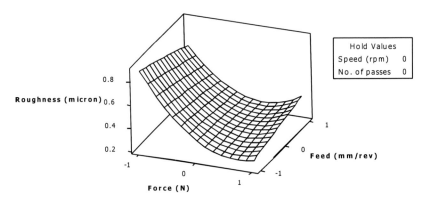

Figure 16. Surface plot of Surface Roughness Vs Coded Burnishing Force, Feed for the mathematical model using RSM.

When the feed rate is high, the distances between the successive burnishing traces will be large and the application of burnishing force will be less, which will cause less improvement in the surface finish of the work piece (as shown in Figure 16). The decrease in hardness with the increase in feed rate, Figure 17, is attributed to the increase of the distances between the successive burnishing traces. This will cause a decrease in the total repeated deformation action on the surface of the work piece, which leads, in turn, to a decrease in hardness.

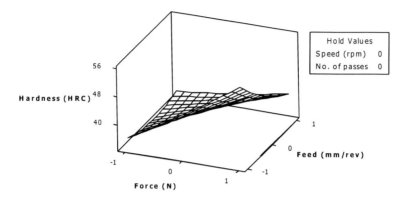

Figure 17. Surface plot of Surface Hardness Vs Coded Burnishing Force, Feed for the mathematical model using RSM.

A combination of low feed with high number of passes leads to a substantial improvement in the surface finish (as shown in Figure 12) compared to combination of low feed and lower number of pass. This may be due to all the asperities of the turned surface may not plastically deform to produce better surface finish. A combination of high feed with lower number of passes deteriorates the surface finish compared to combination of high feed with higher number of passes. This may be due to repeated action of the roller on the surface asperities which deform plastically the asperities. It is better, then, to select high feeds (0.3 mm/ rev) with higher number of passes.

The increase in burnishing force causes an increase in the amount of surface deformation as the tool passes along the surface of the work piece. This will lead to an increase in the work hardening of the surface layers, which have been affected by plastic deformation, so that surface hardness will increase via the increase in burnishing force, as shown in Figures 11, 15 and 16. Hence the corresponding surface finish is improved as shown in Figures 10, 14 and 16.

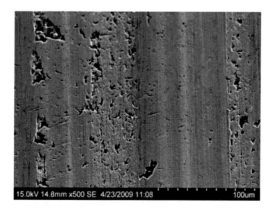

Figure 18. SEM image showing turned surface at 100μm.

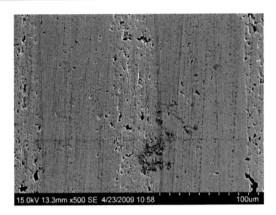

Figure 19. SEM image showing burnished surface at 100μm.

The SEM examinations of the turned surface and burnished surface with resolution of 100 μm are shown in Figures 18 and 19 respectively. The turning marks and surface irregularities are visible in Figure 18. These defects are removed by burnishing process and it is shown in Figure 19.

CONCLUSION

Many factors are affecting the result of the burnishing process and it is quite complicated. Finding the optimum condition and controlling the results is important for the industry. According to this research, the following conclusions may be drawn:

1. A good correlation between the experimental and predicted results derived from the model is exhibited. Thus, using the proposed procedure, the optimum roller burnishing conditions should be obtained to control the surface responses of other materials.
2. It is shown that the spindle speed, burnishing force, burnishing feed and number of passes have the most significant effect on both surface hardness and surface roughness, and there are many interactions between these parameters.
3. The optimized values of the burnishing process is 675.75 N, burnishing feed is 0.3 mm/rev, burnishing speed is 207 rpm and fourth pass using response surface methodology. The optimum surface roughness and surface hardness using response surface model is 0.17 μm and 35.99 HRC respectively.
4. Whenever, the burnishing force increases more than the yield stress of the material, plastic flow of the metal fills the valleys of the surface. Hence the surface roughness is minimized and surface hardness is maximized.
5. At lower feed rates, the more contact time existed between tool and work piece. Hence, the surface roughness is minimized at lower feed compared to higher feed rate.
6. At higher speed, the chattering occurs at the tool. Hence the surface roughness is minimized at higher speed compared to lower speeds.
7. When number of pass is increasing, the surface roughness is gradually decreasing and the surface hardness is increasing. This is due to when roller moves again and

again on the surface, peak of the top surface is deformed and fills the valleys of the surface.

REFERENCES

[1] El-Axir, M. H. (2000) "An investigation into roller burnishing, *International Journal Machine Tools & Manufacture*, Vol. 40, pp. 1603–1617.

[2] El-Khabeery, M. M., and El-Axir, M. H. (2001) "Experimental techniques for studying the effects of milling roller-burnishing parameters on surface integrity", *International Journal of Machine Tools & Manufacture*, 41, 1705–1719.

[3] El-Tayeb, N. S. M., Low, K. O. and Brevern, P. V. (2007) "Influence of roller burnishing contact width and burnishing orientation on surface quality and tribological behaviour of Aluminium 6061", *Journal of Materials Processing Technology*, Vol. 186, pp. 272–278.

[4] El-Taweel, T. A. and Ebeid, S. J. (2008) "Effect of hybrid electrochemical smoothing–roller burnishing process parameters on roundness error and micro-hardness", *International Journal Advance Manufacturing Technology*, DOI 10. 1007/s00170-008-1632-0.

[5] Hamadache, H., Laouar, L., Zeghib, N. E. and Chaoui, K. (2006) "Characteristics of Rb40 steel superficial layer under ball and roller burnishing", *Journal of Materials Processing Technology*, Vol. 180, pp. 130–136.

[6] Hassan, A. M., Al-Jalil, H. F. and Ebied, A. A. (1998) "Burnishing force and number of ball passes for the optimum surface finish of brass components", *Journal of Materials Processing Technology*, Vol. 83, pp. 176–179.

[7] Hudayim Basak and Haldun Goktas, H. (2009) "Burnishing process on al-alloy and optimization of surface roughness and surface hardness by fuzzy logic", *Materials and Design,* Vol. 30, pp. 1275–1281.

[8] Klocke, F. and Liermann, J. (1998) "Roller Burnishing of Hard Turned Surfaces", *International Journal Machine Tools Manufacturing*, Vol. 38, Nos 5-6, pp. 419-423.

[9] Ne´mat, M. and Lyons, A. C. (2000) "An Investigation of the Surface Topography of Ball Burnished Mild Steel and Aluminium", *International Journal Advance Manufacturing Technology*, Vol. 16, pp. 469–473.

[10] Paiva, Anderson P., Jo˜ao Roberto Ferreira., and Pedro P. Balestrassi., (2007) "A multivariate hybrid approach applied to AISI 52100 hardened steel turning optimization", *Journal of Materials Processing Technology*, Vol. 189, pp. 26–35.

[11] Shankar, E., Stalin John, M. R., and Thirumurugan, M., (2008) "Surface characteristics of Al-(SiC)p metal matrix composites by roller burnishing process", *International Journal of Machining and Machinability of Materials*, Vol. 3, No. 3/4, pp. 283–292.

[12] Yeldose, Binu C., and Ramamoorthy, B., (2008) "An investigation into the high performance of TiN-coated rollers in burnishing process", *Journal of Materials Processing Technology*, Vol. 20, No. 7, pp. 350–355.

[13] Yung-Chang Yen. (2004) "Modelling of metal cutting and ball burnishing – prediction of tool wear and surface properties", *Ph D. Thesis*, the Ohio State University.

In: Machining and Forming Technologies, Volume 3 ISBN: 978-1-61324-787-7
Editor: J. Paulo Davim © 2012 Nova Science Publishers, Inc.

Chapter 18

PARAMETRIC OPTIMIZATION OF AFM PROCESS DURING FINISHING OF AL/5WT%SIC-MMC CYLINDRICAL SURFACE

*Harlal Singh Mali[1] and Alakesh Manna[2]**

[1]Mechanical Engineering Department, Chitkara Institute of Engineering & Technology, Rajpura, Punjab; India;
[2]Department of Mechanical Engineering, Punjab Engineering College, (Deemed University); Chandigarh, India

ABSTRACT

The machining of metal matrix composites particularly finishing is a challenge for manufacturing engineers today. In the present study, an abrasive flow machining (AFM) set up has been designed and fabricated to finish the internal cylindrical surfaces of Al/5wt%SiC-MMC components and the results have been encouraging. The influence of AFM process parameters e.g. abrasive mesh size, number of cycles, extrusion pressure, abrasive concentration and media grade on surface finish R_a, R_t, improvement of surface finish ΔR_a, ΔR_t and material removal (MR) have been analyzed. According to the Taguchi quality design concept, a L_{18} (6^1 x 3^7) mixed orthogonal array was used to determine the S/N ratios and based on the determined S/N ratios (dB) optimize the AFM process parameters. The determined optimal parameters are experimentally verified. Analysis of variance (ANOVA) and F-test values indicate the significant AFM parameters affecting the finishing performance. The mathematical models for R_a, R_t, %ΔR_a, %ΔR_t and material removal (MR) are established to investigate the influence of AFM parameters on response characteristics. Conformation test results verify the effectiveness of these models.

Keywords: Abrasive flow machining, Al/5wt%SiC-MMC, Surface finish, Material removal

---

* Email: kgpmanna@rediffmail.com

1. INTRODUCTION

The aluminium alloy reinforced with discontinuous silicon carbide (SiC) particulates (Al/SiC-MMC) is rapidly replacing conventional materials in various industries mainly automotive and aerospace due to considerable weight saving (Allison & Cole, 1993). Despite superior physical and mechanical properties, particulate reinforced metal matrix composites are not widely used in industry because of their poor machinability (Manna and Bhattacharyya, 2005). Wide spread engineering application is resisted due to its poor machining characteristics particularly excessive tool wear and poor surface finish. The hard SiC particles of Al/SiC-MMC, which intermittently come into contact to the tool surface, act as small cutting edges like those of a grinding wheel on the cutting tool edge which in due course is worn out by abrasion, resulting in the formation of poor surface finish during machining (Manna & Bhattacharyya, 2001, 2004). Hence, proper identification of a cost effective surface finishing process to achieve the required quality of surfaces on Al/SiC-MMC jobs is a challenge to the manufacturing engineers. Abrasive flow machining (AFM) is an advanced finishing process that can be used for cleaning and fine finishing of Al/SiC-MMC. This process can also be used for deburring, polishing, radiusing, removing recast layers, producing compressive residual stresses of difficult to reach areas. Application of abrasive flow finishing (AFF) processes has been reported on Al/6063 workpieces and optimization of AFF parameters carried out (Mali and Manna, 2009).

The present work rustically utilized the process and experimentally investigates the internal cylindrical surface characteristic of Al/5wt%SiC-MMC components. The process involves extruding an abrasive-laden semisolid media through a work piece passage. AFM process has three major elements namely the machine, work piece fixture (tooling) and media. The machine used in AFM process hydraulically clamps the work holding fixtures between two vertically opposed media cylinder. These cylinders extrude the media back and forth through the work piece(s). Two strokes are obtained, one from the lower cylinder and the other from the upper cylinder, making up one process cycle. AFM process is an efficient method particularly for the inner surface finishing process. In AFM the media determines the aggressiveness of the action of abrasives, which is resilient enough to act as a self deforming grinding stone when forced through a passageway (Larry Rhoades, 1991). Production of extremely thin chips allows fine surface finish, closer tolerances, and generation of more intricate surface texture. Recently spring collects (J. D. Kim and K.D. Kim, 2004), micro channels (Hsinn-Jyh Tzeng et. al, 2008) and diesel injector nozzles (D. Jung et. al., 2008) have been finished by AFM process and they claimed that the AFM process directly improved the system performances. Materials from soft aluminum to tough nickel alloys, ceramics and carbides can be successfully micro machined by this process (Benedict G.F; 1987).

In view of these above-mentioned finishing problems an abrasive flow machining setup has been fabricated and the process has been explored to finish the internal cylindrical surfaces of Al/5wt%SiC-MMC jobs. The objective of the research is to investigate the effect of AFM parameters such as abrasive mesh size, number of cycles, extrusion pressure, % abrasive concentration and media viscosity grade on surface finish during finishing of Al/5wt%SiC-MMC. According to the Taguchi experimental quality design concept, L_{18} (6^1 x 3^7) mixed orthogonal array is used to determine the S/N ratio, analysis of variance (ANOVA) and F-test values to indicate the significant AFM parameters affecting the finishing performance. Selecting the significant parameters mathematical models have been developed and validated.

2. PLANNING FOR EXPERIMENTS

Figure 1 shows the actual photograph of fabricated AFM setup consisting of a hydraulic power pack and two each hydraulic cylinders and media cylinder. The setup has an extrusion pressure gauge, pressure variation system, cycle counter and limit switches. The work piece along with its tooling is clamped between the two media cylinders. Figure 2 (a) shows the work-pieces used to improve the surface finish. Figure 2 (b) shows the detail dimensions of the work samples used for experiments. Sketch of the tooling used to hold the work samples in AFM is shown in Figure 3.

Figure 1. AFM setup: (1) cylinders for media two nos. (2) Hydraulic cylinder two nos.; (3) Hydraulic Power Pack. (4) Pressure Gauge. (5) Limit switches four nos.

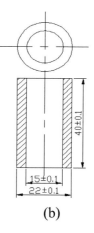

(a)   (b)

Figure 2. Work-piece used for experiment.

Discontinuous Al/5wt%SiC-MMC of 25.4 mm diameter and 50 mm long pieces are casted by liquid stirring method. Stir-cast pieces were turned, faced, centered, drilled, bored and part-off to work-piece samples size is shown in Figure 2(b). Sixty work samples were prepared with inner diameter 15 ± 0.10 mm for eighteen set of experiments, each set of experiments was repeated three times. The bored surface roughness was measured and recorded and the average surface roughness, R_a was found in the range of 1.5 to 6 μm. Table 1 shows the average chemical composition of the Al/5wt%SiC-MMC used for the experimentation.

Table 1. Composition of Al/5wt%SiC-MMC used for experiment

Type of MMC	Type of reinforced particles	% SiC	% Si	% Mg	% Fe	% Cu	% Mn	% Zn	% Ti	Al
Discontinuous MMC	SiC APS: 45μm	5	7.01	0.60	0.12	0.18	0.10	0.10	0.10	Remaining

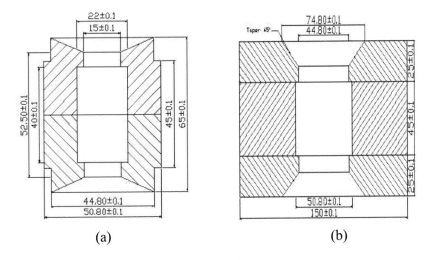

Figure 3. Tooling to house the work-pieces used for experiments.

The rheological properties of AFM media constantly present abrasive particles to the work piece surface in a continuous and uniform manner. Davies (1996) conducted the experiments based on the three qualitative levels of media viscosities by controlling the polymer to gel ratio of media. Similarly, media was prepared of three grades based on the percentage of fluidic element added into the semisolid polymer prepared in bulk just before the use in AFM. Then the prepared media of three different grades were passed through the work sample after thoroughly mixing with abrasive particles.

A Denver Instrument precision weighing balance of least count 0.01 mg was used to measure the weight of specimen before and after each AFM operation. Surfcom 130A surface roughness measuring instrument is used to measure the surface finish. Utilizing the test results percentage of improvement in surface finish is calculated as follows,

$$\text{Percentage improvement in surface finish} = \frac{\text{SR (Boring)} - \text{SR (AFM)}}{\text{SR (Boring)}} \times 100\%, \qquad (1)$$

2.1. AFM Process Parameters and their Levels

There are several factors that effect the quality characteristics of the AFM process such as extrusion pressure; abrasive mesh size; number of working cycle; media flow volume; media rheology; jig & fixture; length of restricting passage etc. In the present study, mainly five AFM parameters namely abrasive mesh size, number of cycles, extrusion pressure, percentage abrasive concentration and media viscosity grade are considered for experimental investigation. Taguchi method based design of experiment (D. C. Montgomery, 1997) and L_{18} (6^1 x 3^7) mixed orthogonal array is utilized to setup the parametric design. Table 2 represents the various parameters considered with their levels for conducting the experiments.

Table 2. AFM parameters and their corresponding values with levels

Symbols	AFM parameters	Levels						Units
		1	2	3	4	5	6	
A	Abrasive mesh size	100	150	200	250	300	350	Nos.
B	No of Cycles	10	20	30	-	-	-	Nos.
C	Extrusion Pressure	3	4.5	6	-	-	-	MPa
D	% Abrasive Concentration	33%	50%	67%	-	-	-	%
E	Media Viscosity Grade*	L	M	H	-	-	-	-

* L= Low, M= Medium and H= High

Other experimental conditions:

Stroke Length   : 50 mm

Volume of media   : 400,000 mm^3 (400 ml)

Average Volume Flow Rate   : 600 mm/min

Room temperature   : 22±0.5^{0}C.

S/N ratio is used to determine the optimal combination of control parameters which will be recognized to gain the surface quality and reduce variability. According to Taguchi method, S/N ratio is the ratio of 'signal' representing the desirable value, i.e. mean for the output characteristics and the 'noise' representing the undesirable value, i.e. the square deviation for the output characteristics. Therefore, S/N ratio is the ratio of mean to square deviation. It is a summery statistics and denoted by η, and the unit is dB. According to quality engineering (Padke M S,1997), the characteristic that the lower observed values represents the better machining performance e.g. surface roughness is known as "lower the better". The summery statistic η (dB) of the lower the better performance characteristic (η_{LB}) is expressed as,

$$\eta_{LB} = -10 \log_{10} \left[\frac{1}{n} \sum_{i=1}^{n} \left(y_i^2 \right) \right]; \quad \text{Where, 'n' is replication;} \qquad (2)$$

This equation is used to determine the S/N ratio (dB) for surface roughness R_a and R_t.

Table 3. Setting of parametric levels and experimental results for R_a and ΔR_a as per L_{18} (6^1 x 3^7)

Exp No.	AFM Parameters					SR (Ra) Before AFM'ing (µm)				SR (Ra) after AFM'ing (µm)					Improvement in SR (ΔRa) after AFM'ing (µm)					% Imp. in Ra based on difference of means.
	A	B	C	D	E	R_{a1}	R_{a2}	R_{a3}	Mean	R_{a1}	R_{a2}	R_{a3}	S/N Ratio (dB)	Mean	ΔR_{a1}	ΔR_{a2}	ΔR_{a3}	S/N Ratio (dB)	Mean	
1	1	1	1	1	1	5.02	5.26	5.68	5.32	3.71	3.77	3.85	-11.543	3.78	1.31	1.49	1.83	3.53	1.54	29.01%
2	1	2	2	2	2	2.65	2.56	2.04	2.42	1.53	1.29	1.54	-3.2747	1.45	1.12	1.27	0.50	-2.57	0.96	39.86%
3	1	3	3	3	3	3.84	3.82	3.72	3.79	2.61	2.42	2.22	-7.6831	2.42	1.23	1.40	1.50	2.69	1.38	36.29%
4	2	1	1	2	2	4.07	3.48	3.61	3.72	3.02	2.93	2.81	-9.3114	2.92	1.05	0.55	0.80	-2.84	0.80	21.51%
5	2	2	3	3	3	5.76	5.12	5.82	5.57	3.18	2.89	3.32	-9.9251	3.13	2.58	2.23	2.50	7.68	2.44	43.77%
6	2	3	3	1	1	3.04	2.88	2.85	2.92	1.96	1.9	1.88	-5.6372	1.91	1.08	0.98	0.97	0.06	1.01	34.55%
7	3	1	2	1	3	3.79	3.37	3.01	3.39	2.51	2.28	2.22	-7.3843	2.34	1.28	1.09	0.79	-0.08	1.05	31.07%
8	3	2	3	2	1	3.97	3.88	3.37	3.74	2.28	2.16	2.11	-6.787	2.18	1.69	1.72	1.26	3.57	1.56	41.62%
9	3	3	1	3	2	3.91	3.79	3.41	3.70	1.97	1.82	1.54	-5.0356	1.78	1.94	1.97	1.87	5.69	1.93	52.03%
10	4	1	3	3	2	2.89	2.95	2.93	2.92	1.03	1.02	1.32	-1.0763	1.12	1.86	1.93	1.61	5.02	1.80	61.57%
11	4	2	1	1	3	2.29	2.14	1.91	2.11	1.07	1.05	0.93	-0.1596	1.02	1.22	1.09	0.98	0.70	1.10	51.89%
12	4	3	2	2	1	3.85	3.44	3.63	3.64	1.34	1.09	1.21	-1.7102	1.21	2.51	2.35	2.42	7.69	2.43	66.67%
13	5	1	2	3	1	3.85	3.5	3.44	3.60	2.18	2.08	2.03	-6.4344	2.10	1.67	1.42	1.41	3.44	1.50	41.71%
14	5	2	3	1	2	2.21	1.89	1.67	1.92	0.71	0.68	0.67	3.2624	0.69	1.50	1.21	1.00	1.49	1.24	64.30%
15	5	3	1	2	3	2.89	2.66	2.78	2.78	0.98	0.93	0.91	0.5332	0.94	1.91	1.73	1.87	5.26	1.84	66.15%
16	6	1	3	2	3	3.66	3.43	3.32	3.47	1.23	1.12	1.09	-1.2007	1.15	2.43	2.31	2.23	7.31	2.32	66.95%
17	6	2	1	3	1	6.29	6.34	6.11	6.25	2.59	3.06	2.98	-9.1999	2.88	3.70	3.28	3.13	10.49	3.37	53.95%
18	6	3	2	1	2	3.87	3.76	3.43	3.69	1.72	1.54	1.61	-4.2172	1.62	2.15	2.22	1.82	6.19	2.06	55.97%

* SR = Surface finish

Table 4. Experimental results for Rt, ΔRt and MR as per $L_{18}\,(6^1 \times 3^7)$

Exp No.	SR (Rt) before AFM'ing (µm)				SR (Rt) after AFM'ing (µm)					Improvement in SR (Rt) after AFM'ing (µm)					% imp. in R_t based on difference of means.	MR 10^{-4} gms	S/N ratio (MR) (dB)
	R_{t1}	R_{t2}	R_{t3}	Mean	R_{t1}	R_{t2}	R_{t3}	S/N Ratio (dB)	Mean	ΔR_{t1}	ΔR_{t2}	ΔR_{t3}	S/N Ratio (dB)	Mean			
1	25.76	23.47	24.72	24.65	15.32	16.55	19.17	-24.654	17.01	10.44	6.92	5.55	16.81	7.64	30.98%	17.7	-55.041
2	10.57	8.28	9.67	9.51	7.04	6.17	6.39	-16.317	6.53	3.53	2.11	3.28	8.77	2.97	31.28%	47.3	-46.503
3	22.83	20.49	21.11	21.48	14.16	13.77	12.79	-22.662	13.57	8.67	6.72	8.32	17.79	7.90	36.80%	9.7	-60.265
4	16.09	15.93	15.37	15.80	12.28	12.06	11.91	-21.644	12.08	3.81	3.87	3.46	11.36	3.71	23.51%	29	-50.752
5	27.19	19.68	27.66	24.84	17.73	16.55	17.86	-24.806	17.38	9.46	3.13	9.80	13.85	7.46	30.04%	28.7	-50.842
6	20.64	14.79	14.43	16.62	10.88	10.55	10.5	-20.543	10.64	9.76	4.24	3.93	13.60	5.98	35.96%	9.6	-60.355
7	16.21	15.81	15.21	15.74	12.68	12.88	10.92	-21.721	12.16	3.53	2.93	4.29	10.77	3.58	22.76%	18.3	-54.751
8	21.13	18.44	17.68	19.08	13.92	12.96	12.89	-22.454	13.26	7.21	5.48	4.79	14.94	5.83	30.53%	8.7	-61.210
9	14.86	14.94	13.96	14.59	8.56	8.19	8.79	-18.606	8.51	6.30	6.75	5.17	15.50	6.07	41.64%	15.9	-55.972
10	12.07	11.25	11.72	11.68	6.12	6.34	6.45	-15.994	6.30	5.95	4.91	5.27	14.53	5.38	46.03%	15.8	-56.027
11	13.13	12.7	12.08	12.64	8.26	7.46	7.34	-17.727	7.69	4.87	5.24	4.74	13.87	4.95	39.17%	8.1	-61.830
12	12.07	12.89	11.92	12.29	6.29	6.34	6.08	-15.901	6.24	5.78	6.55	5.84	15.60	6.06	49.27%	19.1	-54.379
13	18.79	18.33	17.92	18.35	11.9	11.4	11.14	-21.202	11.48	6.89	6.93	6.78	16.73	6.87	37.43%	41.7	-47.597
14	11.46	11.04	10.03	10.84	5.18	4.82	4.08	-13.471	4.69	6.28	6.22	5.95	15.77	6.15	56.72%	7	-63.098
15	11.98	12.04	11.89	11.97	5.34	5.65	4.97	-14.53	5.32	6.64	6.39	6.92	16.44	6.65	55.56%	72	-42.853
16	14.78	15.03	14.69	14.83	6.77	5.23	5.09	-15.189	5.70	8.01	9.80	9.60	19.11	9.14	61.60%	57.8	-44.761
17	26.89	27.05	26.24	26.73	17.33	17.37	15.11	-24.421	16.60	9.56	9.68	11.13	20.05	10.12	37.88%	63.3	-43.972
18	14.48	15.05	13.89	14.47	5.28	5.62	5.2	-14.599	5.37	9.20	9.43	8.69	19.17	9.11	62.92%	48.6	-46.267

* MR = Material removal, SR = Surface finish

Again, according to quality engineering (Padke M S,1997), the quality characteristic for material removal is of higher-the-better type. So, the S/N ratio for higher the better characteristics (η_{HB}) is express as,

$$\eta_{HB} = -10 \log_{10}\left[\frac{1}{n}\sum_{i=1}^{n}\left(1/y_i^2\right)\right]; \quad \text{Where, 'n' is replication;} \tag{3}$$

3. RESULTS AND DISCUSSIONS

According to the L_{18} ($6^1 \times 3^7$) mixed orthogonal array, Table 3 represents the parametric level for different sets of experiments with experimental results. Table 3 also represents the *improvement in surface finish (ΔR_a) and their S/N ratio*. From the Table 3, it is observed that percentage of improvement in R_a based on difference of means is 66.95%. The experimental results for R_t, ΔR_t, MR as per L_{18} ($6^1 \times 3^7$) and their S/N ratios are also represented in Table 4.

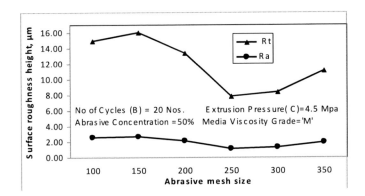

4 (a). Effect of abrasive mesh size on SR, μm.

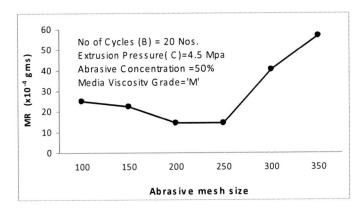

4(b) Effect of abrasive mesh size on MR.

Figure 4. Effect of abrasive mesh size on Ra, Rt and MR keeping other parameters constant.

3.1. Effect of Abrasive Mesh Size on Surface Characteristics

Figure 4 shows the effect of abrasive mesh size on R_a, R_t and material removal keeping other parameters constant. From the Figure 4(a), it is clear that R_a decreases with increase in abrasive mesh size (up to mesh size 250) and beyond this range Ra starts increasing. The bored surface has sharp peaks, when abrasive particles flow over the peaks, cut these peaks and became flatter. Pattern of decrease in R_t is also similar as R_a and least value for both the cases i.e. R_a & R_t is achieved at mesh size 250. During experiments it is also observed that when abrasive-laden semisolid media was passed through a restrictive passage peaks are cut successively and fine finishing took place. The media slug uniformly abraded the walls of the extrusion passage. Material removal has been pronounced both at lower and higher values of abrasive mesh size.

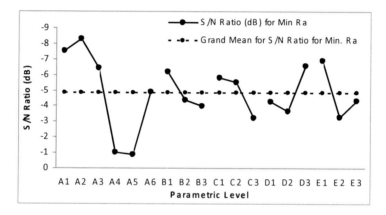

Figure 5(a). S/N Ratio(dB) for surface finish Ra, μm.

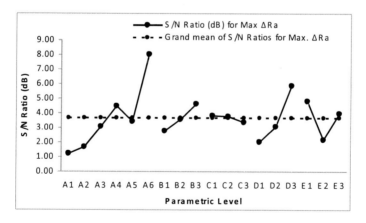

Figure 5(b). S/N Ratio(dB) for improvement of average surface finish ΔRa, μm.

3.2. Parametric Optimization for Surface Characteristics

Figure 5 (a) shows the average S/N ratio (dB) by factor level for R_a. From the graph Figure 5 (a), it is clear that the optimal parametric combination for minimum surface roughness is $A_5B_3C_3D_2E_2$ i.e. at 300 abrasive mesh size, 30 number of cycles, 6 MPa extrusion pressure, 50% abrasive concentration and 'M' grade media viscosity.

Figure 5 (b) shows the average S/N ratio (dB) by factor level for ΔR_a. From the Figure 5(b), it is also concluded that the optimal parametric combination for maximum improvement in average surface finish (ΔR_a) is $A_6B_3C_1D_3E_1$. Figure 6(a) to 6(c) show the S/N ratio for surface finish R_t, ΔR_t and MR. respectively.

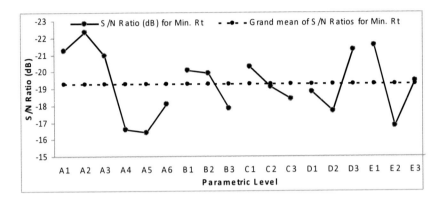

Figure 6(a). S/N Ratio (dB) for R_t, μm.

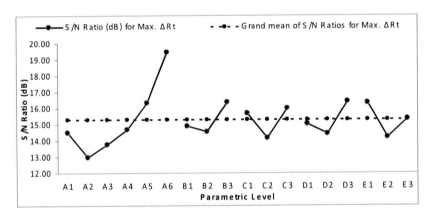

Figure 6(b). S/N Ratio (dB) for ΔR_t, μm.

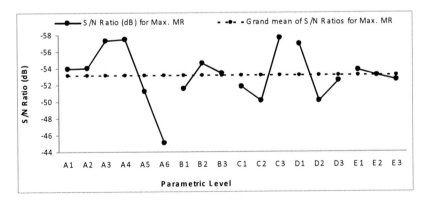

Figure 6(c). S/N Ratio (dB) for material removal (MR).

The plots shown in Figure 6 (a) suggested that the optimal parametric combination for minimum surface roughness height, R_t is $A_5B_3C_3D_2E_2$ i.e. at 300 abrasive mesh size, 30 number of cycles, 6 MPa extrusion pressure, 50% abrasive concentration and 'M' grade media viscosity.

From the Figure 6(b), it is concluded that the optimal parametric combination for maximum ΔR_t is $A_6B_3C_3D_3E_1$.

From S/N ratio graphs Figure 6(c), the optimal parametric combination for maximum material removal is $A_6B_1C_2D_2E_3$.

3.2.1. Experimental Tests for Validation of Optimized Parametric Combinations.

Experiments were conducted on developed AFM setup to verify the optimal parametric combinations for $R_a (\mu m)$, $\Delta R_a (\mu m)$, $R_t (\mu m)$, $\Delta R_t (\mu m)$ and MR, (mg) during AFM of Al/5wt%SiC-MMC and tabulated the results in Table 5. Table 5 represents a comparison of the output performance characteristics of AFM of Al/5wt%SiC-MMC cylindrical surfaces at initial levels setting (i.e. $A_1B_1C_1D_1E_1$). From tabulated results Table 5, while experiments were conducted as per determined optimal parametric combination, it is observed that R_a and R_t decreased by 85.98% and 72.49% respectively; MR and ΔR_a increased by 151.30%, and 264.97% respectively as compared to the initial parametric settings (Exp. No.1) result. The experimental results represented in Table 5 justify the importance of optimal parametric combinations for the developed AFM response characteristics. The identified optimal parameters will help to set the parameters level for effective abrasive flow finishing of Al/5wt%SiC-MMC cylindrical work-pieces.

Table-5. Conformation test results for optimal parametric combinations during AFM of Al/5wt%SiC-MMC

Response characteristics	Exp. No.	Optimized parametric combination	Exp. Results with optimal conditions	Exp. Results with initial levels $(A_1B_1C_1D_1E_1)$	Percentage improvement due to optimal combination over initial level condition.
$R_a (\mu m)$	19	$A_5B_3C_3D_2E_2$	0.53	3.78	+85.98%
$\Delta R_a (\mu m)$	20	$A_6B_3C_1D_3E_1$	3.87	1.54	+151.30%
$R_t (\mu m)$	21	$A_5B_3C_3D_2E_2$	4.68	17.01	+72.49%
$\Delta R_t (\mu m)$	22	$A_6B_3C_3D_3E_1$	11.34	7.64	+48.43%
MR, (mg)	23	$A_6B_1C_2D_2E_3$	6.46	1.77	+264.97%

3.3. ANOVA for Surface Finish R_a, Improvement in Surface Finish ΔR_a and MR

The main purpose of the Analysis of Variance (ANOVA) is to investigate the design parameters and to indicate which parameters are significantly affecting the quality characteristic. In the analysis, the sum of the square deviation is calculated from the value of S/N ratio by separating the total variability of S/N ratio for each control parameter. This analysis helps to find out the relative contribution of machining parameter in controlling the response of the AFM process. Table 6 shows the ANOVA and 'F' test values with percentage of contribution, i.e. effectiveness of the individual AFM parameters on surface finish (R_a), improvement of surface finish *(ΔR_a)* and material removal (MR). From the Table 6, it is concluded that the abrasive mesh size (parameter, A) and media viscosity grade (parameter, E) are the most significant and significant parameters for minimum surface roughness, R_a. From the Table 6, it is found that extrusion pressure (parameter, C), % abrasive concentration (parameter, D) and abrasive mesh size (parameter, A) are the most significant and significant parameters for maximum material removal (MR). It is also concluded that no significant parameters are identified for the maximum surface improvement ΔR_a.

3.4. ANOVA for Surface Finish R_t and Improvement in Surface Finish ΔR_t.

Table 7 shows the ANOVA, 'F' test and 'P' value for surface finish, R_t and improvement in surface finish, ΔR_t. From the Table 7, it is concluded that the abrasive mesh size (parameter, A) and media viscosity grade (parameter, E) are the most significant parameters for minimum surface roughness, Rt. It is also concluded that abrasive mesh size (parameter, A) is the most significant parameter for maximum ΔR_t.

4. DEVELOPMENTAL OF MATHEMATICAL MODELS

The mathematical models for the AFM process performance characteristics (PC) such as R_a; %ΔR_a; Rt, %ΔR_t and MR have been developed in the form of general power factor equation using multivariable curve fitting technique.

$$PC = k \cdot A^{a1} \cdot B^{a2} \cdot C^{a3} \cdot D^{a4} \cdot E^{a5} \tag{4}$$

Where, PC = AFM process performance characteristics,
k = Constant,
and a1, a2, a3, a4, a5 are the exponentials of the regression equation (4) for AFM parameters e.g. abrasive mesh size (A), number of cycles (B), extrusion pressure (C), % abrasive concentration (D) and media viscosity grade (E) respectively. Utilizing the test results the constants k and exponentials a1, a2, a3, a4 and a5 are evaluated. The developed mathematical models represented in equation 5, 6, 7, 8 & 9 for AFM process performance characteristics,

R_a *%ΔR_a, Rt, %ΔR_t and MR respectively.*

$$R_a = 42.2160 \times A^{-0.5613} \times B^{-0.2383} \times C^{-0.4094} \times D^{0.3300} \times E^{-0.3063} \tag{5}$$

$$\% \Delta Ra = 0.484 \times A^{0.533} \times B^{0.2486} \times C^{0.1963} \times D^{0.1647} \times E^{0.0923} \tag{6}$$

$$R_t = 111.7093 \times A^{-0.4964} \times B^{-0.2130} \times C^{-0.3146} \times D^{0.3596} \times E^{-0.2816} \tag{7}$$

Table 6. ANOVA based on S/N Ratio for R_a, ΔR_a and MR

Source of Variation	d.f.	Surface Finish, Ra				Improvement in SR, ΔR_a				MR			
		SS	MS	F Test	P value	SS	MS	F Test	P value	SS	MS	F Test	P value
Abrasive mesh size (A)	5	155.99	31.198	5.94	0.055	3.299	17.98	1.74	0.305	321.23	64.25	3.55	0.122
No. of Cycles (B)	2	16.54	8.271	1.57	0.313	0.295	5.24	0.39	0.700	29.00	14.50	0.80	0.509
Extrusion pressure (C)	2	24.30	12.151	2.31	0.215	0.163	0.34	0.21	0.815	189.24	94.62	5.23	0.076
%Abrasive Concentration(D)	2	28.46	14.231	2.71	0.180	1.63	23.71	2.15	0.232	143.59	71.80	3.97	0.122
Media Viscosity Grade (E)	2	41.51	20.754	3.95	0.113	0.57	10.79	0.75	0.528	4.39	2.20	0.12	0.889
Error	4	21.01	5.254			1.52	14.95			72.31	18.08		
Total	17	287.82				7.47				759.77			
d.f. = degree of freedom, SS = sum of square, MS = mean sum		Smaller is better & R^2= 92.70%				Larger is better & R^2= 80%				Larger is better & R^2= 90.48%			

*SR = Surface finish, MR = Material removal

Table 7. ANOVA based on S/N ratio for surface finish, Rt and improvement ΔR_t

Source of Variation	d.f.	Surface Finish, R_t				Improvement, ΔR_t			
		SS	MS	F Test	P value	SS	MS	F Test	P value
Abrasive mesh size (A)	5	99.005	19.801	5.22	0.067	36.88	7.376	3.50	0.125
No. of Cycles (B)	2	18.787	9.393	2.47	0.200	2.75	1.375	0.65	0.570
Extrusion pressure (C)	2	10.805	5.402	1.42	0.341	1.65	0.825	0.39	0.700
%Abrasive Concent.(D)	2	40.991	20.495	5.40	0.073	7.76	3.877	1.84	0.272
Media Viscosity Grade (E)	2	68.236	34.118	8.99	0.033	7.23	3.615	1.71	0.290
Error	4	15.186	3.796			8.45	2.112	-	-
Total	17	153.009				64.71		-	-
d.f. = degree of freedom, SS = sum of square, MS = mean sum		Smaller is better & R^2= 94.00%				Larger is better & R^2= 87%			

$$\% \Delta R_t = 1.3949 \times A^{0.4539} \times B^{0.2363} \times C^{0.2144} \times D^{-0.0345} \times E^{0.0669} \qquad (8)$$

$$MR = 0.0000344 \times A^{0.5514} \times B^{-0.2292} \times C^{-0.8816} \times D^{0.8008} \times E^{0.1247} \qquad (9)$$

To validate the above developed mathematical models few experiments were conducted (experiment number 19 to 24) and test results are presented in Table 8(a) and Table 8 (b). From the tabulated results (Table 8) it is clear that the experimental results and calculated results using the developed mathematical models bear a good agreement. It is also observed that the percentage of error is less than 6%.

Table 8(a). Confirmation test results for R_a (μm) and % ΔR_a

Exp No.	AFM Parameters					Average surface roughness R_a (μm)			Percentage improvement in average surface roughness %ΔR_a		
	A	B	C	D	E	Exp. Results	Model Results	Error (%)	Exp. Results	Model Results	Error (%)
19	1	3	1	2	1	3.12	3.28251	5.21%	32.02	31.01251	3.15%
20	2	2	3	1	2	1.59	1.52873	3.85%	38.84	39.70945	2.24%
21	3	2	1	3	3	1.87	1.92754	3.08%	50.01	47.13820	5.74%
22	4	3	1	1	2	1.41	1.38384	1.86%	51.11	50.32959	1.53%
23	5	2	3	2	1	1.39	1.46933	5.71%	56.78	57.69770	1.62%
24	6	1	2	3	3	1.38	1.40679	1.94%	58.21	57.89525	0.54%

Table 8(b). Confirmation test results for R_t (μm), %ΔR_t and MR (g)

Exp No.	Maximum peak-valley surface roughness height, R_t (μm)			Percentage improvement in maximum peak-valley surface roughness height %ΔR_t			Material Removal (MṘ)		
	Exp. results	Model Results	Error (%)	Exp. results	Model Results	Error (%)	Exp. Results (10^{-4}gms)	Model Results (gms)	Error (%)
19	15.23	15.904	4.43%	28.62	27.865	2.64%	16.5	0.00174	5.55%
20	8.37	8.076	3.50%	36.98	37.522	1.47%	9.8	0.00101	3.47%
21	9.87	10.021	1.54%	37.89	36.951	2.48%	39.8	0.00406	2.02%
22	6.92	7.150	3.32%	46.24	44.880	2.94%	22.1	0.00226	2.10%
23	7.76	8.081	4.14%	50.82	48.368	4.82%	18.1	0.00190	5.03%
24	7.34	7.744	5.52%	46.76	44.112	5.66%	47.6	0.00453	4.78%

The above developed mathematical model for obtaining the AFM process performance characteristics are of great importance for proper selection of machining parameters during AFM'ing of Al/5wt%SiC-MMC cylindrical surface.

CONCLUSION

Based on the experimental results during abrasive flow machining of Al/5wt%SiC-MMC cylindrical surface the following points are concluded as listed below;

(i) AFM process can be utilized for finishing of Al/5wt%SiC-MMC cylindrical components.

(ii) The surface finish R_a decreases with increase in abrasive mesh size up to 250 mesh and beyond this range R_a starts increasing; this could be attributed to the size of reinforcing particles in MMC.

(iii) The optimal parametric combination for minimum surface roughness, R_a and R_t is $A_5B_3C_3D_2E_2$; *for maximum improvement in average surface finish, ΔR_a is* $A_6B_3C_1D_3E_1$; *for maximum improvement of surface finish, ΔR_t is* $A_6B_3C_3D_3E_1$ and t*he optimal parametric combination for maximum material removal is* $A_6B_1C_2D_2E_3$. *These optimal parametric combinations are experimentally verified.*

(iv) The abrasive mesh size and media viscosity grade are the most significant and significant parameters for minimum surface finish, R_a and R_t. Again the extrusion pressure, percentage abrasive concentration and abrasive mesh size are the most significant and significant parameters for maximum material removal. It is also concluded that no significant parameters are identified for the maximum surface improvement ΔR_a but abrasive mesh size is the most significant parameter for maximum improvement of surface finish ΔR_t.

(v) The maximum percentage of improvement in R_a and R_t based on difference of means are 66.95% and 66.92% respectively

(vi) The developed mathematical models for surface finish criteria and material removal are successfully proposed for proper selection of AFM parameters during finishing of Al/5wt%SiC-MMC work-piece.

REFERENCES

[1] Manna, A. and Bhattacharyya, B., (2001), *"Investigation for effective tooling system to machine Al/SiC-MMC"* Proceedings of the RAMP, Department of Production Engineering, Annamalai University, India, pp. 465–472.

[2] Allison, JE and Cole. GS (1993) "Metal matrix composite in automotive industry: opportunities and challenges". *JOM*, January, pp 19-24.

[3] Manna, A. and Bhattacharyya, B., (2004), "Investigation for optimal parametric combination for achieving better surface finish during turning of Al/SiC-MMC" *Int. J Advanced Mfg. tech.* Vol-23, pp.658-665.

[4] Manna, A. and Bhattacharyya, B., (2005), "Influence of machining parameters on the machinability of particulate reinforced Al/SiC-MMC" *Int. J Advanced Mfg. tech.* Vol-25, pp.850-856.

[5] Mali, H. S. and Manna, A. (2009) "Optimization of AFM Parameters during Finishing of Al-6063 Alloy Cylindrical Surface" Journal *of Machining and Forming Technologies*, Vol. 1, Issue ¾, pp. 237-248 .

[6] Benedict G. F. (1987); *"Nontraditional Manufacturing Processes"*, Marcel Dekker, New York.

[7] Davies, P. J. and Fletcher, A. J. (1996) "Assessment of rheological characteristics of abrasive flow machining process", proc. *Inst. Mech. Engrs.*, vol. 209, pp 409-418.

[8] Douglas, C. Montgomery; (1997) *"Introduction to Statistical quality Control"*, Wiley, 589-601.

[9] Jung, D., Wang, W. L. and Hu, S. J. (2008); "Microscopic geometry changes of a direct-injection diesel injector nozzle due to abrasive flow machining and a numerical investigation of its effects on engine performance and emissions", *Proc. IMechE: Part A: J. Power and Energy;* 222; 241-252.

[10] Hsinn-Jyh Tzeng, Biing-Hwa Yan, Rong-Tzong Hsu and Yan-Cherng Lin. (2007), "Self-modulating abrasive medium and its application to abrasive flow machining for finishing micro channel surfaces", *International Journal Advanced Manufacturing Technology;* 32; 1163–1169

[11] Jeong-Du Kim and Kyung-Duk Kim. (2004), "Deburring of burrs in spring collets by abrasive flow machining" *International Journal Advanced Manufacturing Technology;* 24; 469–473.

[12] Larry Rhoades. (1991), "Abrasive flow machining: a case study" *Journal of Materials Processing Technology,* 28 107-116.

[13] Padke, M. S. (1997); *"Quality Engineering Using Robust Design"* Prentice–Hall, Englewood Cliffs, New jersey.

[14] Williams, R.E. and Rajurkar, K. P. (1992), "Stochastic modeling and analysis of abrasive flow machining", *Trans. ASME, J. Eng. Ind.;* 114, 74–81.

[15] Loveless, T.R., Williams, R.E. and Rajurker, K.P. (1994) "A study of the effects of abrasive flow machining on various machined surfaces", *Journal of Materials Processing Technology;* 47, 133–151.

[16] Jain, V. K. and Adsul, S. G. (2000), "Experimental investigations into abrasive flow machining (AFM)" *International Journal of Machine Tools & Manufacture;* 40; 1003–1021.

Jain, V. K. (2002); *"Advanced Machining Processes",* Allied Pub. Pvt. Ltd., 58-76.

In: Machining and Forming Technologies, Volume 3
Editor: J. Paulo Davim

ISBN: 978-1-61324-787-7
© 2012 Nova Science Publishers, Inc.

Chapter 19

OIM AND EBSD STUDIES ON THE INFLUENCE OF TEXTURE ON THE MECHANICAL PROPERTIES DEVELOPED IN COLD ROLLED DUPLEX STAINLESS STEELS

*K. Manikanda Subramanian1, P. Chandramohan1 and S. Basavarajappa2**

[1]Coimbatore Institute of Engineering Technology, Coimbatore-641109, India
[2]University BDT college of Engineering, Davanagere-577004, India

ABSTRACT

This paper reports an investigation of bulk and micro texture in the cast condition and in the cold rolled conditions (15% and 50% size reduced) of nitrogen alloyed duplex stainless steel. Bulk texture measurements were carried out using orientation imaging microscopy (OIM). The micro texture measurements were carried out using electron backscattered diffraction (EBSD) system for the dual phases, ferrite (alpha) and austenite (gamma). The amount of alpha fibre formation for each reduction is reported under both the conditions of before heat treatment (BHT) and after heat treatment (AHT). Individual texture components for both 15% and 50% reductions under the condition of BHT and AHT are also reported. The results reveal that improvement in the microhardness of 50% cold rolled specimens has a definite correlation with bulk texture, low angle grain boundary and grain size distribution.

Keywords: Rolling, Texture, Orientation, Gamma, Grain boundary

---

[*] Email: basavarajappas@yahoo.com

1. INTRODUCTION

Duplex Stainless Steels (DSS) contain both phases, i.e. ferrite (Alpha-BCC) and austenite (Gamma-FCC) in almost equal proportions. Therefore, to make a study of texture in this cold rolled alloy, it is necessary to report few literatures related to the deformation mechanisms, texturing and orientations that exist in cold rolled DSS alloys.

Fine grained DSS produced by casting, rolling or extrusion and forging followed by quenching in water, yields what are called dual phase steels where ferrite constitutes the matrix while austenite forms islands identified by its turns [Chandramohan, 2006, 2007]. These dual phase steels show superplastic behaviour since the grain growth is effectively retarded at high temperature due to the two-phase-aggregated microstructure. Grain boundary sliding and dynamic recrystallization of the softer phase in a duplex microstructure are considered to be the dominant mechanism for superplasticity in duplex stainless steel [Maehara, 1987]. The role of the dynamic recrystallization is also to transform the low angle grain boundaries into high angle grain boundaries suitable for sliding [Young-Han, 1997].

Cyclic stress-strain behaviour of duplex stainless steels shows that austenite is the harder of the two phases and this higher hardness of the austenite leads to the transfer of plastic deformations from the austenite to the ferrite during cyclic loading. The austenite phase will always show a planar dislocation structure but with different location density. However the ferrite will show dislocation channels, walls and cells depending on the amount of plastic deformation experienced [Lillbacka, 2007]. Depending on the degree of deformation and on the initial cold rolling texture the ferrite phase undergoes recovery or recrystallization, but recovery dominates the softening of the BCC phase. The austenite phase is subjected to constrained nucleation conditions and thus leads to textures different from textures of FCC single phase materials [Keichel, 2003].

Compared to single phase steels, the rolling deformation of duplex steels is much more complex, since dislocation glide, which is the most important mechanism in single phase steel is progressively hindered with increasing nitrogen content [Akdut, 1993]. The rolling deformation mechanisms of duplex structures differs from those of single phase materials, because the phase boundaries (PB) very effectively hinder the deformation on each phase and cause different deformation mechanisms to appear [Chandramohan, 2006, 2007]. A strong grain refinement is observed with the increase in deformation of cold rolled duplex stainless steel and in metastable austenitic steels at room temperature [Tavares, 2006].Akdut et al investigated on the cold rolling texture development of two α/γ duplex stainless steels (DSS) with similar volume fractions of both phases but with totally different microstructures. The austenitic phases of both DSS behave similarly to single phase materials with low stacking fault energy which develop a brass-type rolling texture. In contrast, the texture development of the ferritic phases strongly differs from those of single phase ferrites. Instead of a fibre type texture the α-phase in both DSS exhibits a peak dominated texture regardless of whether it is the matrix phase or not [Akdut, 1996].

The effect of changes in strain path on the flow behaviour in duplex stainless steels was investigated by Johan J. Moverare. It was found that during prestraining in the rolling direction there is a change in microstresses due to different flow behaviour in the two phases. An increase in microstress (tensile in austenite and compressive in ferrite) was observed in the transverse direction. Also, the surface macrostresses increased after prestraining due to the texture gradients observed through the thickness of the sheet. No significant changes in the texture were observed during prestraining up to 5.2% plastic strain and therefore, differences in the anisotropic behaviour between as received and prestrained material do not originate from the crystallographic texture [Johan Moverare, 2002].

From the literature stated above it becomes clear that different type of orientation /texture is formed in the phases which determine the mechanical property of the DSS. Therefore it becomes necessary to carryout research in order to correlate the texture formation of cold rolled duplex stainless steels and the resulting mechanical properties.

2. EXPERIMENTAL WORK

2.1. Melting and Heat treatment

Nitrogen alloyed duplex stainless steel was melted in 100 kg induction furnace (Acid lined), with suitable alloying additions in the ladle. The following materials were used as charge materials: pure iron, low carbon stainless steel scrap, Low carbon Ferro chromium, Low carbon Ferro manganese, Ferro Silicon and nitrogen alloyed Low carbon Ferro chromium. After melting, deoxidization was carried out in the ladle by adding the following deoxidizers: Ca-Si-0.1%, Fe-Si-Zr- 0.05%, Fe-Ti-0.05% and Se-0.02%. CO_2 molding was used for making molds and sprit based zirconium coating was applied on the surface of the mold. The test bar dimensions were followed as per the ASTM A370 specifications. The composition of the material was analyzed using ARL 3460 spectrovac and given in Table 1.

Table 1. Chemical composition in wt%

Alloy	%C	% Si	% Mn	%S	%P	%Cr	%Ni	% MO	% Cu	% V	% W	%N
DSS	0.028	0.94	0.86	0.009	0.035	22.79	5.59	3.42	0.19	-	-	0.18

Based on Fe Cr Ni phase diagram, for Ni-5 to10 % and Cr-20 to 25% the minimum temperature required to achieve dual phase of ferrite and austenite is 1100°C. Hence for safer side a slightly higher range of 1130°C was selected as heat treatment temperature. The temperature before quenching is 1010°C. The test bar of the alloy was solution annealed in the muffle furnace. The water temperature before quenching was 31°C and after quenching 40°C. Heat treatment cycle is as follows: Heated to 1130°C, soaked for 2 hrs, furnace cooled to 1010°C, soaked for 30 minutes and quenched in water.

2.2. Cold Rolling

The casting was subsequently cold rolled at room temperature, to 15% and 50% size through an even reduction of 1 mm in every pass. Immediately after cold rolling, specimens were taken from one half of the casting (BHT-Before Heat Treatment) and the other half of the casting was subjected to solution annealing i.e. heated to 1130°C at the rate of 150°C / hr soaked for 2 hrs furnace cooled to 1010°C and then water quenched (AHT-After Heat Treatment).

2.3. Texture Analysis

Samples were subjected to measurements of bulk crystallographic texture and micro texture. The measurements were obtained at the mid thickness sections of the rolling plane (containing rolling directions (RD) and transverse directions (TD)) and of the long transverse plane (containing RD and normal directions (ND)) for the bulk and micro texture, respectively. All samples were electro polished using the standard [Samajdar, 1998] technique.

For the bulk texture measurement, a PANalytical materials research diffraction (PANalytical, Almelo, The Netherlands) system was used. The orientation distribution functions (ODFs) were obtained through the inversion of four incomplete X-ray pole figures and the software MTM-FHM [Van-Houtte, 1995].The software uses the standard series-expansion technique [Bunge, 1982]. The ODFs were further analyzed to obtain the volume fractions of the ideal FCC and BCC orientations. The pole figures were plotted from the ODF data using orthorhombic symmetry.

The micro texture measurements were obtained in a Tex SEM Laboratories (TSL) energy- dispersive X-ray orientation-imaging microscopy or electron backscattered diffraction (EBSD) system on an FEI Company Quanta 200 high- vacuum scanning electron microscope (SEM). The phenomena like Local property effects, Orientation variations within individual grains, grain size distribution, Phase relationships and Direct ODF measurement which are directly linked to microstructure were studied.

3. RESULTS AND DISCUSSION

Ideal FCC deformation textures are often generalized as α (ND//<110>) and β (<110> tilted 60 deg toward RD) fibers. The α fiber includes goss {011}<100> and brass {011}<211>, while the β starts at brass {011}<211>, goes through S {123}<634>, and finally ends at Cu {112}<111> [Hirsch, 1998]. In addition, cube {001}<100> is often associated with a typical FCC recrystallization [Samajdar, 1998[1], 1999] and different types of rotated cubes (CG {012}<100>, H {001}<110>, and CH {001}<210>) are identified as potential PSN orientations [Samajdar, 1998[1],1998[2]]. In the present study, the results for the bulk crystallographic texture and microtexture are presented separately both for simplicity and for relative ease and clarity in reaching the objectives.

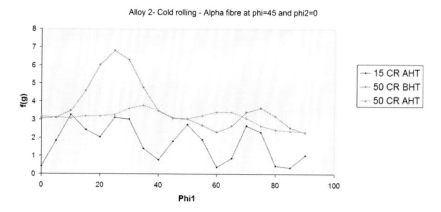

Figure 1. Alpha fiber plot of cold rolled specimen.

Table 2. Texture components of cold rolled specimen

Components	15% Cold AHT	50% Cold BHT	50% Cold AHT
Cube	0.05	0.02	0.03
Taylor	0.13	0.08	0.07
Brass	0.10	0.17	0.13
S	0.22	0.22	0.16
Cu	0.13	0.08	0.06
CG	0.10	0.11	0.10
H	0.03	0.02	0.03
CH	0.08	0.04	0.07
Goss	0.05	0.08	0.07

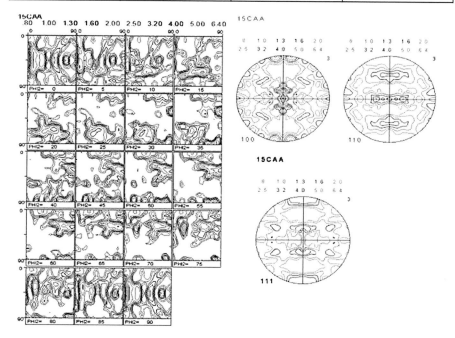

Figure 2. ODF and pole figure of austenite phase in 15% cold rolled AHT (Bulk texture).

3.1. Bulk Texture

3.1.1. 15% Cold rolled

Figure 1 show the alpha fiber plot which is drawn with reductions in x axis and f (g) in y axis. The units of f (g) is X times random the automatic contour levels which were used for plotting the ODFs. It can be observed that for a 15% size cold reduction (AHT), the value of f (g) is 3, which is a mark of weak texture. The same is evidenced with a less preferred orientation in the ODF and pole figure of austenite phase in the 15% cold rolled AHT (Bulk texture) as shown in Figure 2. This is a mark of lower medium texture. The amount of alpha fiber in a 15 % size cold reduction (BHT) is equivalent to the value f (g) = 3.5, which is also mark of weak texture. As given in the Table 2, the bulk texture analysis report reveals that the significant orientations in the bulk texture are Brass and S. Brass orientation is at a medium level of 0.10 and S orientation is predominant to a level of 0.22 which is the maximum among all reductions. Though higher level orientations are achieved, the formations of alpha and beta fibers that improve the mechanical property of a material are lacking. Therefore the specimen was subjected to a higher reduction (50% reduction).

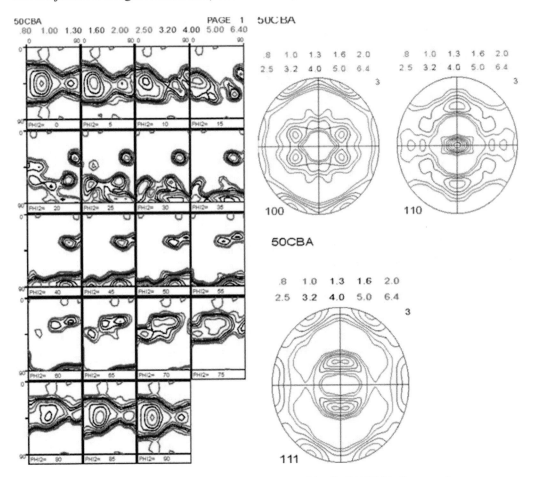

Figure 3. ODF and pole figure of austenite phase in 50% cold rolled BHT (Bulk texture).

3.1.2. 50 % Cold rolled

Alpha fiber plot shown in Figure 1 reveals that for a 50 % size cold reduction (BHT), the value of f (g) (alpha fiber) is 6.9, which is a mark of strong texture. The same is evidenced with a definite and preferred orientation in the ODF and pole figure of austenite phase in the 50% cold rolled BHT (Bulk texture) as shown in Figure 3. This fibre texture is considered to be the contributing factor for deep drawability with higher hardness [Clark, 1991] which is in agreement with the micro hardness result achieved. As given in the Table 2, the bulk texture analysis report reveals that the predominant orientations in the bulk texture are Brass, S and Cube. Brass orientation is present to level of 0.17 (highest among all reductions), S orientation is at a level of 0.22 (highest value is 0.23) and Cube orientation is present in the lowest level of 0.02. This is in line with the findings of C.M. Souza Jr, 2008 that the bulk texture of cold rolled DSS leads to higher amount of brass orientation.

It can also be observed from the Figure 1 (alpha fiber plot) that for a 50 % size cold reduction (AHT), the value of f (g) is 3 which get reduced drastically from value of 6.9 after heat treatment. The same is evidenced with a less preferred orientation in the ODF and pole figure of austenite phase in the 50% cold rolled AHT (Bulk texture) as shown in Figure 4.This is a mark of lower medium texture. The reason for this is the grain refinement due to the presence of two phases, short range atomic ordering of N and lesser stacking fault energy of austenite, retarding dislocation climb between slip planes, and the strain induced by differential contraction of the two phases on cooling from annealing temperatures [Reis, 2000], [Jimenez, 1999]. As given in the table 2, the bulk texture analysis report reveals that the predominant orientations in the bulk texture are Brass, S and CG. Brass orientation is present to level of 0.13 (reduces after heat treatment), S orientation is of level 0.16 (reduces after heat treatment) and Cube orientation is of lowest level 0.03.

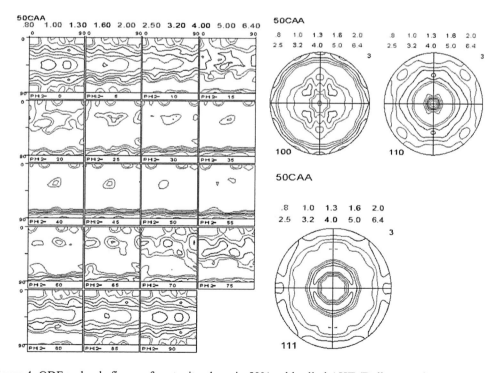

Figure 4. ODF and pole figure of austenite phase in 50% cold rolled AHT (Bulk texture).

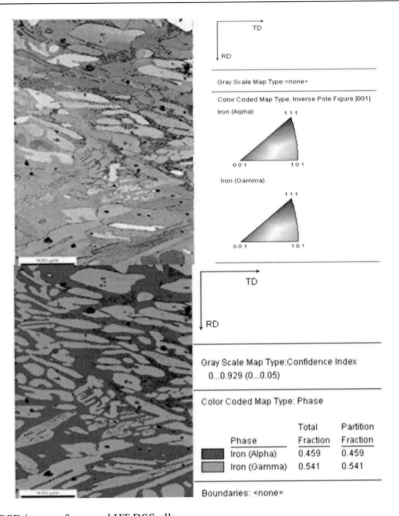

Figure 5. EBSD image of cast and HT DSS alloy.

3.2. Micro Texture

3.2.1. 15% cold rolled

Electron back scattered image of the cast nitrogen alloyed duplex stainless steel alloy is shown in Figure 5. Since the bulk texture of both the 15% cold rolled BHT and AHT specimens are weak, it was decided to make micro texture study on AHT specimen alone. The electron back scattered image of the specimen subjected to cold rolling and size reduced to 15% (AHT) is represented in Figure 6. Rolling direction and transverse direction are shown in the figure for clear understanding. It is observed from the colour coding of the Figure 6 that the ferrite grains are predominantly oriented between <101> and <001> directions. Austenite grains are oriented in <001> direction and in between <111> and <001> directions. It can be observed from the Figure 5 and 6 that the volume fraction of ferrite (alpha) is 46 and 54 percent and that of austenite (gamma) is 54 and 46 percent in the BHT and AHT specimens respectively which means, the rolling process has increased the amount of ferrite by sacrificing austenite, irrespective of the heat treatment process. Therefore, it can be inferred that the orientation which is random as per the micro texture results of this specimen is in agreement with the lower medium texture arrived from the bulk texture analysis.

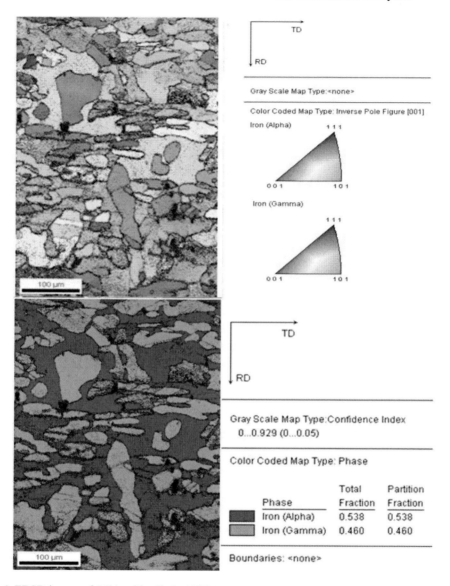

Figure 6. EBSD image of 15% cold rolled- AHT.

3.2.2. *50% Cold rolled*

Electron back scattered image of the specimen subjected to cold rolling and size reduced to 50% (BHT) is represented in Figure 7 and the image of the specimen subjected to cold rolling and size reduced to 50%, followed by heat treatment (AHT) is represented in Figure 8. Rolling direction and transverse direction are shown in all the three figures for clear understanding. It can also be observed from the Figures 7 and 8 that the volume fraction of ferrite (alpha) are 66 and 53 percent and that of austenite (gamma) are 34 and 47 percent which means, the rolling process has increased the amount of ferrite by sacrificing austenite, irrespective of the heat treatment process. It is inferred from the colour coding of the Figure 7 that the material is heavily worked because of 50% cold reduction and hence the probe size fixed in the machine could not provide the scanned image. It is inferred from the colour coding of the Figure 8 that the ferrite grains are predominantly oriented between <101> and <111> directions. Austenite grains are oriented in almost all the <001>, <111> and <101> directions.

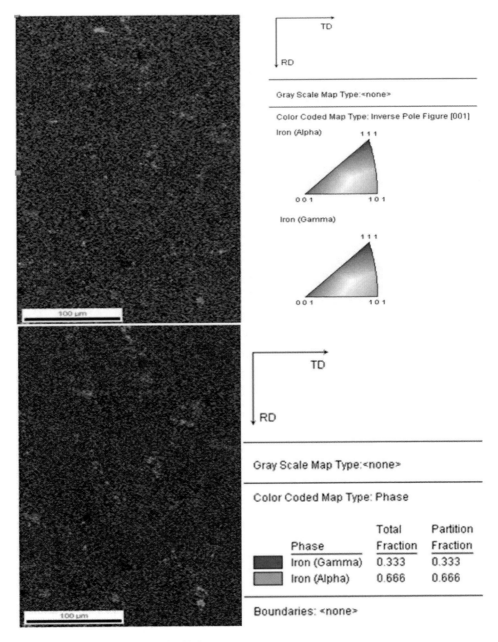

Figure 7. EBSD image of 50% cold rolled-BHT.

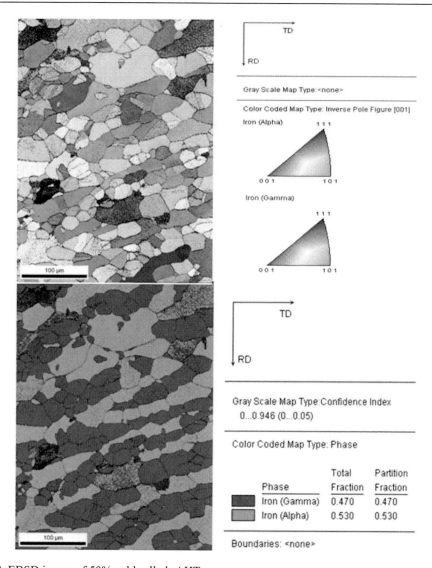

Figure 8. EBSD image of 50% cold rolled- AHT.

3.3. Effect of Cold Rolling on Microhardness and Microstructure

A remarkable improvement is noticed in the microhardness of the alloy from 241.09 VHN to a maximum of 370 VHN, after the cold rolling process as shown in Figure 9. It can be observed that the increase in reduction percentage leads to the increase in hardness at every reduction stage. Also the hardness values of the BHT specimens are higher compared to the AHT specimen in all the stages of reduction which means homogenization of grains has contributed for the hardness reduction after heat treatment. The higher hardness of BHT specimens as compared to that of AHT specimens can be well justified with the strong texture, low angle grain boundaries and grain size distribution as stated below.

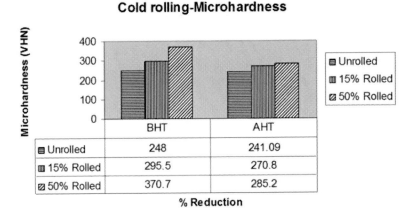

Figure 9. Micro hardness of cold rolled specimens.

Optical microstructures of the cold rolled specimens upon various percentages of reductions are shown in Figure 10. It can be observed that the austenite islands attain an elongated structure and the matrix is free from any detrimental secondary phases.

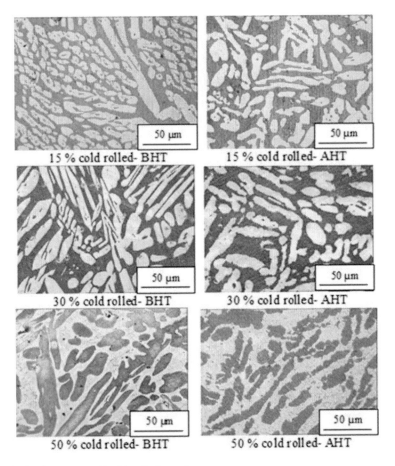

Figure 10. Optical microscopy of cold rolled specimens.

Figures 11-12 represents the grain size distribution of 50% cold rolled BHT and AHT specimens. The average grain size before heat treatment is 1.35 micron whereas, after heat treatment it is 31.6 micron, which means extensive grain growth has occurred during heat treatment. Figures 13-14 represents the misorientation angle of 50% cold rolled BHT and AHT specimens. The misorientation angle is widely distributed over a range upto 60.5^0 in both the specimens. Even though the volume fraction of low angle grain boundary in both the BHT and AHT specimens are same ($\leq 33.5^0$ - 0.06), the BHT specimens contain more amount of low angle misorientations which is considered to be a positive sign as far as mechanical properties are concerned.

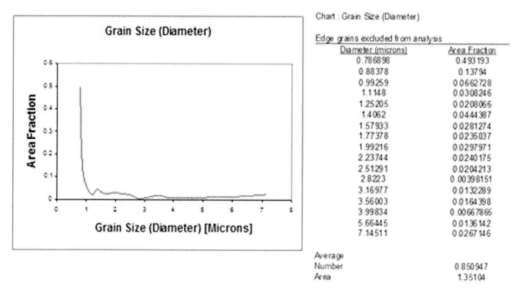

Figure 11. Grain size distribution of 50% cold rolled –BHT.

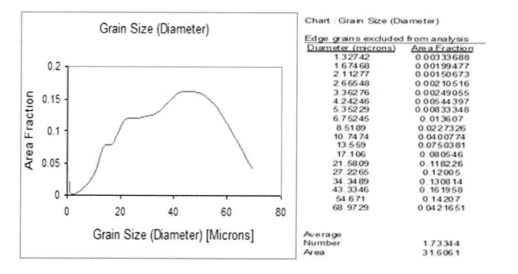

Figure 12. Grain size distribution of 50% cold rolled –AHT.

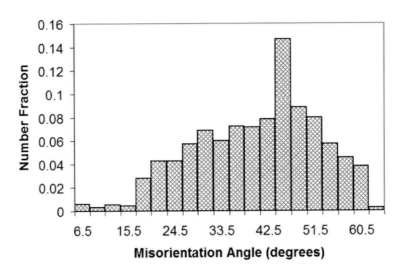

Figure 13. Misorientation angle of 50% cold rolled–BHT.

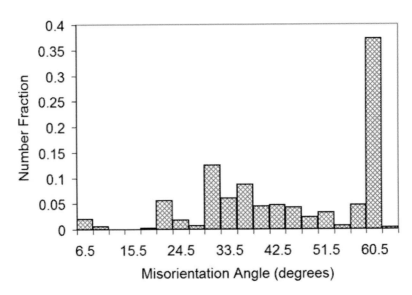

Figure 14. Misorientation angle of 50% cold rolled –AHT.

CONCLUSION

The directions of ferrite and austenite vary with the metal forming process and with the heat treatment procedure practiced. Cold rolling process increases the amount of ferrite (46 to 66) by sacrificing austenite (54 to 34), irrespective of the heat treatment process. In the cast and heat treated specimen, both the ferrite and austenite grains are predominantly oriented between <111> and <001> directions. In the cold rolled (BHT) specimens, due to heavy working (cold reduction) it is not possible to acquire the EBSD image. In the cold rolled (AHT) specimens, the ferrite grains are predominantly oriented between <101> and <111> directions. Austenite grains are oriented in almost all the <001>, <111> and <101> directions.

Bulk texture measurement of 50% cold rolled (BHT) specimen reveals a strong texture with f(g) 6.9 that leads to higher microhardness. It also reveals that the predominant orientations in the bulk texture are Brass, S and Goss.

ACKNOWLEDGMENT

All India Council for Technical Education, New Delhi (RPS grant-8023/BOR/RPS-205/2006-07) and National Facility of Texture and Orientation Imaging Microscopy - Department of Science and Technology, India at IIT-Bombay, supported this work. The authors are grateful to Defence Metallurgical Research Laboratory (DMRL)-Hyderabad, Auto shell cast private ltd and Commando Engineers, Coimbatore for their technical support, material supply and machining process.

REFERENCES

[1]   Akdut, A. and Foct, J, (1993), "Cleavage-like fracture of austenite in duplex stainless steel", *Scripta Metallurgica and Materialia*, 29(2), 153-158.

[2]   Akdut, N., and Foct, J. (1995), "Phase boundaries and deformation in high nitrogen Duplex Stainless Steel-II. Analysis of deformation mechanisms by texture measurements in X2CrNiMo225 (1.4462)", *Scripta Materialia*, 32(1), 109-114.

[3]   Akdut, N., Foct, J. and Gottstein, G. (1996), *"Cold rolling texture development of α/γ duplex stainless steels"*, Steel Research Institute, 67 (10), 450-455.

[4]   Bunge, H. J. (1982). *Texture analysis in material Science*. Butterwoths, London.

[5]   Chandramohan, P, Mohamed Nazirudeen, S. S. and Srivatsavan, R. (2006), "The effect of nitrogen solubility, heat treatment and hot forging on 0.15% N duplex stainless steels", *International Journal of Materials and Product Technology*, 25(4), 281-296.

[6]   Chandramohan, P., Mohamed Nazirudeen, S. S. and Ramakrishnan, S. S. (2007), Studies on hot forging of nitrogen alloyed duplex stainless steels, *Journal of Material Science and Technology,* 23(1), 111-117.

[7]   Clark, R. K., Garrett, T. L., Jungling, R. A. and Vandermeer Vold, C. L, (1991), "Effect of processing variables on texture and texture gradients in tantalum", *Metallurgical and Material Transactions A,* 22, 2039-2048.

[8]   Hirsch, J. and Lucke, K. (1988), "Mechanism of deformation and development of rolling textures in polycrystalline f.c.c. metals—I. Description of rolling texture development in homogeneous CuZn alloys", *Acta Metallurgica,* 36, 2863–2882.

[9]   Jimenez, J. A., Carreno, F. Ruano, O. A. and Carsi, M. (1999), "High Temperature mechanical behaviour of d-g Stainless steel", *Material Science and Technology*, 15, 127-131.

[10] Johan Moverare, J. and Magnus Oden, (2002), "Deformation behaviour of a prestrained duplex stainless steel", *Material Science and Engineering* A, 337(1-2), 25-38.

[11] Lillbacka, R., Chai, G., Ekh, M., Liu, P., Johnson, E. and Runesson, K. (2007), "Stress-strain behaviour and load sharing in duplex stainless steels: Aspects of modeling and experiments", *Acta Materialia*, 55(16), 5359-5368.

[12] Maehara,Y. and Ohmori,Y, (1987), "Microstructural change during superplastic deformation of delta-ferrite/austenite duplex stainless steel". *Metallurgical and Material Transactions A*, 18 (5), 663-672.

[13] Reis, G. S., Jorge Jr, A. M. and Balancin, O. (2000), "Influence of the microstructure of duplex stainless steels on their failure characteristics during hot deformation", *Materials research*, 3(2), 31-35.

[14] Samajdar, B., Verlinden, L., Rabet. and Van-Houtte, P. (1999), "Recrystallization Textures in a Cold Rolled Commercial Purity Aluminum:- An Effort to Define the Probable Macro and Micro Mechanisms Involved", *Material Science and Engineering A*, 266, 146–154.

[15] Samajdar, I. and Doherty, R.D. (1998), "Cube recrystallization texture in warm deformed aluminum", *Acta Materialia*, 46, 3145–3158.

[16] Samajdar, I., Rabet, L. B., Verlinden. and Van-Houtte, P. (1998), "Channel die compression of IF steel: Developments in Macro and Micro Texture", *ISIJ International*, 38, 539–46.

[17] Souza Jr, C. M., Abreu, H. F. G., Tavares, S. S. M. and Rebello, J. M. A. (2008), "The σ phase formation in annealed UNS S31803 duplex stainless steel: Texture aspects" *Materials Characterization*, 59(9), 1301-1306.

[18] Tavares, S. S. M., Silva da, M. R., Paradal, J. M., Abreu, H. F. G. and Gomes, A. M. (2006), "Micro structural changes produced by plastic deformation in UNS S31803 duplex stainless steel", *Journal of Material Processing Technology*, 180(1-3), 318-322.

[19] Van-Houtte, P., (1995). "*MTM-FHM software and manual*", version 2, ed. MTM KULeuven, Belgium.

[20] Young Han, S. and Soon Hong, H. (1997), "The effects of thermo-mechanical treatments on superplasticity of Fe-24Cr-7Ni-3Mo-0.14 N duplex stainless steel", *Scripta Materialia,* 36(5), 557-563.

INDEX

A

abrasive flow finishing (AFF), xiii, 266
abrasive flow machining (AFM), vi, xiii, 265, 266, 267, 268, 269, 270, 271, 275, 276, 278, 279, 280
access, 2, 231
achievement, 116, 132
adaptation, 8, 15, 183, 189
additives, 212
adhesion, 20, 219, 221, 222, 223, 224
adjustment, 6, 9, 10, 55, 90
aerospace, xiii, 18, 19, 20, 148, 266
age, 34, 37, 38, 41
aggressiveness, 266
aircraft structural parts, ix, 18
algorithm, 53, 140, 181, 183, 189, 190, 195, 244
alloys, x, 18, 19, 20, 27, 83, 100, 105, 106, 136
alternatives, ix, x, 18, 86
aluminium, xiii, 17, 18, 19, 20, 26, 83, 157, 191, 266
aluminum, x, 27, 61, 62, 75, 83, 100, 105
aluminum oxide, 61, 62, 164, 168
amplitude, 76, 82, 194
anisotropy, x, 74, 75, 99, 103, 107, 118
annealing, 19, 63, 110, 145, 283, 287
ANOVA, 89, 201, 204, 205, 206, 227, 229, 232, 233, 241, 243, 245, 252, 265, 266, 275, 276, 277
applications, x, 100, 136, 141
asbestos, xi, 199, 200
assessment, 29, 46, 208
authors, ix, 15, 18, 27, 30, 42, 47, 51, 59, 62, 65, 71, 75, 83, 87, 89, 98, 104, 126, 132, 145
automation, 148
automobile parts, 181
automotive industry, x, xi, 100, 136, 144, 148, 162, 279
avoidance, 196

B

Bangladesh, 211, 224
base, x, 25, 106, 116, 132, 133, 182, 225, 231
behavior, 2, 4, 42, 62, 73, 75, 81, 82, 83, 107, 109, 144
Belgium, 296
bending, 5, 30, 87, 100, 103, 107, 185, 186
beneficial effect, ix, 30
benefits, xii, 127, 128, 148, 212, 213
bias, 26
biaxial stretching, 74, 99, 101, 104, 105, 106, 107, 108, 111, 113, 127, 128, 129, 131
Boltzmann constant, 139
bonding, 168
brass, x, 73, 74, 76, 78, 79, 80, 81, 82, 100, 248, 263, 282, 284, 287
Brazil, xv, 1, 15, 16, 17, 29, 42, 43, 73, 83, 99
building blocks, 187
burn, 176

C

calculus, 65
calibration, 58, 214, 215
CAM, 191
Canada, 72
carbides, 147, 212, 266
carbon, 18, 73, 74, 75, 76, 78, 79, 80, 81, 82, 141, 145, 146, 201, 207, 208, 209, 211, 213, 283
case study, 117, 280
casting, 229, 244, 282, 283
catastrophic failure, 224
categorization, 217
cell, 82, 84, 90
ceramic, xi, 19, 199, 200, 201, 202, 204, 205, 206, 207, 208, 225, 248
challenges, 180, 279

channels, 34

chemical, 46, 76, 136, 137, 141, 219, 229, 233, 245, 249, 268

China, 226

chromium, 62, 283

clarity, 284

classification, 148

cleaning, xiii, 266

cleavage, 87

coatings, x, 2, 15, 61, 62, 63, 69, 70

coding, 288, 289

color, 64, 176, 256

combined effect, 217

commercial, 46, 61, 65, 89, 117, 118, 145, 166, 178

common sense, ix, 30

comparative analysis, 97

compensation, xi, 27, 28, 182, 197

competition, 100

complexity, 4, 136, 166

components, ix, 3, 4, 5, 13, 17, 18, 19, 29, 30, 31, 32, 39, 42, 45, 85, 98, 101, 117, 118, 119

composites, ix, xi, xiii, 162, 179, 201, 205, 207, 208, 209, 244, 249, 263, 265, 266

composition, 25, 76, 88, 130, 136, 137, 141, 166, 176, 233, 268, 283

compression, 2, 30, 48, 55, 75, 258, 259, 296

computation, 38

computer, ix, 46, 47, 51, 54, 117, 118, 183, 216

computer software, 118

computing, 183

concentration, 22, 31, 63

conduction, 25, 26, 63, 143

conductivity, xii, 26, 63, 71, 139, 142, 165, 202, 228

confidence, 14, 15, 29, 30, 36, 97

confidence interval, 14

configuration, 37, 38, 164, 167, 181, 185, 192

conflict, 206

congress, 132, 196

construction, 166, 178

consumption, xii, 144, 173, 212, 217, 218, 227, 228, 229, 232, 233, 235, 236, 237, 238, 239, 240, 243

contact time, 19, 262

continuity, 65

contour, 100, 122, 123, 286

contradiction, 154

control, 19, 30, 65, 86, 90, 96, 97, 102, 136

controversial, 46

convergence, 65, 119, 195

convergence criteria, 65

conversion, 78

cooling, x, 24, 26, 135, 136, 137, 138, 140, 141, 143, 144, 145, 146, 165, 171, 177, 178, 212, 219, 223, 224, 225, 287

cooling process, 135, 136, 144

copper, 202

correction factors, 4, 9, 15

correlation, 5, 41, 46, 59, 119, 155, 156, 157, 200, 206, 215, 247, 253, 254, 262, 281

correlation coefficient, 215, 247, 253, 254

correlations, 58, 139

corrosion, xi, xii, 18, 199, 200, 228

cost, x, xii, xiii, 85, 86, 100, 135, 148, 212, 228, 266

costs, x, 30, 61, 100, 136

covering, 73

CPU, 118

crack, 49, 50, 54, 55, 56

cracks, ix, 30, 50, 85, 86, 89, 91, 93, 94, 95, 97

critical value, 105, 106, 108, 113

cubic boron nitride, 62

cutting fluids, xii, 148, 225, 228, 245

cutting force, xii, 1, 2, 3, 4, 5, 8, 9, 10, 11, 12, 13, 14, 18, 19, 25, 45, 46, 47, 51, 54, 55, 57, 58, 59, 147, 148, 149, 151, 152, 157, 158, 166, 177, 182, 183, 184, 185, 188, 189, 191, 195, 197, 211, 212, 213, 216, 217, 218, 220, 224

CVD, 158

cycles, 265, 266, 269, 274, 275, 276

D

damages, 201

damping, 148, 197

database, 152

decision-making process, 228

decomposition, 141, 184

defects, xii, 18, 85, 99, 100, 101, 103, 105, 115, 116, 162, 201, 248, 262

deficiency, 59

definition, 5, 88, 107

deformation, ix, x, xii, xiii, 2, 12, 18, 30, 47, 48, 50, 53, 57, 59, 73, 74, 76, 77, 78, 79, 81, 82, 83, 91, 92, 93, 94, 97, 101, 103, 104, 105, 106, 116, 119, 123, 126, 127, 128, 131, 164, 217, 218, 219, 228, 247, 248, 261, 282, 284, 295, 296

degradation, 20

density, 63, 139, 141, 142

dependent variable, 205

depth, x, xii, 11, 17, 19, 21, 22, 23, 24, 25, 26, 30, 31, 32, 33, 34, 35, 37, 38, 40, 41, 62, 64, 85, 86, 87, 88, 89, 90, 91, 92, 93, 94, 95, 96, 97, 149, 154, 155, 163, 165, 169, 171, 173, 174, 175, 176, 177, 178, 184, 191, 212, 221, 227, 228, 235, 241, 243, 248, 249, 251

developed countries, xi, 200

deviation, 23, 24, 26, 40, 42, 65, 173, 195, 269, 275

diffraction, 281, 284

Index

diffusion, 65, 138, 141, 212, 219, 221, 222, 223, 224
diffusivity, 142, 143
dimensional accuracy, ix, 18
discontinuity, 104, 218
discretization, 53, 65, 118
dislocation, 73, 75, 76, 81, 82, 141, 282, 287
displacement, 6, 10, 37, 39, 49, 50, 51, 90, 102, 119, 123, 124, 126, 129, 166, 185, 186
distribution, x, 17, 20, 30, 31, 42, 45, 49, 54, 55, 61, 62, 63, 66, 83, 87, 103, 110, 121, 123, 124, 126, 130, 138, 140, 141, 162, 171, 174, 180, 183, 188, 281, 284, 291, 293
distribution function, 284
DOI, 263
drawing, 83, 90, 99, 101, 102, 103, 105, 249, 250
ductility, x, 18, 74
Duplex Stainless Steels (DSS), xiii, 282
durability, xi, 199, 200
dyeing, 233

E

elaboration, 5
elastic deformation, 30
elasticity, 32
electrical resistance, 31
electron, 55, 56, 89, 202, 217, 224, 281, 284, 288
electron microscopy, 89
emission, 202
energy, x, xii, 17, 18, 19, 46, 50, 52, 65, 91, 92, 94, 135, 136, 137, 142, 143, 144, 164, 166, 171, 173, 174, 179, 202, 212, 217, 218, 228, 248, 282, 284, 287
energy consumption, 212, 217, 218
energy input, 217
energy transfer, 136
enforcement, 53
engineering, xii, xiii, xv, 2, 42, 47, 100, 103, 148, 228, 245, 266, 269, 272
England, 145
environment, 17, 18, 30, 136, 137, 139, 142, 144, 197, 218, 219
environmental impact, xii, 211
equilibrium, 39, 53, 109, 142, 189
equipment, 33, 34, 100, 151, 233
erosion, 219
ester, 250, 251
estimating, 45, 46
Europe, 100
evidence, 213
evolution, 50, 56, 73, 75, 76, 81, 99, 101, 102, 105, 106, 108, 109, 110, 111, 141, 193
examinations, 262

excitation, 188, 189, 197
exclusion, 37, 38
experimental condition, 213, 217, 269
experimental design, 233, 235, 251
external environment, 30
extraction, 86, 87, 89, 90, 91, 95, 97
extrusion, 265, 266, 267, 269, 273, 274, 275, 276, 279, 282

F

fabrication, x, 86, 99, 100, 115, 116, 117, 123, 126, 128, 129, 130, 131, 132
failure, 57, 105, 124
fatigue, ix, 30, 87, 100
FEM, vi, 39, 58, 59, 62, 132, 147, 149, 152, 154, 157, 161, 162, 166, 176, 178
ferrite, xiii, 137, 138, 140, 141, 281, 282, 283, 288, 289, 294, 296
fiber, xi, 200, 201, 207, 208, 209, 284, 285, 286, 287
films, 71
filters, 35
financial, 15, 27, 59, 71, 98, 113, 145
financial support, 15, 27, 59, 71, 98, 113, 145
finite element method, 30, 32, 42, 60, 149, 158
flank, 55, 62, 200, 211, 214, 217, 220, 221, 222, 223, 224
flexibility, x, 100, 189, 196
flood, 17, 22, 23, 24, 26
flow field, 59
fluid, xii, 17, 18, 22, 23, 24, 26, 33, 148, 165, 171, 211, 212, 214, 217, 225
food, 233
force, 1, 2, 3, 4, 5, 7, 8, 9, 10, 11, 12, 13, 14, 16, 19, 45, 46, 47, 51, 54, 55, 57, 58, 85, 86, 87, 88, 89, 90, 91, 92, 93, 94, 95, 96, 97, 117, 121, 123, 127, 128, 149, 151, 152, 163, 181, 182, 183, 184, 186, 188, 189, 197, 199, 200, 201, 202, 203, 204, 205, 206, 207, 211, 212, 213, 216, 218, 220, 224, 247, 248, 249, 251, 256, 258, 260, 261, 262, 263
formation, x, xiii, 2, 12, 15, 19, 45, 46, 50, 54, 55, 56, 58, 59, 60, 82, 83, 93, 97, 100, 103, 107, 136, 137, 138, 140, 141, 147, 155, 158, 159, 164, 176, 211, 213, 216, 217, 218, 219, 224, 266, 281, 283, 296
formula, 163
fracture toughness, 46, 47, 49, 50, 55, 220
fractures, x, 115, 116, 117, 118, 126, 132
France, 83
free energy, 46
freedom, 150, 165, 166, 205, 206, 235, 242, 254, 277
friction, x, 12, 16, 18, 19, 24, 25, 45, 46, 47, 48, 49, 52, 55, 59, 60, 62, 74, 87, 88, 89, 98, 102, 103,

117, 118, 123, 125, 127, 148, 154, 157, 173, 212, 218, 219, 224
fuzzy membership, 240
fuzzy sets, 239, 240

G

gel, 268
generation, 18, 65, 97
geometrical parameters, 63, 179
geometry, ix, xii, 1, 2, 3, 4, 5, 7, 9, 15, 19, 20, 33, 38, 48, 49, 63, 85, 87, 88, 89, 91, 93, 95, 97, 115, 116, 117, 123, 127, 128, 129, 130, 131, 132, 136, 148, 165, 181, 182, 183, 184, 185, 189, 190, 194, 200, 206, 207, 208, 214, 216, 217, 226, 280
Germany, 16
goals, 30, 130
grades, 141, 240, 268
grain boundaries, 84, 282, 291
grain refinement, 282, 287
grain size, 103, 141, 281, 284, 291, 293
grains, 110
graph, 110, 193, 194, 205, 206, 232, 233, 241, 273
graphite, 215, 249
Greece, 147, 161
groups, 25, 100
growth, xv, 55, 107, 220, 221, 222, 282, 293
guidelines, 164

H

hardness, 20, 31, 33, 77, 89, 90, 94, 136, 138, 140, 141, 158, 171, 176, 220, 221, 225, 226, 247, 248, 249, 250, 251, 252, 253, 254, 255, 256, 258, 260, 261, 262, 263, 282, 287, 291, 292
heat, x, 17, 18, 19, 20, 23, 24, 25, 26, 61, 62, 63, 64, 65, 66, 67, 68, 69, 70, 71, 75, 110, 136, 138, 139, 140, 142, 143, 144, 145, 146
heat capacity, 63, 71, 165
heat conductivity, xii, 202, 228
heat transfer, x, 19, 61, 62, 63, 65, 70, 139, 140, 143, 211
heating, 19
height, 48, 63, 143, 240, 275, 278
heterogeneity, 108
High Speed Machining (HSM), xi, 148
high strength, 117, 119, 145
histogram, 94
homogeneity, 248, 258
HPC, 217, 218, 219, 222, 223
human, ix, 18, 231
Hungary, 180

hybrid, 205, 207, 249, 263
hydrocarbons, 25

I

ideal, 47, 148, 194, 284
identification, xiii, 17, 41, 76, 266
image, 64, 207, 261, 262, 288, 289, 290, 291, 295
image analysis, 207
images, 222, 223
immersion, xi, 181, 182, 183, 184, 189, 195, 196
implementation, 39, 54
improvements, 2, 165, 179
impurities, 110
inclusion, 46
incompatibility, 59
indentation, 31, 85, 86, 87, 88, 89, 90, 91, 92, 93, 94, 95, 96, 97, 98
independent variable, 53
India, 215, 227, 244, 247, 265, 279, 281, 295
indication, 55
indices, 97
indirect effect, 218
induction, 283
industries, x, xii, xiii, 18, 20, 100, 183, 228, 266
industry, ix, x, xi, xii, xiii, xv, 1, 7, 18, 100, 103, 136, 138, 144, 148, 151, 161, 162, 191, 195, 199, 228, 262, 266, 279
inferences, 231
infinite, 37, 139, 140
initiation, 138
insertion, 136
instability, 74, 76, 79, 82, 83, 114
insulation, x, 62, 67
integration, 10, 37, 39, 50, 53, 118, 165, 166
integrity, 29, 149, 173, 196, 212, 213, 217, 225, 263
interaction, 45, 46, 47, 51, 89
interface, 23, 25, 48, 52, 62, 63, 64, 67, 87, 118, 127, 154, 211, 212, 214, 217, 218, 219, 222, 224, 225
interval, 66
intervention, 152
inversion, 284
iron, xi, 146, 159, 199, 200, 207, 283
islands, 282, 292
issues, ix, xv, 30, 45, 148

J

Japan, 217, 224

K

kerosene, 249

L

laboratory tests, 101, 103, 106, 113
language, 140
laws, 120
lead, 5, 6, 9, 10, 20, 45, 47, 48, 59, 74, 82, 136, 176, 178, 261
legend, 79
life expectancy, xi, 200
lifetime, ix, 30, 42
ligament, 50, 55
light, 46, 113, 191
line, 9, 10, 45, 75, 91, 123
linearity, 102
localization, 73, 79, 105, 106, 212
loci, 83
locus, 105, 117
logic rules, 239
lubricants, 249
Luo, 136, 141, 145, 212, 225

M

magnesium, xi, 200, 244
magnetic composites, 179
magnets, 25
magnitude, 2, 30, 31, 74, 124, 127, 142, 155, 189, 192, 194, 218
maintenance, 100
majority, 47
manganese, 283
manufacturing, ix, xi, xii, xiii, xv, 30, 42, 86, 90, 96, 161, 179, 182, 228, 229, 245, 265, 266
market, 86
mass, x, 100, 136, 137, 145, 186
material removal rate (MRR), xii, 228
material surface, 31
materials, ix, x, xi, xii, xiii, 2, 4, 15, 17, 18, 26, 28, 30, 42, 47, 60, 62, 63, 73, 74, 76, 77, 79, 81, 82, 89, 98, 100, 115, 138, 140, 142, 143, 148, 149, 152, 158, 162, 164, 168, 174, 175, 177, 179, 180, 200, 206, 208, 209, 211, 212, 214, 219, 225, 228, 247, 248, 262, 266, 282, 283, 284
matrix, xi, xiii, 32, 33, 53, 70, 188, 199, 200, 201, 202, 204, 205, 206, 207, 239, 240, 244, 249, 263, 265, 266, 279, 282, 292
meanings, 59

measurement, xii, 22, 23, 25, 27, 34, 89, 95, 151, 179, 200, 225, 284, 295
measurements, 21, 22, 29, 35, 36, 37, 39, 40, 42, 47, 50, 51, 54, 57, 58, 115, 117, 121, 151, 169, 281, 284, 295
measures, 51, 142
mechanical properties, x, xi, xiii, 19, 59, 77, 101, 119, 136, 141, 145, 171, 200, 266, 283, 293
media, 37, 265, 266, 267, 268, 269, 273, 274, 275, 276, 279
melting, 19, 283
membership, 231, 233, 239, 240
metal industry, ix, 1
metals, ix, x, xi, 19, 53, 73, 82, 83, 84, 99, 100, 101, 103, 104, 105, 106, 110, 113, 162, 295
meter, 215
methodology, 10, 31, 32, 33, 42, 61, 62, 70, 71, 123, 150, 201, 247, 248, 249, 251, 252, 253, 262
microhardness, 226, 248, 249, 281, 291, 295
micrometer, 34, 217
microscope, 55, 202, 217, 284
microscopy, 89, 281, 284, 292
microstructure, 103, 104, 105, 135, 136, 137, 138, 140, 141, 145, 176, 282, 284, 296
microstructures, x, 83, 136, 282, 292
mixing, 268
model, ix, 1, 2, 4, 5, 7, 8, 9, 12, 14, 15, 19, 36, 37, 38, 39, 42, 45, 46, 47, 49, 54, 57, 59, 60, 62, 63, 70, 77, 86, 87, 88, 89, 99, 100, 101, 104, 105, 106, 107, 108, 109, 110, 111, 112, 113, 115, 116, 117, 126, 152, 157, 162, 165, 171, 180,
models, 2, 3, 4, 29, 31, 36, 37, 39, 41, 42, 46, 60, 65, 89, 99, 101, 103, 106, 123, 132, 133, 147, 149, 152, 154, 156, 157, 161, 162, 164, 165, 166, 167, 171, 178, 179, 182, 184, 196, 207, 228, 244, 248, 249, 265, 267, 276, 278, 279
modifications, 129
modulus, 32, 166, 186
mold, 141, 158, 181, 283
momentum, 65, 104
morphology, 140, 141, 147, 148, 149, 152, 157
motion, 85
movement, 81, 85, 86, 89, 136, 138
multidimensional, 139

N

near linear loading conditions, x, 116
Netherlands, 284
New South Wales, 179
nickel, 63, 225, 266
nitrogen, 281, 282, 283, 288, 295
nodes, 38, 166, 173

noise, 85, 86
non-linear equations, 53
North America, 225
nucleation, 94, 105, 282
null, 186, 187
numerical analysis, 61, 63, 121, 125, 128, 129
numerical computations, 34

O

objectives, 87
observations, 5, 55, 82, 102, 110, 115, 126, 127
oil, 22, 25, 34, 249, 251
omission, 148
operations, ix, x, xiii, 1, 2, 30, 33, 73, 74, 99, 100, 101, 103, 105, 106, 115, 116, 117, 118, 119, 121, 123, 126, 127, 130, 132, 149, 152, 157, 159, 211, 235, 248
opportunities, 279
optimal performance, xii, 228
optimization, xiii, 2, 16, 63, 100, 157, 158, 159, 201, 208, 217, 227, 228, 229, 243, 244, 245, 247, 255, 263, 266
optimization method, 157, 201
optimum output, 201
order, x, 2, 9, 12, 17, 18, 19, 20, 22, 25, 29, 30, 35, 36, 37, 42, 51, 53, 54, 59, 61, 62, 63, 65, 73, 74, 78, 86, 89, 97, 100, 115, 116, 118, 119, 123, 127, 128, 130, 132, 135, 136, 137, 140
orientation, 110
overload, 5
oxidation, 176
oxygen, 25

P

packaging, x, 100
parallel, 10, 152
parameter, 22, 23, 46, 55, 74, 78, 79, 82, 92, 104, 106, 107, 108, 109, 110, 111, 136
parameters, x, 2, 3, 4, 5, 7, 8, 17, 18, 20, 22, 25, 30, 35, 42, 46, 51, 61, 63, 69, 74, 78, 85, 91, 93, 99, 100, 101, 103, 107, 110, 111, 112, 113, 117, 121, 123, 126, 130, 132, 145
particles, 87
partition, 166, 179
passive, 3
peace, 100
pedal, 86
performance, ix, x, 29, 30, 42, 62, 126
permit, x, xi, 136, 182
phase boundaries, 282

phase diagram, 283
phase transformation, 180
phenomenology, 83
Philadelphia, 43
phosphorus, 63
physical and mechanical properties, xiii, 266
physical characteristics, 183
physical chemistry, 225
physics, 45, 55, 59
planning, 71
plastic deformation, ix, x, 18, 30, 47, 50, 53, 57, 92, 97, 101, 105, 116, 127, 128, 164, 219, 247, 248, 261, 282, 296
plasticity, ix, 31, 42, 45, 46, 83, 100, 103, 104, 110, 132
plastics, 208
platform, 151, 203
Poland, 114
polar, 25
polarization, 75
pollution, 148
polymer, 48, 209, 268
polymer composites, 209
poor, 18, 20, 42, 105, 119
porosity, 101, 103, 110
Portugal, xv, 45, 83, 115
power, 4, 19, 33, 116, 117, 120
precipitation, 20
prediction, 1, 4, 28, 57, 74, 99, 101, 103, 110, 111, 112, 113, 114, 123, 124, 126, 132
preparation, 152
pressure, 22, 45, 47, 57, 58, 59, 87, 91, 102, 103, 123
pressure gauge, 267
prevention, 224
principal component analysis, 249
principles, 12, 60, 147, 245
probe, 21, 31, 290
processing variables, 201, 295
product design, 85
product performance, 217
production, ix, x, 18, 19, 100, 101, 136, 137, 144
production costs, 100
productivity, ix, 18, 100
professionals, 103
program, 33, 38, 51, 54, 115, 116, 117, 118, 119, 126
programming, 140
project, 86, 97, 115
propagation, 50, 54, 201
properties, 42, 61, 62, 73, 76, 98, 99, 101, 105, 144, 145
proposition, 12, 45, 231
prototypes, 100
pulse, 51

Index 303

Q

quality control, 30, 42, 100

R

radiation, 65, 137, 139, 143
radius, 5, 9, 10, 12, 20, 33, 36, 63, 87, 91, 127, 173, 182, 191, 214, 216, 225, 227, 235, 236, 237, 241, 243
range, 15, 31, 32, 45, 46, 47, 48, 51, 55, 57, 75, 82, 90, 94, 115, 137
real time, 2
reason, 23, 47, 144
reasoning, 227, 229, 231, 232, 233, 237, 239, 240, 241, 242, 243
recommendations, 42, 78
recovery, 76, 82, 101, 282
recrystallization, 282, 284, 296
redistribution, 31, 141
reengineering, 100
region, 4, 7, 9, 12, 15, 50, 51, 55, 63, 64, 87, 102, 103, 105, 108, 109, 116, 126
regression, 252, 253, 276
regression equation, 252, 276
relationship, 2, 4, 17, 18, 59
relaxation, 29, 30, 31
relevance, xv, 133
reliability, 178
relief, 7, 15, 55, 214
repetitions, 204
replication, 123, 269, 272
reproduction, 117
requirements, xii, 18, 47, 53, 86, 91, 187, 211, 218
researchers, xi, 4, 37, 47, 103, 105, 148, 149, 162, 163, 166, 178, 249
residuals, 93
resistance, xi, xii, 3, 18, 19, 31, 54, 65, 95, 100, 103, 199, 200, 228
resolution, 21, 90, 187, 191, 262
respect, 18, 38, 40, 127
response, 89, 90, 91, 95, 97, 120, 185, 188, 205, 206, 232, 233, 241, 247, 248, 249, 251, 252, 253, 255, 262, 265, 275
restrictions, 126
retardation, 224
rheology, 269
rings, x, 86
risk, 125, 127, 242
rods, 86
rolling, 73, 74, 75, 76, 77, 78, 79, 80, 81, 82, 83, 110, 145

rolls, 77
room temperature, 48, 49, 63, 65, 89, 136, 138, 139, 165, 282, 283
root, 31
roots, 140
roughness, ix, xii, 18, 19, 85, 87, 88, 89, 90, 91, 98, 99, 101, 102, 103, 106, 110, 113, 158, 211, 213, 217, 223, 224, 227, 228, 229, 232, 233, 235, 236, 237, 238, 239, 240, 243, 244, 245, 247, 248, 249, 250, 251, 252, 254, 255, 256, 258, 262, 263, 268, 269, 274, 275, 276, 278, 279
routines, 183
Royal Society, 179
rules, 231, 239, 240

S

scanning electron microscopy, 89
scatter, 229
scholarship, 113
search, x, 86
security, 86
seizure, 219
sensitivity, 31, 37, 82, 83, 99, 103, 107, 113, 191, 224, 228
sensors, 191, 192, 193
separation, x, 45, 46, 50, 54, 55, 59
severity, 115
shape, ix, x, 53, 54, 73, 91, 97, 100, 101, 102, 157, 181, 183, 190, 192, 194, 195, 197, 216, 217, 219
shear, ix, 12, 18, 19, 24, 25, 26, 38, 39, 45, 46, 47, 49, 50, 51, 52, 55, 56, 59, 60, 73, 74, 75, 78, 83, 99, 102, 103, 105, 148, 200, 212
shear deformation, 59, 83
sheet metal blanking, x, 100
sheet metal formability, x, 99, 101, 103, 111, 116
showing, ix, 46, 54, 55, 56, 98, 104, 107, 108, 118, 167, 261, 262
signal-to-noise ratio, 204
significance level, 17, 23, 26, 93, 95, 242
silicon, xiii, 18, 266
silicon carbide (SiC), xiii, 266
simulation, 1, 2, 4, 5, 13, 14, 38, 54, 62, 63, 67, 86, 89, 91, 92, 93, 96, 101, 116, 120, 123, 124, 125, 128, 129, 130, 131, 133, 147, 149, 152, 156, 157, 158, 161, 162, 166, 171, 178, 179, 180, 181, 182, 183, 185, 189, 192, 194, 195, 196, 197, 248
simulations, 46, 51, 62, 70, 92, 101, 136, 138, 144, 147, 152, 154, 157, 161, 179, 197
Singapore, 225, 245
smoothing, 249, 263

software, 33, 34, 35, 38, 51, 62, 63, 65, 66, 69, 85, 88, 89, 100, 120, 147, 152, 166, 191, 203, 252, 255, 284, 296

solubility, 295

solution, 5, 20, 52, 53, 54, 55, 63, 65, 132, 139, 140, 181, 183, 185, 186, 187, 188, 189, 192, 212, 283

space, 39, 52

specific heat, 63, 71, 139, 142, 165

specifications, 283

speed, ix, 1, 2, 5, 7, 8, 9, 12, 17, 19, 20, 22, 23, 25, 26, 27, 28, 33, 34, 47, 51, 64, 77, 89, 90, 91, 119, 140

spindle, xi, 23, 26, 34, 148, 149, 151, 155, 191, 199, 204, 205, 206, 207, 213, 216, 233, 262

stability, 65, 82, 90, 119, 182, 196

stabilization, 82, 91

standard deviation, 195

state, ix, 19, 29, 30, 31, 54, 55, 56, 57, 59, 62, 77, 79, 104, 112, 152, 154, 165, 171

states, 46, 65, 73

statistics, 269

steel, x, xii, 7, 9, 12, 13, 14, 27, 29, 30, 33, 42, 62, 63, 73, 74, 75, 76, 78, 79, 80, 81, 82, 83, 84, 89, 90, 99, 100, 101, 102, 105, 110, 112, 113, 114, 117, 119, 120, 130, 136, 138, 141, 143, 144, 145, 147, 151, 158, 176, 177, 178, 180, 211, 212, 213, 214, 215, 224, 225, 226, 227, 228, 229, 233, 241, 243, 244, 245, 247, 248, 249, 263, 281, 282, 283, 288, 295, 296

strain, ix, x, 24, 29, 30, 31, 33, 34, 35, 36, 37, 38, 39, 40, 41, 46, 47, 48, 49, 51, 52, 53, 54, 59, 73, 74, 75, 76, 77, 78, 79, 80, 81, 82, 83, 84, 97, 98, 99, 100, 101, 102, 103, 104, 105, 106, 107, 108, 109, 110, 111, 112, 113, 115, 116, 117, 118, 119, 120, 123, 125, 126, 127, 130, 132, 133

strategies, 28, 47

strategy, 19, 37, 39, 54, 127

strength, x, 18, 19, 20, 74, 107, 117, 119, 130, 141, 145

stress, ix, x, 12, 29, 30, 31, 32, 33, 34, 35, 36, 38, 39, 42, 47, 49, 52, 53, 54, 55, 58, 59, 73, 74, 75, 76, 79, 80, 81, 82, 87, 92, 97, 98, 105, 107, 109, 112, 114, 115, 116, 117, 118, 119, 120, 123, 124, 125, 128, 129, 130, 131, 132, 133, 162, 165, 166, 177, 212, 226, 248, 262, 282

stress fields, ix, 29, 30, 32, 35, 38, 39, 42

stretching, 74, 99, 101, 102, 103, 104, 105, 106, 107, 108, 111, 113, 127, 128, 129, 131

stroke, 90

structural changes, 296

structural characteristics, 73, 76

structure, 145, 182, 282, 292

substrate, 2, 61, 62, 63, 65, 66, 67, 71

succession, 75

Sun, 162, 180

superplastic forming, 103

superplasticity, 282, 296

supply, x, 86

suppression, 196

surface area, 102, 165

surface hardness, 247, 249, 251, 252, 254, 255, 256, 258, 261, 262, 263

surface layer, ix, xii, 30, 212, 261

surface properties, 263

surface roughness, ix, xii, 18, 19, 85, 87, 88, 90, 98, 101, 103, 106, 158, 211, 213, 217, 224, 227, 228, 229, 232, 233, 235, 237, 238, 239, 240, 243, 244, 245, 247, 248, 249, 251, 252, 254, 255, 256, 258, 262, 263, 268, 269, 274, 275, 276, 278, 279

surface treatment, 248

Sweden, 43

symmetry, 192, 284

T

Taiwan, 199, 207

tantalum, 295

target, 22, 194, 229

technical support, 295

techniques, ix, 2, 19, 31, 65, 145, 152, 180, 185, 188, 207, 231, 244, 245, 263

technology, ix, x, xi, xv, 18, 100, 173, 182

teeth, 20, 25, 189, 195

temperature, x, xi, 17, 18, 19, 20, 22, 23, 24, 25, 26, 27, 47, 48, 49, 61, 62, 63, 65, 66, 67, 68, 69, 70, 71, 73, 74, 89, 103, 135, 136, 137, 138, 139, 140, 141, 142, 143, 145, 162, 165, 166, 167, 171, 173, 174, 175, 176, 177, 178, 179, 199, 200, 202, 211, 212, 213, 214, 215, 217, 218, 219, 221, 222, 223, 224, 269, 282, 283

tension, ix, 30, 73, 74, 75, 76, 78, 79, 80, 81, 82, 201

testing, 42, 47, 49, 57, 78, 79, 99, 101, 110, 111, 136, 152, 250

texture, xiii, 75, 76, 83, 101, 110, 266, 281, 282, 283, 284, 285, 286, 287, 288, 291, 295, 296

thermal analysis, 61, 179

thermal energy, 17, 18, 135, 136, 137, 142

thermal expansion, 17, 18, 23, 25, 26, 166

thermal properties, 62, 65, 71, 136

thermal resistance, 65

thermal treatment, 135, 138

thermosets, 208

thinning, 103, 123, 129, 131, 132

time increment, 183, 184, 189

titanium, x, xi, 62, 100, 148, 158

tooth, 19, 155, 183, 190

Index

topology, 189, 194

torsion, 5

trajectory, 192

transducer, 51, 90

transformation, 136, 137, 138, 141, 146, 176, 180, 183, 185, 251

transformations, 38, 39, 136, 176, 177, 178

translation, 10

transmission, 86

transport, 65

trapezoidal membership, 239

treatment, x, 20, 104, 110, 136, 138, 145, 149, 212, 248, 281, 283, 287, 288, 289, 291, 293, 294, 295

trends, 29, 39

tungsten, 19, 26, 34, 202, 214, 215, 247, 249, 251

tungsten carbide, 19, 26, 34, 202, 214, 215, 247, 249, 251

twist, xi, 1, 2, 4, 7, 13, 199, 200, 201, 202, 205, 206, 207, 208

U

uniform, 19, 31, 43, 61, 65, 70, 92, 103, 110, 165, 248, 268

united, 196

United States, 196

USA, 27, 43, 72, 145, 157, 158, 179, 180, 196

V

vacuum, 85, 86, 89, 91, 284

validation, 37, 39, 42, 147, 149, 162, 178, 179, 227, 248

valve, 85, 86, 214

variability, 85

variables, x, 2, 19, 45, 46, 52, 53, 54, 63, 65, 74, 75, 86, 88, 89, 91, 93, 94, 95, 97, 99, 101, 187, 189, 201, 205, 228, 231, 233, 239, 252, 253, 295

variance, 17, 23, 89

variations, 1, 2, 5, 9, 23, 61, 66, 87, 95, 99, 101, 106, 110, 111, 113, 126, 136, 176, 284

vector, 184

vegetable oil, 25

vehicles, xi, 200

velocity, xii, 52, 53, 104, 148, 200, 217, 218

vessels, 233

vibration, 20, 23, 181, 183, 185, 187, 188, 189, 195, 196, 216

viscosity, 266, 269, 274, 275, 276, 279

W

Wales, 179

Washington, 196

waste, 140

water, 22, 225, 282, 283, 284

wear, x, xi, xii, xiii, 2, 19, 20, 33, 62, 63, 100, 146, 148, 149, 155, 158, 164, 199, 200, 201, 211, 212, 213, 217, 218, 219, 220, 221, 222, 223, 224, 225, 228, 245, 263, 266

windows, 51

workers, xii, 248

working conditions, 2, 4

Y

yield, 12, 31, 54, 105, 107, 118, 130, 177, 248, 262

Z

zinc, 17, 18

zirconium, 283